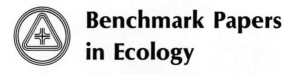

Benchmark Papers
in Ecology

Series Editor: **Frank B. Golley**
University of Georgia

PUBLISHED VOLUMES

CYCLES OF ESSENTIAL ELEMENTS / *Lawrence R. Pomeroy*
BEHAVIOR AS AN ECOLOGICAL FACTOR / *David E. Davis*
NICHE: THEORY AND APPLICATION / *Robert H. Whittaker and
  Simon A. Levin*
ECOLOGICAL ENERGETICS / *Richard G. Wiegert*
ECOLOGICAL SUCCESSION / *Frank B. Golley*
PHYTOSOCIOLOGY / *Robert P. McIntosh*

Benchmark Papers
in Ecology / 6

A BENCHMARK® Books Series

PHYTOSOCIOLOGY

Edited by

ROBERT P. McINTOSH
University of Notre Dame

Dowden, Hutchinson & Ross, Inc.
STROUDSBURG, PENNSYLVANIA

80 79 78 1 2 3 4 5
Manufactured in the United States of America.

LIBRARY OF CONGRESS CATALOGING IN PUBLICATION DATA

Main entry under title:
Phytosociology.
  (Benchmark papers in ecology ; 6)
  Includes indexes.
  1. Plant communities—Addresses, essays, lectures. I. McIntosh,
Robert Patrick.
QK911.P47   582'.05'24   77-20258
ISBN 0-87933-312-X

Distributed world wide by Academic Press,
a subsidiary of Harcourt Brace Jovanovich,
Publishers.

SERIES EDITOR'S FOREWORD

Ecology—the study of interactions and relationships between living systems and environment—is an extremely active and dynamic field of science. The great variety of possible interactions in even the most simple ecological system makes the study of ecology compelling but difficult to discuss in simple terms. Further, living systems include individual organisms, populations, communities and ultimately the entire biosphere; there are thus numerous subspecialties in ecology. Some ecologists are interested in wildlife and natural history, others are intrigued by the complexity and apparently intractable problems of ecological systems, and still others apply ecological principles to the problems of man and the environment. This means that a Benchmark Series in Ecology would be subdivided into innumerable subvolumes that represented these diverse interests. However, rather than take this approach, I have tried to focus on general patterns or concepts that are applicable to two particularly important levels of ecological understanding: the population and the community. I have taken the dichotomy between these two as my major organizing concept in the series.

In a field that is rapidly changing and evolving, it is often difficult to chart the transition of single ideas into cohesive theories and principles. In addition, it is not easy to make judgments as to the benchmarks of the subject when the theoretical features of a field are relatively young. These twin problems—the relationship between interweaving ideas and the elucidation of theory, and the youth of the subject itself—make development of a benchmark series in the field of ecology difficult. Each of the volume editors has recognized this inherent problem, and each has acted to solve it in his or her unique way. Their collective efforts will, we anticipate, provide a survey of the most important concepts in the field.

The Benchmark series is especially designed for libraries of colleges, universities, and research organizations that cannot purchase the older literature of ecology because of cost, lack of staff to select from the hundreds of thousands of journals and volumes, or unavailability of the reference materials. For example, in developing countries where a science library must be developed *de novo*, I have seen where the Benchmark series can provide the only background literature available to the students and staff. Thus, the intent of the series is to provide an authoritative selection of literature that can be read in the original form, but which is cast in a matrix of thought provided by the editor. The volumes are designed to explore the historical development of a concept in ecology and point the way toward new developments, without being a historical study. We hope that even though the Benchmark Series in Ecology is a library-oriented series and bears an appropriate cost it will also be a sufficient utility so that many professionals will place it in their personal library. In a few cases the volumes have even been used as textbooks for advanced courses. Thus we expect that the Benchmark Series in Ecology will be useful not only to the student who seeks an authoritative selection of original literature but also to the professional who wants to quickly and efficiently expand his or her background in an area of ecology outside his or her special competence.

This volume deals with the subject of phytosociology or the interrelationships of species of plants in space. It is one of the oldest interests of ecology and the literature is enormous. Dr. Robert McIntosh of the University of Notre Dame is editor of the volume. Dr. McIntosh has been concerned with phytosociological problems throughout his professional career and brings to the subject substantial field experience. However, Dr. McIntosh has also been interested in the history of ecology and could well be considered one of our few ecological scientist-historians. As a consequence he also brings to the discussion of phytosociology impartiality and breadth of knowledge. Given the serious constraint of a limitation on length, Dr. McIntosh has provided us a selection that balances contrasting and often contentious views, captures the essence of an often complicated terminology and numerical analysis, and shows how this old topic relates to the most modern concerns of ecologists.

FRANK B. GOLLEY

PREFACE

The present volume brings together a selection of articles from an extremely extensive literature on vegetation and the plant communities which comprise it. The title of the book, *Phytosociology,* is one of several words which have been used to describe the study of aspects of the communal relations of plants. Phytosociology was coined by Paczoski (Mayock 1967),* and unfortunately there has been confusion of terminology among ecologists. In Continental Europe, field botany has traditionally been called geobotany; and the study of plant communities was sociological geobotany—phytosociology, plant sociology, and phytocoenology being common synonyms. In the English-speaking countries, the study of communities was commonly called synecology, phytosociology or, even more inclusively, plant ecology. Mueller-Dombois and Ellenberg (1974, p. 7) provide a convenient table of these terms and their several uses and propose "vegetation ecology" as an alternate descriptor for phytosociology. Egler (1942) proposed "vegetation science" for the study of aggregates of plant communities, in an area, acting as a whole.

Some traditional schools of ecology using variations of these terms may circumscribe their content differently. For the purposes of the present volume, I have construed phytosociology broadly as including structure and distribution in space, both vertically and horizontally, of plant communities at all scales of distribution, species composition, and social interactions among species which may occur together as members of any plant community, and the reciprocal relations of plant community and environment that are related to organization and distribution of communities. Phytosociology in this sense is not restricted to the tenets of any school or to any concept of community or method of study. It does not imply a particular classification scheme or even classification as a method for the study of communities. The pioneer figures in the early history of phytosociology considered physiognomy, composition, and habitat relations of communities; and it does no violence to tradition to include these, as well as recent quantitative methods for the study of vegetation, in the logical concept of phytosociology.

*Preface citations are listed in the References for the Introduction.

An important adjunct of phytosociology, particularly in some traditions, has been the attempt to develop a hierarchical classification of communities and to establish a consistent nomenclature. Mapping of vegetation also depends upon a system of recognizing and classifying communities (Küchler 1967, Küchler and McCormick 1965). Vegetation maps may be useful in relating communities to environmental variables (site or habitat conditions) or as a basis for management studies or land planning. Neither the taxonomy of vegetation or mapping are considered in detail in this volume. The process of community change or development, *succession,* is the subject of a separate volume, edited by Dr. Frank B. Golley, of the Benchmark Papers in Ecology. It will not be a primary concern of this volume except as it is inherent in some of the conceptualization of the nature of plant communities, their distribution and classification.

It has become fashionable in an era of the "new ecology," which has placed major emphasis on community function, to apply the somewhat pejorative term "mere description" to much of the phytosociological work which has attempted to sort out the bewildering complex of plant communities, or more comprehensive ecosystems, which covers the earth's surface. It should be noted that the extensive functional studies of recent years were related to each other within the mold of community organization (the biome) provided by traditional "mere description." Functional attributes are necessarily of something; and in community ecology they are of a community defined and related to other communities by structural, compositional, or habitat characteristics. Definition or delimitation of communities on functional attributes has not been done, and numerous current efforts to arrive at useful bases for classification for scientific and management purposes still face the problems of "mere description."

The rigorous selection from the literature of a century of vegetation studies, dictated by limits of space, inevitably omits many significant developments in phytosociology and, lamentably, selections from the work of eminent contributors to phytosociology in diverse places. It is hoped that the papers in this volume, which reflect the long tradition of phytosociological work, will introduce the reader to some of the significant traditions and personalities in the study of vegetation and identify some of the salient current trends in this earliest root of formal ecology.

The assistance of R. H. Whittaker, G. Cottam, and E. Van der Maarel in identifying appropriate papers for inclusion is gratefully acknowledged. Edward J. Kormondy graciously made available translations of Paper 2.

ROBERT P. McINTOSH

CONTENTS

Contents

PART III: ORDINATION AND NUMERICAL CLASSIFICATION

PART IV: RECENT PERSPECTIVES

CONTENTS BY AUTHOR

PHYTOSOCIOLOGY

INTRODUCTION

The landscape is given much of its character by its vegetative cover; and human evolution and history have been substantially influenced by vegetation. Animal and human life still depend on, and in many ways are related to, vegetation, either natural or as influenced by human actions. Vegetation is not equivalent to the list of the plant species, *the flora,* of an area but is made up of aggregations of species, *communities,* of plants whose visual and biological impact is due largely to the growth form and relative abundance or *dominance* of the largest and most conspicuous plants (Egler 1942). Much of the early development of formal ecology is associated with the work of botanists who transformed the traditional treatment of the plant cover of a region, primarily in terms of its floristics (classical phytogeography), to the description and analysis of vegetation in terms of communities, their gross appearance (physiognomy), species composition, and relation to the environment (Egerton 1976, Whittaker 1962).

Although humans had always lived in and exploited plant communities, the earliest systematic description of vegetation was by the geographer-explorer, Alexander von Humboldt (1805), who classified areas dominated by plants of similar growth form into vegetation types. More detailed and better defined descriptions and classifications were developed by Schouw (1832), Kerner (1863), and Grisebach (1872) (see Egerton 1976). In the last decade of the nineteenth century a number of European botanists, notably Drude (1890), Warming (1895, 1909) (*see* Goodland 1975), and Schimper (1898, 1903), published works which profoundly influenced the development of plant ecology (Egerton 1976, Whittaker 1962). The books by Warming and Schimper were widely known and stimulated American botanists, particularly Charles E. Bessey and John M. Coulter and their students (Bessey's—Conway MacMillan, Roscoe Pound, and Frederick E. Clements; Coulter's—Henry C. Cowles) to some of the pioneer work on American vegetation (McIntosh 1976).

HISTORY AND SCHOOLS OF PHYTOSOCIOLOGY

Two aspects of vegetation have been widely used in characterizing it:

1. *physiognomy,* or gross appearance due to the growth form of the plants;

2. *composition,* the species present and their relative proportions.

Plant geographers and pioneer plant ecologists, such as Grisebach, Drude, Warming, and Schimper, largely described vegetation according to its physiognomy. Grisebach, in 1838, had applied the term *formation* to a community of plants distinguished by its physiognomy. Consensus on terminology and concept was not, however, achieved among ecologists concerned with vegetation. The confusion about the term *formation* caused Warming to recommend its disuse, a fate not uncommonly wished on other confounded ecological terms (e.g., association, climax, stability). A major concept of plant geography and early phytosociology was that of plants growing together; i.e., *associated,* and of a group of associated plants, an *association.* There was an extended period of confusion about the proper use of formation and association, but by the early twentieth century there was general consensus that formation applied to communities distinguished by physiognomy and association to communities, included within a formation, which were defined by composition. However, no unanimity was achieved then, or since, on the precise scope or nature of the formation or association (Paper 1).

In addition to characteristics of the vegetation itself, habitat or site characteristics were used to classify areas (site-types, biotope-types, or habitat-types) which may have specific vegetation. Phytosociologists recognized a definite relation between vegetation and environment, notably water, heat, or *edaphic* (soil) factors. The complex of environment (e.g., climate, soil) and community may also be used to identify landscape-types, or biogeo-coenose-types, in an effort to deal with the multifaceted relationships of organisms and their physical or abiotic environment.

Because of variation in vegetation, environment, history of human impact, and the several bases for describing vegetation, diverse traditions and schools of phytosociology developed. The proliferation of schools in the early stages of the development of a science is familiar and phytosociology was no exception (Kuhn 1970). The several traditions in phytosociology have been reviewed in detail by Whittaker (1962) (*see also* Trass 1976) and in summary by Whittaker (1973b). Whittaker recognizes six major traditions in the history of scientific approaches to vegetational study and separate schools within these. It has not been possible in this volume to include representative articles from each of these traditions or to represent many of the major contributors to them.

The physiognomic tradition is parental in the emergence of vegetation studies from the tradition of biogeography and continues as an aspect of all of the schools of phytosociology and as a continuing approach in its own right. It is particularly used in large-scale regional descriptions of vegetation and its relation to climate or environmental gradients and also in regions where the floristics of the vegetation formations is not well known (e.g., the tropics). It is essential in efforts to classify and compare vegetations of very different areas when there is no floristic identity. Paper 10 by Beard is an example of this tradition and Beard (1973) provides an excellent review of the physiognomic approach.

The northern tradition is seen in the work of many plant ecologists of northwestern Europe working in species-poor vegetation of the Scandinavian and Baltic countries. This tradition made major contributions to the study of vegetation in the work of Du Rietz (1921) and a series of distinguished predecessors in the Uppsala school (Sweden), Lippmaa (1939) in Estonia, and Cajander (1926) in Finland. These are, respectively, reviewed in articles by Trass and Malmer (1973), Barkman (1973), and Frey (1973). The most distinctive contributions of these schools are: (1) the study of vegetation as a composite of its several vertical *strata* (layer societies, synusia) as semi-independent vegetational entities of species belonging to the same layer or growth form and occupying the same microhabitat, and (2) the use of undergrowth strata or layers as a basis of classification and indication of site quality by the Finnish school, notably in the work of Cajander. This approach is seen in North America in Heimberger's (1934) studies of the Adirondacks.

The southern European tradition has converged from diverse roots on the school stemming from J. Braun (Paper 2) and variously described as the Zurich-Montpellier school, French-Swiss school, or by the acronym SIGMA (Weadock and Dansereau 1960). The Braun-Blanquet school is the largest, most persistent, and most influential existing school of phytosociology and has produced an enormous body of work on the vegetation of Europe within a coherent scheme of vegetational study. Part of its success is that it seems elastic enough to accommodate diverse approaches to the study of vegetation. The essence of the Braun-Blanquet school is recognition of vegetational entities, or associations, on the basis of characteristic combinations of species and the elaboration of a classification hierarchy, with the association as basic unit, for a taxonomy of communities. Extended commentaries on the Braun-Blanquet school at various times in its history are available (Egler 1954, Poore 1955 a,b, Becking 1957, Weadock and Dansereau 1960, Moore 1962, Braun-Blanquet 1968, Werger 1974, Westhoff and Van der Maarel 1973, Van der Maarel [Paper 18]).

In addition to the diversity of vegetation and botanical tradition encountered in various areas, the discreteness of the several phytosociological traditions was accentuated and perpetuated by differences in language. This was particularly true in the case of the Slavic languages. Few ecologists, not publishing in these, read Russian or Polish and relatively little of the work of phytosociologists written in these languages was translated. It is not surprising, therefore, that much work was published in these languages which parallels that of other students of vegetation. These parallels are seen in the discussions of Russian phytosociology by Ponyatovskaya (1961), Aleksandrova (1973), and Sobolev and Utekhin (1973). The relative ignorance of most English-speaking ecologists of Russian vegetation studies has been recently combated by the reviews of much current Russian work in recent volumes of the journal *Ecology* by Jack Major.

Classical Russian phytosociology is represented here by V. N. Sukachev (Paper 7) and characteristically recognizes major vegetational communities on the basis of the physiognomic dominant subdivided by stratal dominants. There is in the Slavic language literature a striking parallel to phytosociology in English-speaking countries in the development of a divergent view treating vegetation as continuously varying rather than as a series of distinct vegetation types (*see* Curtis 1959, Ponyatovskaya 1961, McIntosh 1967, 1975).

The British and American traditions, although not identical,

4

are sufficiently similar to be combined. Both stressed recognition of vegetational units by their dominant species and were more concerned with successional processes than most of the Continental European traditions with the exception of the Russian, where the concept of "zonal" vegetation is explicit (Aleksandrova 1973). American phytosociology, and ecology generally, were heavily influenced by the ideas of Frederick E. Clements (see Cain, Paper 6). Clements recognized that basic, stable, vegetational units or "associations," which were climax or mature and self-perpetuating barring disturbance, were controlled by the regional climate, hence the "climatic climax" (Clements 1936). According to Clements, the "association" was a "supra-organism" which had equivalent characteristics to an individual and included all of the successional communities as part of its development. Clement's ideas stimulated H. A. Gleason to codify his own thoughts about the "individualistic concept" of the plant association (see Gleason, Paper 4). This set the stage for a persistent controversy (Ponyatovskaya 1961, McIntosh 1967, 1974, 1975). Both Clements's and Gleason's concepts were questioned by the preeminent, British, plant ecologist of the early twentieth century, Sir Arthur Tansley (1920, 1935, Godwin 1977). Tansley, most British plant ecologists, and some American plant ecologists recognized multiple stable communities in an area, the *polyclimax* view, in contrast to the *"monoclimax"* concept of a single climatic climax of Clements. Much confusion and disputation about the nature of the plant association occurred and has persisted in phytosociology because of the differences in concept, scale, and methods of study of the association between most Continental schools and most Anglo-American phytosociologists, particularly those following Clementsian ideas.

APPROACHES TO VEGETATION STUDY

The several ecological traditions have all used a limited number of approaches to vegetation study, and the sometimes exaggerated differences between them are essentially due to the different emphasis placed on particular aspects of vegetation study. They share, however, a common conviction, namely, that the mantle of vegetation on the earth is susceptible of scientific study, that such study will reveal relationships among species, allow analysis of characteristics of aggregation of species distinguished as communities, demonstrate relationships between these and the environment, and prove useful for management and conservation purposes.

The essence of much classical phytosociology was the assumption that individual instances or *stands* of vegetation were subject to classification as abstract *vegetation types,* and a related issue was whether such abstract vegetation types were "natural." That a set of objects such as stands of vegetation can be classified in diverse ways is indisputable. That any such classification is "natural" in the sense of reflecting relatively independent, internally homogenous groups of species brought together by interactions among themselves and between them and the habitat is the subject of much dispute. Much of the contention among phytosociologists derived from the effort to develop a standardized method of classifying vegetation and an agreed upon system of nomenclature, at least analogous to that of plant taxonomists, which surfaced in the Botanical Congresses early in this century and continues to the present (Barkman et al. 1976). There are parallels between systematics, taxonomy, and botanical nomenclature and their counterparts in vegetation study; and assertions about genetic parallels, or lack of them, appear frequently in phytosociological disputes (cf., Nichols 1926, Gleason 1926). Phytosociology has the problem of finding in, or imposing on, vegetation an order which is the basis for understanding. It suffers the same problems of taxonomy in dealing with things which may differ in diverse ways and for very different reasons, but its entities lack the genetic connection of taxonomic entities. Some of the differences of recent decades between classical, or alpha, taxonomy and numerical taxonomy have parallels in the distinctions between classical phytosociology and recent numerical and multivariate approaches to plant ecology using similar, even identical, techniques to those of numerical taxonomy.

The characteristics of vegetation may be assessed in diverse ways, and the resulting description may be a consequence of the preconceptions of the phytosociologist. The scientific study of vegetation grew out of classical phytogeography, which was based primarily on floristics. Much of the transition literature was the recognition of plant communities largely on the basis of physiognomy and particular habitats. The detailed description of the community was largely floristic still, and the species list was the most characteristic descriptor of the plant community in most of the early work.

Among the early and continuing sources of contention among phytosociologists were the divergent views which developed concerning the nature of the association and its definition. One of the early definitions of association was that of Flauhault and Schroter (1910), "An association is a plant community of definite floristic composition." While phytosociologists commonly recognized with Warming (1909) that, ". . . certain species group themselves into communities which are met with more or less frequently," it was also widespread belief as Raunkaier (1934) noted that, ". . . No case has ever been established in which two or more species are completely alike from the point of view of their adaptations to conditions and there is no reason to suppose that such a thing ever exists . . ." Hence, many observed with Braun-Blanquet (1932), ". . . with few exceptions no two bits of vegetation have precisely identical floristic composition." The difficulty of dealing with vegetation was expressed in Raunkaier's (1934) words,

> . . . Wherever we travel we find in nature that the environment alters from place to place often from step to step. Correspondingly the vegetation alters qualitatively, i.e. in its species composition, and quantitatively, i.e. in the frequency of the component species. The qualitative floristic changes can be fairly easily observed and followed, but this is not true of the quantitative relationships giving no fixed points for our observations.

In spite of the problems posed by gradual change, Raunkaier asserted, "But in order to recognize and account for anything so that we can compare it with something else, we are obliged to define it, in other words delimit it . . . ."

Thus, a major dilemma faced in phytosociology was that of identifying and delimiting the units with which it hoped to deal. Some plant ecologists followed Clements in conceiving the plant association as a supra-organismic unit of species integrated closely with each other and developing from a pioneer stage to a stable or mature climax paralleling the life history of an individual organism. Others, like Tansley, tempered this extreme view by recognizing the community as a quasi-organism but still one which had a high degree of integration and which behaved as a whole. Tansley was responsible for coining the word *ecosystem*, which is the key entity of modern ecology with conceptual problems of its own. Many ecologists recognized with Braun-Blanquet (1932) that ". . . the possible combinations of plant species are indeed endless . . ." but argued that similar pieces, or stands, of vegetation could be grouped into abstract *association-types*. These, by analogy, were species and could be classified into larger units in a hierarchical system of progressively more inclusive entities. The early

7

phytosociological literature made much of the difference between the individual stand, or concrete instance of the association, and the abstract association, the aggregate of several stands, and it is important to recognize that distinction.

The degree of integration inherent in a plant community and the relations among the component species were subject to very different interpretations. In contrast to Clement's supra-organism, or even Tansley's quasi-organism, Flauhault and Schroter (1910) are quoted by Mueller-Dombois and Ellenberg (1974) as stating that,

> . . . the plant association does not imply a harmonious concurrence of diverse activities working toward a common end, as in every society founded on a division of labor. It is applied to the coexistence of forms which specifically and morphologically are foreign to one another, each having its object, its own exclusive profit. They live side by side according to the similarity or diversity of their environment or determined by the presence of other organisms.

This and other assertions of species individuality were similar to the individualistic species concept of H. A. Gleason which he elaborated into the individualistic concept of the plant association (Gleason 1926). This concept was developed independently by several persons in diverse places and holds that because the environment varies constantly from place to place and selects from the available propagules of species in the area no two individual stands can be considered alike and vegetation cannot be classified into discrete associations except on arbitrary grounds (Ponyatovskaya 1961, McIntosh 1967, 1974, 1975).

COMMUNITY ANALYSIS AND SYNTHESIS

The question, "What is a plant community?", ("Was ist eine Pflanzengesellschaft?") has been asked and answered in various languages, and by various precepts, as suggested above, and pervades the history of phytosociology. Subjective or intuitive preconceptions of the nature of the community and the recognition of specific entities as communities, based on particular characteristics in common, have heavily weighted the various traditional views of plant communities. Probably all studies of vegetation are predicated upon some familiarity with the general vegetation of an area based on a more or less thorough reconnaissance of the area—the recognition of vegetation segments on the basis of physiognomy, the occurrence of a particular species or species combinations, or juxtaposition of species and habitat uses characteristics which are immediately perceivable. The earliest stages of phytosociology were concerned with subjective and qualitative descriptions of vegetation using either its physiognomy or a floristic list to describe individual stands. The validity and reproducibility of the subjectively determined plant community depended largely on the experience and the "soziologischer Blick," of the phytosociologist. Some still urge that recognition of vegetational units is dependent on the virtuosity of the phytosociologist and that one sufficiently familiar with an area can recognize "iterative groups" of species and hence define associations (Guinochet 1968).

The complexity of vegetation is such that a complete survey of even a relatively small area is prohibitively time consuming. Plant ecologists, early in the twentieth century (ca. 1900–1905), began to elaborate the methods of characterizing their objects of study—the stands or individual communities. They began to use small areas, called quadrats, to represent the communities comprising vegetation. The use of samples introduced into phytosociology many problems including delimiting the area to be sampled, the number, size, shape, and placement of the sample(s), what information to record from the samples, and how the data should be analyzed. A large portion of phytosociological work in the subsequent three-quarters of a century has revolved around these questions and more abstract conceptions of the community based on the sample data.

Perhaps the most crucial choice made by the phytosociologist is placing the sample or samples. There is, inevitably, some subjective choice of the area, habitat, or type of vegetation to be sampled. However, within such an area one or more samples may be placed subjectively or objectively and taken as representative of the area which was usually presumed to be homogeneous. The concept of *homogeneity* was crucial to much of classical phytosociology but was most difficult to define and apply in an operational way. The usually subjective choice of presumably homogeneous areas was designed to avoid mixtures (*ecotones*) between different communities or fragments of a community. An important corollary of homogeneity was the concept of *minimal area* which was the size of the sample area required to adequately represent the characteristic species composition of the community. *Richness* or a characteristic number of species necessary for stabilization of a community was an important community attribute (Gleason 1936). These much debated concepts, the associated idea of

9

the species-area curve and their descendent, the concept of diversity, have been the focus of continued discourse in phytosociology and, more recently, ecology in general.

The introduction and development of samples and sampling methods early in the history of phytosociology were paralleled by the development of statistics, and the introduction of quantitative statistical approaches in phytosociology opened up a Pandora's Box which is still producing surprises for plant ecologists.

Among the earliest explicit quantitative approaches to phytosociology was valency or frequency analysis introduced notably by Raunkaier (1934). This method simply recorded the presence or absence of a species in each of a series of *quadrats* or sample areas within a stand of a community and the number of samples in which the species occurred was termed its *frequency*.

In addition to frequency, phytosociologists soon began to record other quantitative characters: (1) Number of individuals, usually called *density*, equivalent to *abundance* of animal ecologists although abundance also had a meaning peculiar to some phytosociologists. (2) Size of individual, usually as an indirect measure of its *cover* or *basal area* and, more recently, as direct measures of weight (*biomass*).

These quantities were variously measured, or estimated according to a variety of scales, in an effort to provide useful methods of describing the occurrence, distribution, and importance of the several species in the stand (Greig-Smith 1964). The Braun-Blanquet sampling, or *relevé*, method commonly describes each species in a single quadrat by numbers based on two rating scales, one representing a combination of number (density) and size (cover), the other an estimate of its *sociability*, a five-point scale, indicating degree of clumping *(gregariousness)* of individuals.

In addition to quantitative characters used to analyze the plant community, qualitative characters were used describing its vertical structure in *layers* or *strata*, the ability of a species to grow and complete its life cycle, *vitality*, and the seasonal course of development of species groups *periodicity (aspect)*, brought about by the struggle for existence in the community (Braun-Blanquet 1928).

Development of the concept of community-types led to a number of synthetic community attributes such as: *presence*— the number of individual stands of a community-type in which a species occurred; *constance*—presence based on equal-sized samples of individual stands; *fidelity*—the degree to which a species was restricted to one or a few community-types. The substance of much work of classical phytosociology was that these properties, deriving from comparison of large numbers of individual stands from many plant associations, would demonstrate the aggregation of species into communities.

The literature of phytosociology, with a tradition of nearly two centuries if von Humboldt is credited as its initiator, is extremely diverse and extensive. Many of its early nineteenth-century classics are verbal and qualitative descriptions of vegetation formations of specific geographic and habitat areas. In the twentieth century, extended studies of methods of description and analysis of vegetation were done in an effort to increase accuracy and to develop more effective quantitative methods that would be precise and provide a standard basis for analysis of vegetation. Increasingly detailed studies of vegetation continued as these methods were incorporated into phytosociology.

Both the pioneers of phytosociology and its recent students extended their efforts in description and analysis of plant communities into considerations of their relations to the environment and the evolution and distribution of morphological and autecological properties of component species that could be interpreted in terms of their physiological function and thus serve to elucidate the causal relationships between environment and vegetation. The early emphasis on water, temperature, and soil relations, seen in the work of Warming, Schimper, and others, continues in much more detailed and sophisticated environmental analysis (physiological ecology) and experimental studies in recent work on phytosociology. Phytosociology is sometimes associated solely with the systematic description and classification of plant communities. There is an extended literature on classification, nomenclature, and the development of a hierarchy of systematic units for this purpose. However, the proper study of phytosociology, as J. Braun-Blanquet stated, ". . . includes all phenomena which touch upon the life of plants in social units." The investigation of composition is simply the beginning of phytosociology.

The many facets of phytosociology are seen with varying emphasis in a number of general works which will serve to introduce the reader to the phytosociological literature. Classical sources for all aspects of phytosociology are the three editions of J. Braun-Blanquet's *Pflanzenoziologie,* only the first (1928) translated into English to date (1932). Other general references are Dansereau (1957), Daubenmire (1968), Ellenberg (1957), Mueller-Dombois and Ellenberg (1974), Oosting (1956), Tansley (1946), Weaver and Clements (1938), and Sukachev and Dylis (1964). The increasing emphasis on quantitative methods and statistics in post-World War II phytosociology (McIntosh 1972) is seen in Cain and Oliviera-Castro (1959), Gounot (1969), Greig-Smith (1964), Kershaw (1964), Orloci (1975), Shimwell (1971), and Vasilevich (1969). Classification of vegetation is emphasized in Aleksandrova (1969) and Tüxen (1968). A most useful general summary of the history, methods, and current state of phytosociology is found in Whittaker (1973a).

REFERENCES

Aleksandrova, V. D. 1969. *Classification of Vegetation: Principles of Classification and Classification Systems of Various Phytocoenological Schools* (in Russian). Leningrad: Nauka, 275 pp.

Aleksandrova, V. D. 1973. Russian Approaches to Classification of Vegetation. In *Ordination and Classification of Vegetation,* ed. R. H. Whittaker, pp. 493–527. The Hague, The Netherlands: W. Junk Publ.

Barkman, J. J. 1973. Synusial Approaches to Classification. In *Ordination and Classification of Vegetation,* ed. R. H. Whittaker, pp. 437–491. The Hague, The Netherlands: W. Junk Publ.

Barkman, J., J. Moravec, and S. Rauschert. 1976. Code of Phytosociological Nomenclature. *Vegetatio* **32**:131–185.

Beard, J. S. 1973. The Physiognomic Approach. In *Ordination and Classification of Vegetation,* ed. R. H. Whittaker, pp. 355–433. The Hague, The Netherlands: W. Junk Publ.

Becking, R. 1957. The Zurich-Montpellier School of Phytosociology. *Bot. Rev.* **23**:411–488.

Braun-Blanquet, J. 1928. *Pflanzenziologie: Grundzuge der Vegetationskunde.* Berlin: Springer Verlag. Second edition 1951, Vienna: Springer Verlag, 631 pp.; third edition 1964, Vienna: Springer Verlag. First edition translated by G. D. Fuller and H. S. Conard, *Plant Sociology,* New York: McGraw-Hill, 1932, 438 pp.

Braun-Blanquet, J. 1968. L'ecole phytosociologique Zuricho-Montpellieraine et la S.I.G.M.A. *Vegetatio* **16**:1–78.

Cain, S. A., and G. M. de Oliveira-Castro. 1959. Manual of Vegetation Analysis. New York: Harper, 325 pp.

Cajander, A. K. 1926. The Theory of Forest Types. *Acta for. Fenn.* **29**:1–108.

Clements, F. E. 1936. Nature and Structure of the Climax. *J. Ecol.* **24**:252–284.

Curtis, J. T. 1959. *The Vegetation of Wisconsin: An Ordination of Plant Communities.* Madison: University of Wisconsin Press, 657 pp.

Dansereau, P. 1957. *Biogeography: An Ecological Perspective.* New York: Ronald Press, 394 pp.

Daubenmire, R. F. 1968. *Plant Communities: A Textbook of Plant Synecology.* New York: Harper and Row, 300 pp.

Drude, O. 1890. *Handbuch der pflanzengeographie.* Stuttgart: Engelhorn, 582 pp.

Du Rietz, G. E. 1921. *Zur methodologischen Grundlage der modernen Pflanzensoziologie.* Vienna: Holzhausen, 267 pp.

Egerton, F. E. 1976. Ecological Studies and Observations Before 1900. In *Issues and Ideas in America*, ed. B. J. Taylor and T. J. White, pp. 311–351. Norman: University of Oklahoma Press.

Egler, F. E. 1942. Vegetation as an Object of Study. *Philosophy of Science* **9**:245–260.

Egler, F. E. 1954. Philosophical and Practical Considerations of the Braun-Blanquet System of Phytosociology. *Castanea* **19**:54–60.

Ellenberg, H. 1957. *Aufgaben und Methoden der Vegetationskunde.* Stuttgart: Eugen Ulmer, 136 pp.

Flauhault, C., and C. Schroter. 1910. Rapport sur la nomenclature phyto-geographie. *Actes III Int. Bot. Congr. Bruxelles* **1**:131–164.

Frey, T. E. A. 1973. The Finnish School and Forest Site-Types. In *Ordination and Classification of Vegetation*, ed. R. H. Whittaker, pp. 403–434. The Hague, The Netherlands: W. Junk Publ.

Gleason, H. A. 1926. Plant Associations and Their Classification: A Reply to Dr. Nichols. In *Proc. International Congress of Plant Sciences*, pp. 643–644, Ithaca, New York.

Gleason, H. A. 1936. Is a Synusia an Association? *Ecology* **17**:444–451.

Godwin, H. 1977. Sir Arthur Tansley: The Man and the Subject, The Tansley Lecture, 1976. *J. Ecol.* **65**:1–26.

Goodland, R. J. 1975. The Tropical Origin of Ecology: Eugen Warming's Jubilee. *Oikos* **25**:240–245.

Gounot, M. 1969. Methodes d'etude quantitative de la vegetation. Paris: Masson et Cie, 314 pp.

Greig-Smith, P. 1964. *Quantitative Plant Ecology*, 2nd ed., London: Butterworths, 256 pp.

Grisebach, A. 1872. *Die Vegetation der Erde nach ihrer klimatischen Anordnung. Ein Abrise der verghleichenden Geographie der Pflanzen.* Leipzig: Englemann, 2 volumes.

Guinochet, M. 1968. Continu ou discontinu en phytosociologie. *Bot. Rev.* **34**:273–290.

Heimberger, C. C. 1934. Forest-type Studies in the Adirondack Region. *Mem. Cornell Univ. (N.Y.) Agric. Exp. Stn.* **165**:1–22.

Humboldt, A. von. 1805. *Essai sur la geographie des plantes.* Paris: Levrantet, Schovell et Cie, 155 pp.

Kerner, A. 1863. *Das Pflanzenleben der Donaulander.* Innsbruck: University Verlag, 452 pp.

Kershaw, K. A. 1964. Quantitative and Dynamic Ecology. London: Edward Arnold Publishing Co., 183 pp.

Küchler, A. W. 1967. *Vegetation Mapping.* New York: Ronald Press, 472 pp.

Küchler, A. W., and J. McCormick. 1965. *International Bibliography of Vegetation Maps, Vol. 1: North America.* University of Kansas Library Series, 453 pp. Subsequent volumes by A. W. Küchler: *Vol. II: Europe*, 1967; *Vol. III: USSR, Asia and Australia*, 1968; *Vol. IV: Africa, South America and the World* (General), 1970.

Kuhn, T. S. 1970. *The Structure of Scientific Revolutions*, 2nd ed., Chicago: University of Chicago Press, 210 pp.

Lippmaa, T. 1939. The Unistratal Concept of Plant Communities. *Am. Midl. Nat.* **21**:111–145.

Maarel, E. van der. 1975. The Braun-Blanquet Approach in Perspective. *Vegetatio* **30**:213–219.

Maycock, P. F. 1967. Josef Paczoski: Founder of the Science of Phytosociology. *Ecology* **48**:1031–1034.

McIntosh, R. P. 1967. The Continuum Concept of Vegetation. *Bot. Rev.* **33**: 130–187.

McIntosh, R. P. 1974. Plant Ecology 1947–1972. *Ann. Mo. Bot. Gard.* **61**: 132–165.

McIntosh, R. P. 1975. H. A. Gleason—"Individualistic Ecologist," 1882–1975. *Bull. Torrey Bot. Club* **102**:252–273.

McIntosh, R. P. 1976. Ecology since 1900. In *Issues and Ideas in America*, eds. B. J. Taylor and T. J. White, pp. 353–372. Norman, Oklahoma: University of Oklahoma Press.

Moore, J. J. 1962. The Braun-Blanquet System: A Reassessment. *J. Ecol.* **50**: 761–769.

Mueller-Dombois, D., and H. Ellenberg. 1974. *Aims and Methods of Vegetation Ecology.* New York: John Wiley and Sons, 547 pp.

Nichols, G. 1926. Plant Associations and Their Classification. In *Proc. International Congress of Plant Sciences*, pp. 629–641, Ithaca, New York.

Oosting, H. J. 1956. *The Study of Plant Communities: An Introduction to Plant Ecology*, 2nd ed. San Francisco: W. H. Freeman and Company, 440 pp.

Orloci, L. 1975. *Multivariate Analysis in Vegetation Research.* The Hague, The Netherlands: W. Junk Publ., 276 pp.

Ponyatovskaya, V. M. 1961. On Two Trends in Phytocenology. *Vegetatio* **10**: 373–385.

Poore, M. E. D. 1955a. The Use of Phytosociological Methods in Ecological Investigations. I. The Braun-Blanquet System. *J. Ecol.* **43**:226–244.

Poore, M. E. D. 1955b. The Use of Phytosociological Methods in Ecological Investigations. II. Practical Issues Involved in an Attempt to Apply the Braun-Blanquet System. *J. Ecol.* **43**:245–269.

Raunkaier, C. 1934. *The Life Forms of Plants and Statistical Plant Geography.* Oxford: Clarendon Press, 632 pp.

Schimper, A. F. W. 1898. *Pflanzeogeographie auf physiologischer Grundlage.* Jena: Fisher, 870 pp.

Schimper, A. F. W. 1903. *Plant Geography upon a Physiological Basis*, trans. W. A. Fischer. Oxford: Clarendon Press, 839 pp.

Schouw, J. F. 1823. *Grundzuge einer allgeminen Pflanzenogeographie.* Berlin: Reimer, 524 pp.

Shimwell, D. W. 1971. *Description and Classification of Vegetation.* London: Sedgewick and Jackson, 322 pp.

Sobolev, L. N., and V. D. Utekhin. 1973. Russian (Ramensky) Approaches to Community Systematization. In *Ordination and Classification of Vegetation*, ed. R. H. Whittaker, pp. 75–104. The Hague, The Netherlands: W. Junk Publ.

Sukachev, V., and N. Dylis. 1964. *Fundamentals of Forest Biogeocoenology*, trans. J. M. Maclennon, 1968. Edinburgh: Oliver and Boyd, 672 pp.

Tansley, A. G. 1920. The Classification of Vegetation and the Concept of Development. *J. Ecol.* **8**:118–149.

Tansley, A. G. 1935. The Use and Abuse of Vegetational Concepts and Terms. *Ecology* **16**:284–307.

Tansley, A. G. 1946. *Introduction to Plant Ecology.* London: Unwin Bros., 260 pp.

Trass, H. 1976. *Vegetation Science: History and Contemporary Trends of Development* (in Russian). Academy of Sciences USSR, 248 pp.

Trass, H., and N. Malmer. 1973. Northern European Approaches to Classification. In *Ordination and Classification of Vegetation,* ed. R. H. Whittaker, pp. 531–574. The Hague, The Netherlands: W. Junk Publ.

Tüxen, R. (ed.). 1968. *Pflanzensoziologische Systematik.* Ber. Symp. Int. Ver. Vegetationskunde. The Hague, The Netherlands: Stolzensau/Weser, Junk, 346 pp.

Vasilevich, V. I. 1969. *Statistical Methods in Geobotany* (in Russian). Leningrad: Nauka, 231 pp.

Warming, E. 1895. *Plantesamfund. Grundtrak af den okologiska Plantegeograf.* Kjobenhaven: Philipsen.

Warming, E. 1909. *Oecology of Plants: An Introduction to the Study of Plant Communities.* London: Oxford University Press, 422 pp.

Weadock, V., and P. Dansereau. 1960. The SIGMA Papers: A Short History and a Bibliographic Overview. *Sarracenia* **3**:1–47.

Weaver, J. E., and F. E. Clements. 1938. *Plant Ecology.* New York: McGraw-Hill, 601 pp.

Werger, M. J. A. 1974. On Concepts and Techniques Applied in the Zurich-Montpellier Method of Vegetation Survey. *Bothalia* **11**:309–323.

Westhoff, V., and E. van der Maarel. 1973. The Braun-Blanquet Approach. In *Ordination and Classification of Vegetation,* ed. R. H. Whittaker, pp. 614–726. The Hague, The Netherlands: W. Junk Publ.

Whittaker, R. H. 1962. Classification of Natural Communities. *Bot. Rev.* **28**:1–239.

Whittaker, R. H. (ed.). 1973a. *Ordination and Classification of Vegetation.* The Hague, The Netherlands: W. Junk Publ., 737 pp.

Whittaker, R. H. 1973b. Approaches to Classifying Vegetation. In *Ordination and Classification of Vegetation,* ed. R. H. Whittaker, pp. 323–354. The Hague, The Netherlands: W. Junk Publ.

Part I

HISTORY, CONCEPTS, AND TERMINOLOGY

Editor's Comments
on Papers 1 Through 6

The British phytosociologist Moss (Paper 1) commented in 1910 that ecological plant geography suffered and still suffers from a lack of uniformity in use of its principal terms. This difficulty persisted through much of the history of phytosociology, and the lack of consensus on terminology and concepts had led a pioneer American plant ecologist, H. C. Cowles, to describe plant ecology in 1904 as chaos (McIntosh 1976). Nineteenth-century plant geography developed the fundamental concepts of growth forms of plants and, particularly, the ideas of aggregations of plants into communities or formations, which were recognized by the growth form of the dominant species and associations, which were determined by their species composition. These concepts provided the background for the development of morphological and physiological plant geography that metamorphosed into "oecological" plant geography in the late nineteenth century under the influence of European plant geographers (Pound 1896, Du Rietz 1921, Tansley 1924, Whittaker 1962, Egerton 1976).

Linked to conceptual problems of the new science were terminological problems, which were recognized by both the International Congress of Geographers in 1899 and a succession of International Botanical Congresses in 1900, 1905, and 1910. It is hardly surprising that such difficulties should arise; as Moss noted, they were common especially in the biological sciences. At the turn of the century the very idea of "oecology" as a science was in dispute as well as what name to apply to it (McIntosh 1976); the history of and relations among ecology, plant geography, and geobotany were still being considered by Rubel in 1927, and complete consensus on concept and terminology is not evident even today, which is, perhaps, surprising.

Much of the early decades of the twentieth century was occupied by intensive studies of vegetation and by the efforts of pioneer phytosociologists to develop an effective conceptual framework. The papers in Part I are representative, if not comprehensive, of this period. Moss traces the history of the concepts of plant formation and association and early efforts to establish a consistent and international hierarchical classification and terminology of plant communities. Schouw (1823) had invented a method of naming plant associations by appending the suffix -*etum* to the stem of the generic name of the dominant species. The concept of the abstract plant association as a reasonably natural and consistent unit, or community-type, comprised of association-individuals, or actual stands, was one of the fundamental concepts of traditional phytosociologists although they disagreed on the details. Given a widespread consensus on the community-type concept as a basis for phytosociology, it was expected that these should be subject to hierachical classification and nomenclature. Various classification schemes were produced (Flauhault 1901), most notably that of the Braun-Blanquet school, which retained the ending -*etum* for the association. The term phytosociology is sometimes limited to the identification and classification of plant communities by floristic methods, but this seems an unwarranted restriction of the word.

Of the many schools of phytosociology founded in this era, the most influential and persistent is represented in an early statement of the major features of the Braun-Blanquet school (Braun and Furrer [Paper 2]). The concept of the association as the basic vegetational unit, based on character species, which can be grouped

19

into higher units, was elaborated by Braun-Blanquet (1921) and established the core ideas of this most productive and long-lived patriarch of phytosociology, who is still active. In the 1913 paper, the concept of characteristic species was introduced. The methods of establishing the characteristic species have become more elaborate with time (Mueller-Dombois and Ellenberg 1974) and, even more recently, have been automated, as discussed in Parts III and IV of this volume. The characteristic species were deemed better indicators of the ecological conditions than the more ubiquitous dominants, although a dominant could be a character species. In spite of many criticisms, the Braun-Blanquet system has flourished and remains today as the most widely established system of phytosociology.

While Braun and Furrer warned against interpreting the successional possibilities of the plant community, noting that local conditions might impede its development, the role of succession was deemed of paramount importance by Frederick E. Clements in North America. According to Clements emphatically holistic viewpoint, the community developed from a pioneer stage, in a progressive sequence, to a relatively stable *climax* community under the control of the, also relatively stable, regional climate. This "supra-organism," like an individual organism, he said, is born, grows, matures, reproduces, and dies. Although not alone in holding such a view of the community, Clements was its greatest expositor (Clements 1936). Cain's 1939 paper (Paper 6) was one of many (e.g., Phillips 1934–35) which attempted to clarify the often obscure complexities of the climax community concept of Clements and the elaborate system of terminology associated with it.

Nowhere are the conceptual and terminological problems of phytosociology better (or worse) represented than in the word "association" and the different concepts of community incorporated within it. The only obvious consensus among the many diverse ideas of association that appear in the phytosociological literature is that is should be based on compositional criteria. The greatest distinction was between the association of Braun-Blanquet and that of Clements. The Braun-Blanquet "association" was traditionally derived by an inductive comparison of tables of samples (relevés, aufnahmen) by identifying groups of species that differentiated associations (cf. Mueller-Dombois and Ellenberg 1974, Chaps. 5 and 9). The differential species were not necessarily dominants in the community and the successional status of the community was not a consideration. Braun-Blanquet recognized several possible relatively stable communities in an area, each controlled by local conditions (Braun-Blanquet 1921, Du Rietz 1921). The Clementsian "association" was derived by a largely deductive method predicated on the concept of a climax association recognized on the basis of a few dominant species. It was, by definition, climax or stable and included all developmental or successional communities (Cain, Paper 6). Clements devised an elaborate logical system of relating all of the vegetation in an area to the single putative climax *(monoclimax)* which was given the class name association, the *-ion* ending designating the climax. According to Clement's classification, the physiognomic formation was divided into two or more floristically defined associations. Braun-Blanquet's association is the fourth below the formation in a series of floristically defined classes. Between these two concepts of association there was little possibility of rapprochment, and the European and North American traditions of phytosociology remained substantially separate (Egler 1951) in spite of efforts such as that of Du Rietz (Paper 5) to bring order out of the chaos of these and other concepts of community and terminological systems (Cain 1932, Tansley 1920, 1935, Braun 1956).

A striking counterdevelopment to the concepts of the unity of the plant community which pervaded early twentieth-century phytosociology, in spite of differences in detail, was the concept of the individualistic plant community. Based on the widespread intuition that each species had individualistic ecological properties, the individualistic community concept was asserted nearly simultaneously by different individuals in Russia, France, Italy, and the United States (Ponyatovskaya 1961, McIntosh 1975). Its substance was that the plant community was a coincidence of the local environment, available plant species and their individualistic properties, and, therefore, that vegetational change was gradual and continuous making any system of classification an arbitrary exercise. These ideas were, in the United States at least, expressed specifically in opposition to the community-unit concept. H. A. Gleason expressed the individualistic concept in three successive papers (McIntosh 1975) and his individualistic dissent is represented by the second of these (Paper 4). The best known European proponent of the individualistic concept, the Russian phytosociologist Ramensky, is represented by an abstract (Paper 3). Gleason's efforts to influence the course of phytosociology were unsuccessful and he abandoned them for plant taxonomy. Ramensky persisted and pioneered quantitative studies of a type that later appeared in North America and elsewhere in Europe although he does not appear to have been in the mainstream of Russian phytosociology (Ponyatovskaya 1962, Sobolev and Utekhin 1973). Similar techniques and ideas were discovered in other early and unheralded work (Kulczynski 1928). These ideas had little impact on the mainstream of phytosociological thought being substantially ignored until they were given new support in the late 1940s and the 1950s, a development that is considered in Part III of this volume.

REFERENCES

Braun-Blanquet, J. 1921. Prinzipien einer Systematik der Pflanzengesellschaften auf florististicher Grundlage. *Jahrb. St. Gallen Nature. Ges.* **57**:305–351.

Braun, E. L. 1956. The Development of Association and Climax Concepts: Their Use in Interpretation of the Deciduous Forest. *Am. J. Bot.* **43**: 906–911.

Cain, S. A. 1932. Concerning Certain Phytosociological Concepts. *Ecol. Monog.* **2**:476–508.

Clements, F. E. 1936. Nature and Structure of the Climax. *J. Ecol.* **24**:252–284.

Du Rietz, G. E. 1921. *Zur methodologischen Grundlage der modernen Pflanzensoziologie.* Vienna: Holzhansen, 267 pp.

Egerton, F. E. 1976. Ecological Studies and Observations Before 1900. In *Issues and Ideas in America*, eds. B. J. Taylor and T. J. White, pp. 311–351. Norman: University of Oklahoma Press.

Egler, F. E. 1951. A Commentary on American Plant Ecology Based on the Textbooks of 1947–1949. *Ecology* **32**:673–695.

Flauhault, C. 1901. A Project for Phytogeographic Nomenclature. *Bull. Torrey Bot. Club* **28**:391–409.

Kulczynski, S. T. 1928. Die Pflanzenassociation der Pieninen. *Bull. Int. Acad. Pol. Sci. Lett. Cl. Sci. Math.* Ser. B. Supple. **2**:57–203.

McIntosh, R. P. 1975. H. A. Gleason—"Individualistic Ecologist," 1882–1975. *Bull. Torrey Bot. Club* **102**:252–273.

McIntosh, R. P. 1976. Ecology since 1900. In *Issues and Ideas in America*, eds. B. J. Taylor and T. J. White, pp. 353–372. Norman: University of Oklahoma Press.

Mueller-Dombois, D., and H. Ellenberg. 1974. *Aims and Methods of Vegetation Ecology.* New York: John Wiley and Sons, 547 pp.

Phillips, J. F. V. 1934–35. Succession, Development, the Climax and the Complex Organism: An Analysis of Concepts. Parts I–III. *J. Ecol.* **22**: 554–571, **23**:210–246, **23**:488–508.

Ponyatovskaya, V. M. 1961. On Two Trends in Phytocenology. *Vegetatio* **10**:373–385.

Pound, R. 1896. The Plant-Geography of Germany. *Am. Nat.* **30**:465–468.

Rübel, E. 1927. Ecology, Plant Geography and Geobotany: Their History and Aim. *Bot. Gaz.* **84**:428–439.

Schouw, J. F. 1823. *Grundzuge einer allgeminen Pflanzenogeographie.* Berlin: Reimer, 523 pp.

Sobolev, L. N., and V. D. Utekhin. 1973. Russian (Ramensky) Approaches to Community Systematization. In *Ordination and Classification of Vegetation*, ed. R. H. Whittaker, pp. 75–104. The Hague, The Netherlands: W. Junk Publ.

Tansley, A. G. 1920. The Classification of Vegetation and the Concept of Development. *J. Ecol.* **8**:118–149.

Tansley, A. G. 1924. Eug. Warming in Memoriam. *Botan. Tidssk.* **39**:54–56.

Tansley, A. G. 1935. The Use and Abuse of Vegetational Concepts and Terms. *Ecology* **16**:284–307.

Whittaker, R. H. 1962. Classification of Natural Communities. *Bot. Rev.* **28**:1–239.

1

Reprinted from *New Phytologist* 9:18–22, 26–41, 44–53 (1910)

THE FUNDAMENTAL UNITS OF VEGETATION:

HISTORICAL DEVELOPMENT OF THE CONCEPTS OF THE PLANT ASSOCIATION AND THE PLANT FORMATION.

BY C. E. MOSS, D.Sc.

INTRODUCTION.

THE subject of ecological plant geography has suffered and still suffers very considerably from a lack of uniformity in the use of its principal terms. This defect need not be a source of great wonder or surprise, as it obtains to a greater or lesser degree in all branches of knowledge, and more especially in the biological sciences. Many biological concepts possess an inevitable vagueness; and this is a reflection of the lack of sharply defined images in nature herself. So marked indeed is this the case that one suspects as artificial any classification of biological units which is capable of a rigid and determinate application. Taxonomy, the oldest of the branches of biology, has despaired of giving logically perfect definitions of the genus, species, and variety. It cannot be expected, therefore, that ecology, one of the youngest branches, shall do better with regard to the formation, association, and society; for the units of the plant-geographer agree with those of the systematist in being entities which are not invariably sharply marked off from each other in nature. Much, however, of the prevailing confusion in phytogeographical nomenclature lies less deeply seated than this; and to some extent, therefore, the malady is one of which hopes may be entertained of an ultimate, if not of a speedy cure.

The need of greater harmony in the use of terms in ecological plant geography was recognized at the International Congress of Geographers held at Berlin in 1899, and by the International Congresses of Botanists held at Paris in 1900 and at Vienna in 1905; and the matter is also to be considered at the forthcoming International Congress of Botanists to be held at Brussels in May, 1910. Warburg (1900), Flahault (1900-1901), Clements (1902), Olsson-Seffer (1905), and Gradmann (1909) have published papers dealing specially with matters of phytogeographical nomenclature; but their differing schemes and recommendations have not brought about the desired uniformity.

Warburg's suggestions were of a tentative and general nature, and referred to the nomenclature of groups of formations rather than to formations themselves. Flahault's suggestions were twofold, relating, on the one hand, to " geographical and topographical units," such as regions and zones, and, on the other hand, to vegetational or " biological units," such as associations and formations. Clements' suggestions received a severe handicap in meeting with the immediate disapproval of Engler. Whilst it must be confessed that Clements' system was perhaps in some respects too elaborate to be workable, and that parts of it were undesirable on general grounds, it is to be feared that the prompt displeasure expressed by such an eminent plant geographer as Engler had the unfortunate and doubtless undesired effect of preventing certain other portions of Clements' scheme from receiving a fair and unbiassed consideration.

Recently, Warming (1909) has issued a book in English on the *Œcology of Plants*; and Professors Flahault and Schröter, in a leaflet circulated in connection with the forthcoming Congress at Brussels, have stated their opinion that this book solves nearly all the difficulties connected with phytogeographical nomenclature. It seems difficult, however, to support this opinion in the matter of Warming's use of the term formation; although, in certain other particulars, especially with regard to the use of the term association and with regard to the grouping of allied associations into formations, the opinion can be fully justified.

Perhaps too much stress has been laid by critics on the particular *terms* employed by plant geographers, and too little on the *concepts* which the terms were intended to convey; and the object of the present communication is to endeavour to trace the historical development of the harmonious elements now existing in the concepts relating to plant associations and formations, and to give to these general ideas those terms which historical and present day usage would seem to indicate as right and proper.

It must be remembered that, in the development of every branch of knowledge, a time occasionally arrives when it becomes necessary to expand a concept which for long has been attached to a particular term. A few years ago, for example, it was held to be desirable to extend the signification of the term seed so that it should be applicable to integumented megasporangia destitute of embryos. So long as the expansion of a concept is due to a real advance in knowledge, and so long as the broader conception is in harmony with the more fundamental aspects of the previous use of the term, the new use is not merely legitimate and desirable but absolutely essential, unless all branches of learning are to be overburdened with new terms whenever concepts undergo a necessary, logical, and developmental expansion.

The Plant Association.

The concept of the plant association is one of the oldest in plant geography. Without referring to the literature of the ancients, in which doubtless the germs of these concepts may be found, it is sufficient to say that, with regard to the plant association, the concept and the term were used by the illustrious Humboldt more than a century ago; whilst the term formation dates from the time of Grisebach (1838). Humboldt showed that he had a well-defined concept of the plant association even when he did not use the term, as the following passages illustrate (Humboldt, ?1806 ; Sabine's English tr., 1849 : 264) :—" In the temperate zone, and especially in Europe and northern Asia, forests may be named from particular genera or species, which, growing together as social plants, form separate and distinct woods. In the northern forests of oaks, pines, and birches, and in the eastern forests of limes or linden-trees, usually only one species of Amentaceæ, Coniferæ, or Tiliaceæ prevails or is predominant." After thus referring to *pure associations*, Humboldt continues with reference to *mixed associations:*—"Tropical forests, on the other hand, decked with thousands of flowers, are strangers to such uniformity of association ..... A countless number of families are here crowded together; and, even in small spaces, individuals of the same species are rarely associated." "The existence of a heath," according to another passage from Humboldt (1819: 295), "always supposes an association of plants of the family of *ericæ.*" It is perhaps scarcely to be expected that the originators of great conceptions should always have kept them within consistent and logical bounds, especially when these conceptions were closely related to wider but, at the time, unformulated ideas; and it is not surprising, therefore, that Humboldt should occasionally comprise groups of associations within his term association. Thus it was, perhaps, when he wrote as follows (1807: 17):—" Les bruyéres, cette association de l'*erica* [*Calluna*] *vulgaris*, de l'*erica tetralix*, des *lichen icmadophila* et *hæmatomma* se répandent depuis l'extrémité la plus septentrionale du Jutland, par le Holstein et le Luneberg, jusqu'au 52ᵉ degré de latitude." In the same essay, Humboldt (1807: 15), after mentioning such plants as *Vaccinium Myrtillus*, wrote:—" Ces plantes associées sont plus communes dans les zones tempérées que sous les tropiques, dont la végétation moins uniforme est par cela même plus pittoresque." The modern concept of the plant association as well as the term may thus be traced to the works of Humboldt; and this concept and this term have, after many vicissitudes, come into very general use in the writings of modern ecologists and plant geographers.

Schouw (1822; German tr., 1823: 165) briefly mentioned a few plant associations, and invented a method of denoting them. To the stem of the generic name of the dominant plant of a pure association, he applied the suffix *-etum*. Thus a plant association whose chief constituent is the beech (*Fagus sylvatica*) he termed a Fagetum. Similarly Schouw referred to Palmeta, Pineta, Querceta, etc. Meyen (1836; English tr., 1846: 78, 80) spoke of Ericeta, Fageta, Oliveta (*sic*), Palmeta, and Pineta. This matter is important in more than its nomenclatorial aspect, a matter which is discussed later on in this paper, for it proves that the concept of the association, had become firmly established in the minds of plant geographers of different nationalities before Grisebach had enunciated his famous definition of the formation.

This definition was put forward in 1838 ; and, although

Grisebach's formation was wider than Humboldt's association, yet, from this date onwards, the term "formation" began to enter into direct competition with the term association. It is most interesting, however, to observe that the modern tendency is to retain the usage of the two great pioneers in so far as the term formation now denotes a vegetation unit more comprehensive than the association.

[*Editor's Note:* Material has been omitted at this point.]

Moss (1907) was the first among British ecologists to speak both of formations and associations. Thus, the sand dune formation of Somerset may be subdivided into progressive associations of strand plants (*Atriplex* spp., etc.), of sea-couch grass (*Agropyrum junceum*), of marram grass (*Ammophila arenaria*), of dune sward-forming plants, of dune pasture plants, and of dune marsh plants. Similarly, in the acidic peaty moorland formation of the Pennines, stable associations of *Calluna vulgaris* and of *Eriophorum vaginatum* may be distinguished, and also retrogressive associations of such species as *Vaccinium Myrtillus*, *Empetrum nigrum*, and *Rubus Chamaemorus* in varying proportions.

Warming (1909 : 139, *et seq.*) followed what has become the general practice of subdividing the formation into associations. The latter he defined (p. 145) as follows :—"An association is a community of definite floristic composition within a formation." This point of view may now be regarded as having become all but universal ; but it would be better to refer to the minor differences of habitat as well as to the differences of floristic composition in all definitions of the association. By so doing, the main object of the study of vegetation would be emphasized ; and a tendency—by no means an imaginary one—to regard plant geography as a branch of floristic botany would be checked. It is perhaps overbold for a science that is still only in its infancy to state that it depends upon " mere accident " whether one or other of these associations prevails at a given spot (Warming 1909: 140), especially as, in many instances, these minor differences in habitat may be actually demonstrated.

It will be seen that the concept of the plant association, and also the term, have been long established, and that they have come to be very widely accepted among ecologists and plant geographers of several nations. Although for a time the term was threatened by the term "formation," and although many other terms have been used to convey the same general idea, and the term has been used to denote other concepts, yet the original term of Humboldt, and in the general sense used by Humboldt, meets with wide acceptance at the present time. It is true that Humboldt's concept has been expanded, as he appears to have had no notion of uniting allied associations into a group of wider significance; but such a classification of associations was ultimately inevitable. It appears likely that the term association, in the general sense here indicated, will be submitted for adoption at the International Congress of Botanists shortly to be held at Brussels; and one hopes that its adoption will be effected. A firm if small step will then have been taken in the direction of the unification of the use of phytogeographical terms; although the precise degree of relationship between the formation and its associations may not be fully realized for some time to come.

THE PLANT FORMATION.

The concept of the plant formation is, in a general way, much older than the term; for whilst the latter only dates from Grisebach (1838: 160), the former may be found in the writings of several of the pre-Linnæan botanists. The foundation of the idea of the formation, as understood by many recent writers, is the habitat; and from this point of view therefore it is fair to trace, as was done by Clements (1904: 12), the modern concept of the formation back to the old idea of the general habitat, even though the early systematists failed to realize fully the vital relations of habitat and vegetation. Linnæus, in his *Philosophia Botanica* (1751: 263-270), distinguished twenty-five habitats, and mentioned their characteristic genera; whilst in the *Stationes Plantarum* (1754; 1760: 64-87) the characteristic species were enumerated. Many of these habitats and accompanying lists of species might easily pass muster as a summary of the corresponding modern plant formations; and one is therefore tempted to speculate as to what would have been the relationships of the branches of modern botany had Linneæus developed this concept on the vegetational rather than on the floristic side. As it is, whilst the general habitat of the early systematists has become the formation of the plant geographer, it has long been the species-station of the taxonomist.

Even earlier than Linnæus, Tournefort (1717, iii.: 178, *et. seq.*) had more than an inkling of ecology when he distinguished the vertical belts of vegetation in Armenia, thus proving, according to Clements (1904: 150), that "the concept of zonation is the oldest in phytogeography.'

It was shown in the preceding section of this paper that the term "formation" has frequently been used to signify an association. Clements (1905: 292) has stated that "there have been as many different opinions in regard to the application of the term formation as there are concerning the group which is to be called a species;" and Flahault (1901: 405) also wrote: "It is not to be marvelled at that several botanists, who had doubtless lost their way in this confusion, acknowledge having employed the word without thinking of its definition."

It is not surprising, therefore, that the development of the concept has proceeded along somewhat divergent lines. One such line has already been indicated: this was the use of the term by Hult (1881, etc.) and others to signify a minor though important unit of vegetation, a unit which is now regarded as a subdivision of a formation, and termed an association. A second line of development may now be traced.

An oft-quoted passage of Grisebach (1838: 160), introducing the term "formation," is thus rendered by Clements (1905: 3):—
"I would term a group of plants which bears a definite ["abge-schlossenen"] physiognomic character, such as a meadow, a forest, etc., a *phytogeographical formation*. The latter may be characterized by a single social species, by a complex of dominant species belonging to one family, or, finally, it may show an aggregate of species, which, though of various taxonomic character, have a common peculiarity; thus, the Alpine meadows consist almost exclusively of perennial herbs." It is obvious that this general statement of Grisebach's early idea of the formation does not exclude all associations, for many of the latter " bear a definite physiognomic character" and are " characterized by a single social species." Hence those writers, such as Hult, who have used the term " formation " in the narrow sense, *i.e.*, as equivalent to an association, are able to find some historical support for their procedure. On the other hand, it must be admitted that the particular units of vegetation to which Grisebach applied his term were almost invariably of greater extent than single associations.

A review by Grisebach (1849: 339, *et seq.*) of a work by Duchartre (1844) enables one to gather the meaning which the eminent plant-geographer attached to the term in his earlier writings. Duchartre himself does not appear to have used the term formation, although his divisions of the vegetation prove that he held the concept quite sharply. Grisebach in his review applies the term to Duchartre's divisions, thus speaking of the formation of the dunes, of the salt-water marshes, of siliceous soils, of calcareous soils, of argillaceous soils, etc. Such a use of the term was frequently adopted by Grisebach (cf. 1846: 73; etc.) about this time; and one sees that here Grisebach regarded the formation and the habitat as indissolubly connected. However, in his classical work *Die Vege-tation der Erde* (1872), the term is used to denote physiognomical units, such as tundra, and the forests of definite geographical regions, which most later authors agree in subdividing into smaller units. Hence, practically every author, however he may use the term, may find some support for his procedure in one or other of the writings of Grisebach. The greatness of Grisebach must not lead us to regard his masterly concept as a fetish; and the analysis of his " phytogeographical formations" into more definite vegetation units is a necessity demanded by the more rigorous methods of modern phytogeography.

Perhaps the most usual application of the term formation by present-day investigators is closely akin to the one which Grisebach held to in 1846 and 1849; and from this point of view a plant formation may be regarded as the vegetation co-existent with a definite habitat.

Several of the " associations " of Lecoq (1854, i: 291, *et seq.*) are either groups of associations or formations.

Kurz (1870), in a report on the vegetation of the Andaman Islands, classified the vegetation into divisions which, to some extent, correspond to formations, and subdivided these, in some cases, into associations, although Kurz does not use these or any other special terms. He described the vegetation (1) of the shores, (2) of the coasts, (3) of the high forests of the interior parts, (4) of the woodless spots, (5) of the cleared lands, and (6) of the sea. Some of these he subdivided into smaller units. Thus, the vege-tation of the coasts is subdivided into (a) evergreen forests, (b) deciduous forests, and (c) bamboo jungles. The evergreen forests are further subdivided into (1) " kuppalee " (*Mimusops indica*) forests and (2) mixed forests; and these are associations.

In his account of the forest and other vegetation of Pegu, Kurz (1875) divided the vegetation of that region into evergreen forests, deciduous forests, bamboo jungles, savannahs, natural pastures,

riparian vegetation, freshwater vegetation, saltwater vegetation,
etc. Several of these groups of plant-communities are further
subdivided; for example, the evergreen forests are subdivided into
littoral and other forests, and the littoral forests themselves into
mangrove jungles and tidal jungles.

Two years later, Kurz (1877) described the forests of British
Burma in a masterly way which has elicited the commendation of
Drude (1889: 27). In none of his works did Kurz use the special
terms formation and association, though he held the concepts clearly.
His forests were separated mainly by habitat. Thus he distinguished
(1) littoral forests, (2) swamp forests, (3) tropical forests, (4) hill
forests, (5) open forests, (6) dry forests, (7) mixed forests, (8) dune
forests, and adds, as an appendix, (9) bamboo jungles and savannahs,
and (10) deserted clearings. These units are in some cases subdivided. Thus, the hill forests he subdivided into (a) deciduous
forests and (b) coniferous forests; and these he further subdivided.
Thus, the coniferous forests provide pure associations (though Kurz
does not use the term) of *Pinus Kasya* and of *P. Merkusii.*

Although Kurz avoided the use of special terms in the description
of his vegetation units, it is impossible to deny that he had clear
concepts not only of formations and associations, but that he also
understood, in some measure, the relations of these vegetation units
to each other.

In the interesting account of the vegetation of northern France
by the abbé Masclef (1888), the concepts of the formation and its
associations may also be found. This account is like those by Kurz
in that the terms formation and association are not used; but
Masclef evidently found some terms necessary, or at least desirable.
He accordingly termed his vegetation units "zones" and "zones
secondaires"; and these, in a general way, correspond to formations
and associations. Thus, Masclef, described a "zone marine," a
"zone des vases," a "zone des falaises," a "zone des eaux sumâtres,"
and a "zone des dunes." Masclef had a follower in de Lamarlière
(1894), who, in a short account of the maritime vegetation of de
Quineville (Manche) stated (p. 137):—"On y retrouve les mêmes
zones établies par M. Masclef et si bien étudiées par lui, et elles son
caractérisées par les mêmes espèces communes."

To Drude (1889) modern plant geographers are greatly indebted
for a clear (and perhaps the earliest) formulation of the concepts of
the formation and the association, as well as the application of these
general ideas to the vegetation of definite districts. He expressed

(p. 29) dissatisfaction with Grisebach's physiognomical definition of the formation, and held that whilst this might serve to distinguish classes of formations, a precise idea of the formation can only be obtained by taking into account the floristic composition. As there is a definite relation between habitat and the total floristic composition, this aspect of the formation may be regarded as in accordance with much of the later work of ecological plant geographers. Drude applied his general ideas to the vegetation of the Hercynian hill-country of Central Europe, which he subdivided into twenty-seven formations. These he then subdivided into associations; for example, the halophyte formation (p. 48) he subdivided into "Salzsumpf-Bestand," "Salztrift-Bestand," and "Trockne *Salsolaceen*-Flur." In connection with the descriptions of British moorland associations, the "montane Grassmoor-Formation (Warming)" is especially interesting. This formation (p. 44) Drude subdivided into the following associations:—"Sumpfwiesen-Bestand," "Binsenmoor-Bestand," "Wollgras- und Riedmoor-Bestand ("Grünmoore ")," and "Torfsumpf-Bestand"; and the first three of these appear to be very closely related to some of the associations which occur on British upland peat moors. Although Drude at the outset of this communication stated that his object was the deepening of floristic studies, he by no means overlooked the importance of the habitat. He laid down (p. 28) the general principles that formations should be based on all conditions of existence (soil, water, atmospheric moisture) and on the associated species, and that the alternation of various principal and subordinate species within the same formation gives rise to the various associations (Bestände).

Warming (1895: 3-10) was so impressed with the confusion which had overtaken the term "formation" that he recommended its disuse. The wider meaning Warming thought might be indicated by the word vegetation; and for a plant-community of no assigned rank he used the term *Plantesamfund* (Pflanzenverein). The term "formation," however, had become so firmly rooted in ecological and phytogeographical literature that even the weight of Warming's authority was insufficient to cause its extirpation; and the trenchant treatment of the "formation" by Drude in the following year (coincident with the appearance of Knoblauch's German translation of the *Plantesamfund*) gave the term a new lease of vitality.

In his *Deutschlands Pflanzengeographie* (1896), Drude again emphasized the habitat, and stated (p. 281) that the plant-covering of a country is expressed in the arrangement according to definite habitats and coincides with the alternation of the principal plant associations (" Pflanzenbestände "). On p. 286, Drude defined the conception of the "formation" which he held at that time as follows:—" Any independent principal association, which has found its natural termination in itself, which consists of similar or of biologically connected plant-forms, and which is confined to a locally determined substratum of similar conditions of maintenance (altitude, exposure, substratum, water supply) has the value of a vegetation formation,—it being assumed that no actual change of association would occur on the site of such a principal association without external changes : the association has "reached its term " [" abgeschlossen "]" (cf. Warming, 1909: 143).

As a definition of a closed, ultimate, or *chief association* of a formation this statement of Drude's is excellent ; and there can be no doubt that Drude's arrangement of the different forests of Germany into fourteen "formations" is a great advance on the merely physiognomical view ; though, as his "formation" is essentially only a particular kind of association, it is not quite consistent with the views of those authors who regard the formation as related to the association as the genus is to the species. There may be associations which are also formations, just as certain genera only contain a single species ; but probably such formations are less frequent than monotypic genera.

Schimper (1898, 1903: 160, *et seq.*) laid the proper emphasis on the fundamental relation between the habitat and the formation. He stated (p. 161) that "every formation is in its floristic and ecological character a product of climate and soil."

Pound and Clements (1898 ; 1900 : 314) wished to preserve Drude's concept of the formation ; but their subdivision of " the river bluff *formation* " into " (1) the red oak hickory *formation*" and "(2) the burr oak elm *formation* " (pp. 324-333 [italics mine]) does not tend to clearness of thought.

Cowles (1899: 111) propounded a view which has become widely adopted when he compared the formation with the genus and the association (" society ") with the species. Cowles stated his view as follows :—" One might refer to particular sedge swamp societies near Chicago, or to the sedge swamp formation as a whole : by this application, formation becomes a term of generic value, plant-society of specific value."

Ganong (1903 : 300, *et.seq.*) regarded the salt marsh associations with which he was dealing as subdivisions of a larger unit, the wild salt marsh formation ; and his treatment therefore of the formation which he so closely studied was strictly in accord with the usage which was becoming more and more general. His formation was obviously determined by habitat, his associations by floristic composition and minor differences of the general habitat.

Harshberger (1904), in a phytogeographical sketch of a portion of Pennsylvania, subdivided formations into associations. He determined his formations and associations by the character of the land on which they occur (p. 133), *i.e.*, by habitat. However, many of his "associations" (cf. p. 145) are undoubtedly units which British authors would term only societies.

Although many earlier writers regarded the formation and the habitat as vitally connected, it is to Clements (1905) that ecologists owe the most emphatic expression of this view. Clements (1905: 292) stated unequivocally that " the connection between formation and habitat is so close that any application of the term to a division greater or smaller than the habitat is both illogical and unfortunate. As effect and cause, it is inevitable that the unit of the vegetative covering, the formation, should correspond to the unit of the earth's surface, the habitat." This view, as has been shown, was by no means new; but no one had previously stated with sufficient emphasis and in general terms what must be regarded as the foundation of the modern treatment of vegetation. The concept is much more stimulating and much more scientific than a merely physiognomical view of the formation ; and this latter view, useful enough in the early days of plant geography, has now been quite outgrown.   It is no longer possible to regard a forest as a " formation," nor even a coniferous forest. Such complex pieces of vegetation must be resolved into separate associations, and the latter rearranged into formations on a basis which shall commend itself to those who search after real affinities and under-lying causes. The rearrangement of associations into formations will not be accomplished at once, except in the case of well-marked habitats.   Where the habitats are less sharply defined, much exact and quantitative experimental work remains to be done; and here again Clements, in his *Research Methods in Ecology*, has performed useful and pioneer work. Until much work of this character has been performed, until certain habitats have been more closely investigated, ecologists and plant-geographers must be content to refer certain communities simply to their associations, rather than hastily build up formations on flimsy foundations.

35

Clements' treatment of the subdivision of the formation is also on the main line of development, for his primary division of this unit is the association, termed by him the " consocies."

The floristic view of the formation put forward by Brockmann-Jerosch (1907 : 237, *et seq.*) has recently found an exponent in Gradmann (1909) ; and their views may therefore be considered together. Whilst these authors recognize that formations should be subdivided into associations, they think that the best way of delimiting formations is by their floristic composition. " One begins to record," wrote Gradmann (1909 : 99), " as completely as possible, and in as many places as possible, the natural closed[1] plant-communities as one finds them in the country. It soon appears that certain single associations have much more in common with each other than with all the others. They can be united into groups, and each of these groups makes a formation of closely allied single associations. The formation comprises in this case all the species of the single associations belonging to it, and thus appears as an abstraction, in that it is not easily realized in a single narrowly limited locality, but only by considering the whole extent of all the associations belonging to it."

Whilst nearly, if not quite all modern ecologists and plant geographers will agree most cordially with this ultimate concept of the formation as composed of allied associations, it does appear that, in cases where formations are characterized by well-marked habitats, the method of delimiting the formation by the method outlined by Gradmann is, in some cases, unnecessarily circuitous. It is surely superfluous to have to draw up full lists of species in order to decide whether one is examining a sand-dune, or, peradventure, a peat moor ! Even Clements, than whom no one has more strongly insisted on the necessity of examining the factors of the habitat, states (1905 : 293) that " this test of a formation is superfluous in a great many cases, where the physiognomy of the contiguous areas is conclusive evidence of their difference."

Gradmann does not overlook the possibility of determining formations by habitat ; for he says (p. 96) that " since necessarily each formation must correspond with definite habitat conditions, it is quite conceivable that we can give adequate diagnoses simply with the help of habitat conditions for all the formations of a region . . . . . This would be an ecological diagnosis of which one certainly could not deny the scientific value." Thus, as Gradmann agreed, this

[1] Why should the investigation be limited to *closed* associations ?

aspect of the formation is on a very different level from the physiognomical one or even the "ecological" one on the basis of water-content alone.

Whilst the goal reached by the two methods—by the *floristic* method advocated by Gradmann and by the *habitat* method advocated in this paper—must in all cases be the same, the latter method would appear to be the more appropriate, and indeed the more fundamental, from the point of view of the study of vegetation as distinct from the study of the flora. The study of vegetation is not a department of taxonomy. Each is a separate department of science, although it is true that a knowledge of the latter is essential to a study of the former. The view advocated by Gradmann should not, in my judgment, supersede the view that the formation must be determined primarily by an investigation of the habitat; but Gradmann's method furnishes an auxiliary and confirmatory test of the formation in all cases of doubtful habitats. I fully agree that no formation can be properly described without giving a full list of its species. I would go further, and insist that such lists must include the characteristic "elementary species" of the formation, for these are frequently much more characteristic of formations than the well-defined species; but to insist that the floristic composition of the formation is more important than the habitat is to maintain that effect is more fundamental than cause.

Moss (1907) followed many previous authors in delimiting formations primarily by habitat, and then subdividing the formations into associations. Applying the test of habitat to the natural and semi-natural woods of Somerset, he distinguished habitats which are characterized respectively by woods of oak (*Quercus Robur* = *Q. pedunculata*), of ash (*Fraxinus excelsior*), and of oak and hazel (*Q. Robur* and *Corylus Avellana*). More extended observations in other parts of the country (cf. Moss, Rankin, and Tansley, 1910) prove that the "oak-hazel woods" of southern England belong to two different communities, one characteristic of non-calcareous clays and derived from woods dominated by *Q. Robur*, and another characteristic of calcareous clays and marls, and related to woods whose dominant tree is *Fraxinus excelsior*. The latter woods are characteristic of many kinds of calcareous soils in England, but do not appear to have been described by continental writers. In the Peak District of Derbyshire, woods of *Quercus sessiliflora* are sharply marked off from woods of *Fraxinus excelsior*: the former are strictly confined to siliceous hill slopes, and the latter, with equal strictness, to hill slopes on the Carboniferous Limestone. Whatever view be taken of the influence of calcium on plants, a matter on which physiologists have no definite information to offer, there can be little doubt that the woods of *Quercus sessiliflora* and those of *Fraxinus excelsior* are related to special habitats and are characterized by distinctive species, and that they should therefore be placed in separate formations. To place these communities in the same formation because they possess the same " definite physiognomical character," or because they are characterized by the same plant form, is to obtain a blurred concept of vegetation and to obscure the relations of vegetation and habitat.

Following Warming, Graebner, Cowles, Clements, and others, this writer laid stress on the succession of plant associations, especially on *the succession of associations within the same formation.* Clements (1904, 1905) had previously discussed the phenomena of succession at some length, and had brought out most clearly its importance in vegetation. It is necessary, however, to distinguish the series of associations within a whole succession, that is, the succession from one formation to another, and the succession of associations within one and the same formation; and Moss (1907 : 12) enunciated a statement of the formation from the latter point of view.

Succession of associations within a formation may be either progressive or retrogressive. In the salt marshes of the south of England, for example, a succession of *progressive associations* of Zostera, of Spartina, of Salicornia, etc., culminates in a comparatively stable association of close turf formed of *Glyceria maritima.* The latter association, however, may be attacked by the waves and ultimately destroyed; and thus *retrogressive associations* are produced. In the case of established woods, we do not know the progressive associations which culminated in the woodland associations; but we can determine retrogressive stages through scrub to grassland. Similarly, the retrogressive associations which are seen in denuding peat moors are recognizable.

A plant formation, then, comprises the progressive associations which culminate in one or more stable or *chief associations,* and the retrogressive associations which result from the decay of the chief associations, so long as these changes occur on the same habitat.

It sometimes happens, as in the case of the peat moors on the Pennine watershed, that the original habitat is wholly denuded and a new rock or rock-soil surface laid bare. In other cases, as when sand dunes are built up on the site of a pre-existing salt marsh, a habitat may be overwhelmed by a new one. In such cases, the succession passes from one formation to another formation. Again, a new habitat is created when an open sheet of water becomes choked up with silt and peat. It is absurd to dignify every stage in this succession by the term "formation"; and it is equally absurd to regard the vegetation of the open water as belonging to the same "formation" as the vegetation of the peat moor. Taking a wide view of this succession, it is obvious that the series of associations has passed from one formation to another. There is less difficulty than might be expected in delimiting the formations in such a succession, as is proved by the success of the existing Ordnance maps of such localities as the Norfolk Broads. Without any special ecological or botanical training, the makers of these maps distinguish, and distinguish successfully, between an aquatic formation (which they map as water) in which is included the reed swamps, and the fen formation (which they map as land) which begins where the reed swamps end. The succession, in such cases as these, passes from one formation to another.

The above examples of succession are given in order to show the importance of regarding the formation from the point of view of its developmental activities; but this point of view, like that of Brockmann-Jerosch and Gradmann, cannot supersede the view that the formation must be primarily determined by habitat. It cannot be successfully maintained that the view which I here put forward introduces into the concept of the formation any more subjectivity than it at present contains; for, since the stages of the conversion of a shallow "mere," "broad," or lake into a "moss," "fen," or peat moor are gradual changes, and since no one would include the earliest of such stages in the same formation as the last, there is therefore, at present, a certain degree of arbitrariness in deciding which of the various stages shall be included in one formation and which shall be included in another. It is, of course, quite obvious that no system of scientific nomenclature is possible which is not, in some of its aspects, more or less subjective.

From the point of view of succession, the "formation" of Drude (1896 : 286), variously termed by him "Formation," "Hauptformation," and "Haupbestand," must be regarded as a *chief association* of a formation. The chief associations of a district, however, do not comprise the whole of the vegetation of that

39

district: they serve to give a vivid, but to some extent an impressionist picture of such vegetation; and the complete picture requires the addition of the details provided by the progressive and retrogressive associations, or, as these may be collectively termed, the *subordinate associations*.

Every formation has at least one chief association : it may have more ; and they may be regarded (*cf.* Drude, *loc. cit.*) as equivalent to one another in their vegetational rank. They are more distinct and more fixed than progressive or retrogressive associations. They are usually, but not invariably, closed associations. They always represent the highest limit that can be attained in the particular formation in which they occur, a limit determined by the general life conditions of the formation. In desert and sub-glacial regions, the chief associations are open ; and, in such cases, it is legitimate to speak of open formations. Open progressive and retrogressive associations, however, frequently occur in formations whose chief associations are closed. On a British salt marsh, for example, open progressive associations of *Salicornia annua* frequently occur, while the chief association of *Glyceria maritima* is closed ; and similarly on a Pennine peat moor, open retrogressive associations of *Vaccinium Myrtillus* occur, whilst the chief associations of *Calluna vulgaris* and of *Eriophorum vaginatum* are closed. Unless, however, the progressive and retrogressive associations are included in the same formation as the related chief associations, an incomplete or an unbalanced picture of the vegetation results. Cowles (1910) has used the term " climatic formations " for those vegetation units which I regard as chief associations. Cowles' term is peculiarly unfortunate, as it is used in a very different sense from the same term of Schimper's.

It is maintained that the view of the formation which is here advocated is more likely to be productive of valuable results than any view which allows of the delimitation of plant formations merely by physiognomy or by plant form. It concentrates attention on the habitat and on the relation of this to the plant covering ; and it cannot be denied that the superficiality of much ecological work is due to the neglect of investigation of habitat conditions. At the same time, the view is essentially in harmony with the floristic view of the formation of Brockmann-Jerosch(1907) and of Gradmann (1909), since, in general, the floristic composition is a result of the habitat conditions ; but it differs from their view in regarding the plant covering of a habitat more from the vegetational standpoint than from that of the flora, and in thus keeping more in line with the historical development of ecological research.

Warming (1909) has recently adopted the term formation; but he has given the concept an unfortunate bias. However, in sub-dividing his formation into associations, he has taken up a view which has forced itself on the minds of nearly all close students of of vegetation. Warming (p. 131, *et cet.*) compares the formation with the taxonomic genus and the association with the taxonomic species. It follows, as many others had previously maintained, that a formation includes a number of allied associations.

Cowles (1909: 150) in reviewing Warming's (1909) book has again emphasized his view that the " conception of a formation as an ecological genus and an association as an ecological species is now becoming generally accepted in principle ; but this concrete state-ment by the father of modern ecology should make its acceptance universal." If therefore the Congress at Brussels should adopt this view, a second firm step will have been taken towards unifying the concepts and terms of ecological plant geography.

Warming's view of the formation itself, however, is sufficiently at variance with historical and present-day usage to demand some examination of his treatment of this unit of vegetation. Confusion is apparent even in Warming's summary statement of the formation. The latter, he stated (1909: 140), may " be defined as a community of species all belonging to definite growth-forms, which have become associated together by definite (edaphic or climatic) characters of the habitat to which they are adapted." Thus, instead of a single *fundamentum divisionis*, Warming puts forward two tests, namely, definite plant forms (" growth-forms ") and definite characters of the habitat, of the formation. It is not clear, either from his definition or from his general treatment of formations, what Warming precisely means by the term " definite growth-forms." Does he mean that a plant formation is characterized by a single plant form, by a single dominant plant form, or by a single set of different but biologically connected plant forms ? In any case, the definition is defective, as plant form is not necessarily related to habitat; and therefore the two tests put for-ward in the one definition will frequently yield contradictory results. It would appear that Warming himself felt this difficulty; for, in spite of his definition, he also stated (p. 142) that " it is nature of locality that must be represented by formations." Yet Warming

insisted (p. 232) that a salt marsh characterized by suffruticose Salicornias " must be set apart from " salt marshes
characterized by herbaceous Salicornias " as a separate formation," merely because the plant form in the two cases is
different. Such paradoxes occur throughout the whole of
Warming's book; and indeed this Janus-like " formation " is
inevitable if plant form be allowed to enter into competition with
habitat in the determination of formation. Warming's view might
find some justification if definite plant forms were invariably
related to definite habitats; but it is quite certain that this is not
the case. For example, on salt marshes in the south of England, it
is no unusual thing to find associations characterized (a) by herbaceous species of Salicornia, (b) by suffruticose species (*S. radicans*
and *S. lignosa*), and (c) by a mixture of these. To place these
associations in separate " formations," however, simply because of
the different nature of the plant forms, is to reduce the study of
formations to an absurdity.

This criticism is not intended to convey the meaning that the
study of plant form is not desirable or even essential. On the
contrary, such a study in the hands of ecologists and plant geographers
has led to most useful results. It must, however, be emphasized in
Warming's own words (1909: 5, 6) that " we are yet far distant "
from the " ecological interpretation of the various growth forms,"
that " it is an intricate task to arrange the growth forms of plants
in a genetic system," and that among growth forms " it is difficult
to discover guiding principles that are really natural."

It is to be regretted that Warming has made so unfortunate an
experiment at a time when the delimitation of the formation by the
habitat was becoming generally adopted; and it seems clear that
the Brussels Congress should not recommend a view of the plant
formation that has not yet been shown to be capable of a logical or
consistent application. Apart from Warming's new book, a clear
lead, however, was distinctly indicated; and, in my judgment, the
Congress would be justified on the grounds both of historical
and present-day usage, in ignoring that element of Warming's
definition of the formation which refers to plant form, and in
recommending for adoption the view that the formation should be
determined primarily by habitat. If the Congress cannot undertake this step, it seems clear that the matter should for the present
remain in abeyance.

METHODS OF DENOTING ASSOCIATIONS AND FORMATIONS.

Whilst Schouw's plan (1822, 1823: 165) of referring to an an association by the suffix *-etum* is simple and brief, it has its imperfections. In the first place, it does not distinguish the species of the dominant plant, but only the genus to which the species belongs. The plan is particularly ambiguous in the case of such genera as Juncus, Quercus, and Pinus, each of which contains several species which are dominant in unrelated associations.

Cajander (1903 : 23) has a modification of Schouw's plan, which, to a large extent, overcomes this difficulty; and Cajander's plan appears to meet with Warming's approval (1909: 145). By this method, an association of *Juncus effusus* is not referred to merely as a Juncetum, a term which might refer to an association of *J. glaucus*, or of *J. maritimus*, or of *J. obtusiflorus*, or of *J. squarrosus*, and so on: Cajander's plan is to add as a genitive the specific name of the dominant plant of the association. Thus an association of *J. effusus* would be a Juncetum Junci effusi, or more briefly, a Juncetum effusi; and similarly an association of *J. maritimus* would be a Juncetum maritimi.

A second objection applies equally to Schouw's plan and to Cajander's modification of it. Neither method is applicable to mixed associations, which have no single dominant species. Of course, some characteristic species might, in certain cases, be utilized; but this would be arbitrary and unscientific, and in some cases, perhaps, impracticable.

The designation of a pure association by its dominant species is a very different matter from determining the association by such a species. Against the latter superficial method of determining associations Drude has expressed himself on several occasions; and recently Gradmann has emphatically condemned such a procedure. The naming of a pure association, however, by its dominant species is comparable with the plan of naming a systematic group after an easily recognizable character; and in neither case does such a name exhaust the characters of the group it denotes.

[*Editor's Note:* Material has been omitted at this point.]

Groups of Formations.

It is obvious that allied formations may be arranged into still more comprehensive classes. Every phytogeographical mono- grapher, in fact, has arranged his formations in some order; but very few have claimed for these higher groups the value of natural divisions. For example, the majority of such authors have arranged their formations or associations into groups which followed the ancient division of the vegetable kingdom into trees, shrubs and herbs ; but it is scarcely possible that such a basis can ever yield a scientific or natural classification of plant communities; although Warming (1909 : 140-141) makes this " the prime basis of classification " in subdividing the groups of hydrophilous, xerophilous, and mesophilous plants. One finds too that authors have adopted points of view which differ according to the size and extent of the earth's surface which they happened to be investigating ; and this perhaps is inevitable. It is the same in taxonomy. Those systematists who have dealt with continental areas or with whole phyla of plants have invariably given a wide significance to the species; and those who have studied intensively the flora of small districts, or who have monographed special genera, have generally used the term species in a much narrower sense.

Many of the physiognomical "formations," *e.g.*, tundra, of Grisebach and others, are to be regarded either as groups of formations or groups of associations ; and these may or may not be natural groups. Kerner (1863) arranged his minor "formations" into groups (Gruppen von Genossenschaften). Drude (1890) in his *Handbuch*, in which he dealt with the whole world, adopted a physiognomical classification on the whole : in his *Deutschlands Pflanzengeographie* (1896), his "formations" (chief associations) were arranged into groups some of which depend largely on physiognomy, *e.g.*, wood "formations" and scrub "formations," and others largely on habitat, *e.g.*, moor "formations" and aquatic "formations ; " whilst in his still more detailed study of the formations of the Hercynian region (1899), he adopted a more natural classification and one based more completely on habitat.

Warming's (1895) classification of plant communities, to some extent based on the work of Schouw (1822, 1823), was undoubtedly the most fundamental which had at that time been put forward. Warming based his major divisions, hydrophyte communities, xerophyte communities, halophyte communities, and mesophyte communities, on the nature of the habitat, partly on the water content and partly on the mineral content of the soil. This classification, however, met with some adverse criticism, particularly with regard to the inclusion of " bog xerophytes " among hydrophytes, with regard to the separation of halophytes from xerophytes, with regard to the inclusion of all conifers among xerophytes and of all deciduous trees among mesophytes (cf. Stopes, 1907 ; Moss, 1907b : 6), and with regard to the group of mesophytes in general. This appears a somewhat formidable indictment; but the fact remains that, at the time, Warming's grouping was the best constructive effort which had been put forward in the direction of a natural and scientific classification of plant communities.

Schimper's (1898) division into hygrophytes, xerophytes, and tropophytes was in several respects superior to the earlier one put forward by Warming.

Schimper's "climatic formations" are doubtless, in general, groups of formations; whilst others may really be formations. As Clements (1904: 27), Smith (1904: 620), and Warming (1909: 132) have maintained, it is undesirable to separate those habitat factors depending on climatic conditions from those depending on edaphic conditions. Still, with regard to certain districts, such as the rainier portions of several tropical countries, whose vegetation and flora, partly from lack of exploration and partly owing to certain inherent difficulties, are little known, such "climatic formations" as tropical rain forest are at present the only practicable vegetation units. Kurz (*op. cit.*) has, however, indicated that the classification of even tropical forests will yield ultimately to more intensive treatment; and it seems clear that many "climatic formations" must ultimately be analysed into formations. With regard to well-worked countries, there is, from the standpoint of an ultimate and natural grouping of plant communities, nothing whatever to be said in favour of regarding as "formations" such complex groups of associations as deciduous forests, coniferous forests, grassland, and so on. For example, it can scarcely be contended that Warming's "formation of deciduous dicotyledonous forest" (1909 : 144 and 329) is of the same rank as his "low moor formation" (p. 196) or his "high moor formation" (p. 200); and the real difficulty is only temporarily shelved by subdividing highly heterogeneous "formations" into "subformations" of little less heterogeneity.

The new classification of formations into higher groups by Warming and Vahl (1909) is much more involved, and much less simple and clear, than Warming's earlier one; and it is very doubtful if the changes have resulted in any gain in accuracy. Already it has met with some adverse criticism by Cowles (1909: 151), Tansley (1909: 221), and Moss (1909 : 351) in their respective reviews of Warming's new book (1909). In particular, the "ecological class" of coniferous formations has been singled out for criticism. "Whilst this group," says Cowles (*loc. cit.*), "is a floristic unit, ... it is far from being a geographical unit of any sort. It would seem better to put many conifers with the lithophytes and psammophytes, whilst others are certainly oxylophytes, and others still pronounced mesophytes, the most mesophytic forests of the United States being dominated by conifers." Several closely allied formations are, in my judgment, placed apart by the first divisions of Warming and Vahl's new system. It must be remembered too that many plants, *e.g.*, "bog xerophytes," possess some organs which, considered by themselves, would result in these plants being termed hydrophytes, whilst other organs would result in their being termed xerophytes; and certain plants, such as deciduous trees, are extreme xerophytes during winter, yet possess leaves which are "mesophytic" during summer. Hence it is to be feared that any attempt to place every plant in a verbal class with the termination *-phyte* is either impossible or premature.

Graebner's (1901, 1909) classification of plant formations into larger groups has much to recommend it. Graebner (1909) makes three main divisions :—

(*a*) Formations where the soil-water is rich in nutritive salts.

(*b*) Formations where the soil-water is poor in nutritive salts.

(*c*) Formations where the soil-water is saline.

These main divisions are further subdivided chiefly on the basis of the water-content of the soil, and then still further subdivided by other physical factors. Thus the formations themselves are ultimately reached. This classification, although its author, of course, makes no claim that it is final, is, on the whole, and within an area of fairly constant climatic conditions, a natural one because it takes into account the whole of the factors comprising the habitat, and because no conflicting factor, such as plant form, is introduced. Doubtless future research will suggest some emendations; but it may be regarded as the most successful of the existing systems, and as planned on a general and scientific basis which will lead to ultimate success.

Gradmann (1909 : 93) has maintained that, in the matter of classifying formations, plant geographers should be given full liberty. It is sufficient to point out that plant geographers, have, in the past, exercised such liberty ; and the result is chaos.

CONCLUSION.

I should have liked, had space permitted, to have also traced the development of the concepts of those minor units of vegetation which are subdivisions of associations. These are important from two points of view. First, they are the outstanding vegetation units to the student of vegetation who is investigating the detailed structure of a very limited area. That such intensive studies may yield results of great value is seen from a perusal of such works as those of Woodhead (1906) and Brockmann-Jerosch (1907), and from unpublished results by several British ecologists. It is becoming usual in this country to speak of the subdivisions of the association as plant societies (cf. Clements, 1905: 296). Graebner (1895, 1901) and some other continental ecologists have termed them "facies"; but the term "facies," like the term "society," has been used in several other senses. In fact, so many terms have been used by ecologists and plant geographers with so many different significations that it would appear to be impossible to find any term to which the above objection does not apply. The intensive study of plant societies permits of a rigorous and quantitative determination of habitat factors, as has been done by Hedgcock (1902), Livingston (1906), Yapp (1909), and others, and in unpublished work by Professor Oliver, Mr. Crump, and Mr. Adamson. Woodhead's useful generalisation with regard to "competitive" and "complementary" communities (1906: 396, 7, etc.) is also an outcome of this kind of work. Such work also lends itself to the investigation of the physiological responses of plants to definite habitat factors. The latter work has scarcely been begun by ecologists; but it is certain that a rich reward of results awaits investigators who undertake such research.

Secondly, the investigation of plant societies is necessary to the study of the formation; for several writers have displayed an unfortunate facility in mistaking the one vegetation unit for the other. Pound and Clements, Drude. Clements, and others have protested against the tendency of certain writers to elevate to the rank of a formation any phase of a succession or any slightly different patch of vegetation. This tendency has to be guarded against in all intensive studies of vegetation; and the *quasi* definiteness of the local patches of vegetation in a restricted area can only be corrected by comparative observations in other districts.

The study of vegetation needs workers from all the departments of ecology and plant geography indicated in this paper. The monographers of the wider areas have still to concentrate attention on a natural grouping of formations, the investigators of the very restricted areas on the factors of the habitat and on the responses made by the vegetation to these factors. Formations themselves require to be carefully delimited; and the associations of each formation need detailed study from the points of view of zonation, alternation, succession, and floristic composition. The study of vegetation touches on physiology at many points, but at present there is not and cannot be any essential identity between them: it needs the systematist, but it is in no sense a branch of taxonomy. Vegetation must be studied from the standpoints of its development, its structure, and its activities; and the object of this study is the elucidation of the relationships of vegetation to the factors of the habitat.

LITERATURE REFERRED TO.

Beck von M., G., 1901. " Die Vegetationsverhältnisse der illyrischen Länder"; in Engler u. Drude, *Veg. d. Erde*, IV; Leipzig.

„ 1902. " Über die Umgrenzung der Pflanzenformationen "; in *Österr. Bot. Zeitschr.*, LII.

Blackman, F. F., and Tansley, A. G., 1905. "Ecology in its Physiological and Phytogeographical Aspects"; in *New Phytologist*, *IV*. (Review of Clements' " Research Methods in Ecology.")

Bonpland, A., 1807. See Humboldt and Bonpland.

Brockmann-Jerosch, H., 1907. " Die Flora des Puschlav (Bezirk Bernina, Kanton Graubünden) und ihre Pflanzengesellschaften "; Leipzig.

Brown, F. B. H., 1905. " A Botanical Survey of the Huron River Valley, III: The Plant Societies of the Bayon at Ypsilanti "; in *Bot. Gaz.*, XL.

Cajander, A. K., 1903. " Beiträge zur Kenntniss der Vegetation der Alluvionen des nördlichen Eurasiens. Die Alluvionen des unteren, Lena-Thales "; in *Act. Soc. Sc. Fenn.*, XXXII.

Celakovsky, L., 1869-1874. "Prodromus der Flora von Böhmen "; Prag. (Cited in Loew, 1879).

Clements, F. E., 1898; 1900. See Pound and Clements.

„ 1902. " A System of Nomenclature for Phytogeography "; in Engler's *Bot. Jahrb.*, XXXI., Beibl., 70.

„ 1904. " The Development and Structure of Vegetation "; Lincoln, Neb., U.S.A.

„ 1905. "Research Methods in Ecology "; Lincoln, Neb., U.S.A.

Cowles, H. C., 1899. "The Ecological Relations of the Vegetation on the Sand Dunes of Lake Michigan "; in *Botan. Gaz.*, XXVII.

„ 1901. "The Physiographic Ecology of Chicago and vicinity . a Study of the Origin, Development, and Classification of Plant Societies "; in *Botan. Gaz.*, XXXI.

Cowles, H. C., 1909.   Review of Warming's Ecology of Plants; in *Botan. Gaz.*, XLVIII., pp. 149-152.

„   1910.   " The Fundamental Causes of Succession among Plant Associations " ; in *Report Brit. Assoc.* (Winnipeg, 1909); London: (see also New Phytologist, 1909 : 367).

Drude, O., 1889.   " Über die Principien in der Unterscheidung von Vegetationsformationen, erläutert an der centraleuropäischen Flora"; in Engler's *Bot. Jahrb.*, XI.

„   1890.   " Handbuch der Pflanzengeographie"; Stuttgart. (French tr. by Poirault, Man. de Geog. Bot., 1897 ; Paris.)

„   1896.   " Deutschlands Pflanzengeographie, I."; Stuttgart.

„   1902.   " Der Hercynische Florenbezirk " ; in Engler u. Drude, *Veg. d. Erde*, VI ; Leipzig.

Duchartre, P., 1844.   " Sur la géographie botanique des environs de Beziers (Hérault) " ; in *Compt. Rend. de l'Acad. des Sciences*, XVIII. ; Paris. (" Extrait par l'auteur.")

Flahault, C., 1896.   " Essai d'une carte botanique et forestiére de la France "; in *Ann. de Geogr.*, V.

„   1896.   " Au sujet de la carte botanique, forestière, et agricole de la France '' ; in *Ann. de Géogr.*

„   1897.   " La flore de la vallée de Barcelonette "; Montpellier. (Soc. Bot. de France.)

„   1900-1901.   " Premier essai de nomenclature phytogéographique"; in *Bull. de la Soc. Languedocienne de Géographie*, 1901 ; Montpellier. (See also "A Project for Phytogeographic Nomenclature"; in *Bull. Torr. Bot. Club*, 1901, XXVIII., an English tr. of Flahault's paper read at the International Botanical Congress, Paris, 1900.)

Früh, J., und Schröter, C., 1904.   " Die Moore der Schweiz "; Bern.

Ganong, W. F., 1903.   " The Vegetation of the Bay of Fundy Salt and Diked Marshes : an Ecological Study"; in *Botan. Gaz.*, XXXVI.

Gradmann, R., 1909.   " Über Begriffsbildung in der Lehre von den Pflanzenformationen"; in Engler's *Botan. Jahrb.*, XXXVIII., Beibl., 99.

Graebner, P., 1895.   " Studien über die norddeutsche Heide. Versuch einer Formationsgliederung "; in Engler's *Botan. Jahrb.*, XX.

„   1901.   " Die Heide Norddeutschlands "; in Engler u. Drude, *Veg. d. Erde*, V ; Leipzig.

„   1909.   " Die Pflanzenwelt Deutschlands: Lehrbuch der Formationsbiologie," etc. ; Leipzig.

Grisebach, H. R. A., 1838.   " Ueber den Einfluss des Climas auf die Begränzung der natürlichen Floren "; in *Linnaea*, XII.

„   1846.   " Report on Botanical Geography during the year 1842"; " Report on Botanical Geography during the year 1843 "; in *Reports and Papers on Botany*, London (Ray Soc.)

„   1849.   " Report on the Progress of Geographical Botany during the year 1844 " ; " Report on the Progress of Geographical and Systematic Botany during the year 1845 "; in *Reports and Papers on Botany*, London (Ray Soc.)

„   1872.   " Die Vegetation der Erde "; Leipzig. (French tr. by Tchihatchef, La Vég. du Globe ; 1877-8, Paris).

„   1880.   " Gesammelte Abhandlungen und kleinere Schriften zur Pflanzengeographie "; Leipzig (2nd ed., 1884).

Hardy, M., 1905. "Esquisse de la géographie et de la végétation des Highlands D'Ecosse"; Paris. (See also "La Végétation des Highlands D'Ecosse": in *Ann. de Géogr.*, XV., 1905, with vegetation map.)

Harshberger, J. W., 1904. "A Phyto-geographic Sketch of extreme southeastern Pennsylvania"; in *Bull. Torr. Bot. Club*, XXXI.

Hedgcock, G. G., 1902. "The Relation of the Water Content of the Soil to certain Plants, principally mesophytes"; Studies in the Vegetation of the State, II; Lincoln, Neb., U.S.A.

Höck, F., 1895. "Genossenschaften in unserer Kiefernwaldflora"; in *Nat. Wochenschrift.*

Hult, R., 1881. "Försök till analytisk Behandling af Växtformationerna"; in *Meddel. af Soc. pro Fauna et Flora Fennica*, VIII.

Humboldt, A., ?1806. English tr. by Sabine, "Aspects of Nature," 1849; London.

    ,, et Bonpland, A., 1807. "Essai sur la géographie des plantes," in "Voyage de Humboldt et Bonpland"; Paris.

    ,,   ,, 1819. English tr. by Williams, "Personal Narrative of Travels to the Equinoctial Regions of the New Continent during the years 1799-1804"; 7 vols., 1814-1829; London.

Kerner, von M. A., 1863. "Das Pflanzenleben der Donauländer"; Innsbruck.

    ,, 1887-1891. English tr. by F. W. Oliver, "The Natural History of Plants," 1895; London.

Kirchner, O., 1896-1902. See Schröter and Kirchner.

Kurz, S., 1870. "Report on the Vegetation of the Andaman Islands"; Calcutta.

    ,, 1875. "Preliminary Report on the Forests and other Vegetation of Pegu"; Calcutta.

    ,, 1877. "Forest Flora of British Burma"; Calcutta. (See also: "Burma, its People and Productions," vol. 2, 1883, Hertford).

Lamarliére, L. G. de, 1894. "Note sur la flore maritime des environs de Quinéville (Manche)"; in *Bull. Soc. Bot. de France*, XLI.

Lecoq, H., 1854-8. "Études sur la géographie botanique de l'Europe, et en particulier sur la végétation du plateau central de la France"; 9 volumes; Paris.

Lewis, F. J., 1904a. "Geographical Distribution of Vegetation of the Basins of the Rivers Eden, Tees, Wear, and Tyne, Part I."; in *Geogr. Journ.*, XXIII., with vegetation map.

    ,, 1904b. "Geographical Distribution of Vegetation of the Basins of the Rivers Eden, Tees, Wear, and Tyne, Part II."; in *Geogr. Journ.*, XXIV., with vegetation map.

Linnaeus, C., 1751. "Philosophia Botanica"; Stockholm.

    ,, 1754. "Stationes Plantarum." (Diss. in "Amoenitates Academicae" IV, pp. 64-87, 1760.)

Livingston B. E., 1901. "The Distribution of the Plant Societies of Kent County, Michigan"; in *Ann. Rep. Mich. St. Bd. Geol. Surv.*

    1903. "The Distribution of the Upland Plant Societies of Kent County, Michigan"; in *Botan. Gaz.*, XXXV.

    1905. "The Relation of Soils to Natural Vegetation in Roscommon and Crawford Counties, Michigan"; *ib.*, XXXIX.

    906. "The Relation of Desert Plants to Soil Moisture and to Evaporation"; Carnegie Inst., Washington.

Loew, E., 1879. "Ueber Perioden und Wege ehemaliger Pflanzenwanderungen im norddeutschen Tieflande"; in *Linnaea*, XLII.

Masclef, (M. l'abbé), 1888. "Études sur la géographie botanique du Nord de la France"; in *Journ. de Bot.*, II.

Meyen, F. J. F., 1836. "Grundriss der Pflanzengeographie"; Berlin. English tr. by Johnson, "Outlines of the Geography of Plants," 1846; London (Ray Soc).

Moss, C. E., 1903. See Smith and Moss.

" 1907. "Geographical Distribution of Vegetation in Somerset: Bath and Bridgwater District," with vegetation map; Royal Geographical Society (also Edward Stanford); London.

" 1907b. "Xerophily and the Deciduous Habit"; in *New Phyt,*, VI

" 1909. Review of Warming's Œcology of Plants; in *Science Progress*, IV., pp. 350-2.

" Rankin, W. M., and Tansley, A. G. (1910, in the press). "The Woodlands of England."

Olsson-Seffer, P., 1905. "The Principles of Phytogeographic Nomenclature"; in *Botan. Gaz.*, XXXIX.

Pavillard, J., 1905. "Recherches sur la flore pélagique de l'étang de Thau." (Cited in Hardy, 1905).

Pethybridge, G. H., and Praeger, R. Ll., 1905. "The Vegetation of the District lying south of Dublin"; in *Proc. Roy. Irish Acad.*, XXV., B., 6, with vegetation map; Dublin.

Pound, R., and Clements, F. E., 1898; 2nd ed., 1900. "The Phytogeography of Nebraska"; Lincoln, Neb., U.S.A.

Praeger, R. Ll., 1905. See Pethybridge and Praeger.

Rankin, W. M., 1903. See Smith and Rankin.

" 1909. See Moss, Rankin, and Tansley.

Schimper, A. F. W., 1898. "Pflanzengeographie auf physiologischer Grundlage"; Berlin. English tr. by Fisher; "Plant Geography upon a Physiological Basis," 1903-4; Oxford.

Schouw, J. F., 1822. "Grundtraek til en almindelig Plantegeografie"; Kjobenhavn. German tr. "Grundzüge einer allegemeinen Pflanzengeographie," 1823; Berlin.

Schröter, C., 1894. "Notes sur quelques associations de plantes rencontrées pendant les excursions dans le Valais"; in *Bull. Soc. Bot. Fr.*, XLI.

" und Kirchner, O., 1896-1902. "Die Vegetation des Bodensees"; Lindau.

" 1904. See Früh and Schröter.

Smith, R., 1898. "Plant Associations of the Tay Basin"; in *Proc. Perthshire Soc. Nat. Sc.*, II.; Perth.

" 1899. "On the Study of Plant Associations"; in *Natural Science*, XIV.; Edinburgh and London.

" 1900a. "Botanical Survey of Scotland: I., Edinburgh District," with vegetation map; in *Scott. Geogr. Mag.*, XVI. Also Bartholomew, Edinburgh.

" 1900b. "Botanical Survey of Scotland: II., Northern Perthshire District," with vegetation map; in *Scott. Geogr. Mag.*, XVI. (See also "Plant Associations of the Tay Basin," Part II.; in *Proc. Perthshire Soc. Nat. Sc.*, III., with vegetation map.) Also Bartholomew, Edinburgh.

Smith, R. (the late), and Smith, W. G., 1904-5. "Botanical Survey of Scotland; III. and IV., Forfar and Fife"; in *Scott. Geogr. Mag.*, XX. and XXI., with two vegetation maps.

Smith, W. G., and Moss, C. E., 1903. "Geographical Distribution of Vegetation in Yorkshire: Part I., Leeds and Halifax District," with vegetation map; in *Geogr. Journ.*, XXI. Also Bartholomew, Edinburgh.

,, and Rankin, W. M., 1903. "Geographical Distribution of Vegetation in Yorkshire: Part II., Harrogate and Skipton District," with vegetation map; in *Geogr. Journ.* XXII. Also Bartholomew, Edinburgh.

,, 1904-5. See Smith and Smith.

Stopes, M., 1907. "The Xerophytic Character of the Gymnosperms: is it an ecological adaptation?" in *New Phytologist*, VI.

Tansley, A. G., 1904. "The Problems of Ecology"; in *New Phytologist*, III.

,, 1905. See Blackman and Tansley.

,, 1909. Review of Warming's Œcology of Plants; in *New Phytologist*, VIII., pp. 218-227.

,, 1910. See Moss, Rankin, and Tansley.

Tournefort, J. P. de, 1717. "Rélation d'un Voyage du Levant"; Lyon. (Amsterdam, 1718.)

Transeau, E. N., 1905-6. "The Bogs and Bog Flora of the Huron River Valley"; in *Botan. Gaz.*, XL. and XLI.

Vahl, M. See Warming (1909).

Warburg, O., 1900. "Einführung einer gleichmässigen Nomenclatur in die Pflanzengeographie"; in Engler's *Botan. Jahrb.*, XXIX., Beibl., 66.

Warming, E., 1887. "Om Grönlands Vegetation." (Synopsis, "Über Grönlands Vegetation," in Engler's *Botan. Jahrb.*, X.)

,, 1895. "Plantesamfund." (German tr. by Knoblauch, "Lehrbuch der Ökologischen Pflanzengeographie: eine Einführung in die Kentniss der Pflanzenvereine," 1896; Berlin. New German ed. by Graebner, 1902; Berlin).

,, 1899. "On the Vegetation of Tropical America"; in *Botan. Gaz.*, XXVII.

,, assisted by Vahl, M., 1909. "Œcology of Plants: an introduction to the study of plant-communities"; Oxford.

Whitford, H. N., 1901. "The Genetic Development of the Forests of Northern Michigan: a study in physiographic ecology"; in *Botan. Gaz.*, XXXI.

,, 1905. "The Forest of the Flathead Valley, Montana"; *ib.*, XXXIX.

Woodhead, T. W., 1906. "Ecology of Woodland Plants in the Neighbourhood of Huddersfield"; in *Linn. Journ.*, Botany, XXXVII.

Yapp, R. H., 1909. "On Stratification in the Vegetation of a Marsh, and its Relations to Evaporation and Temperature"; in *Ann. of Botany*, XXIII.

Young, R. T., 1907. "The Forest Formations of Boulder County, Colorado"; in *Botan. Gaz.*, XLIV.

2

REMARQUES

SUR L'ÉTUDE DES GROUPEMENTS DE PLANTES

Par Josias BRAUN et Ernst FURRER

On s'est beaucoup occupé, depuis quelques années, surtout dans les pays de langue allemande, de travaux phytogéographiques, et en particulier du groupement méthodique des végétaux. Plusieurs siècles de travail constant ont été nécessaires pour résoudre une quantité de problèmes systématiques et floristiques, relatifs aux végétaux supérieurs de nos pays. Ainsi armés, les botanistes, observant directement la nature, ont pu s'intéresser à l'ensemble de la végétation, rechercher les conditions suivant lesquelles elle se groupe et se répartit.

L'étude des groupements de plantes (la *synécologie*) envisage quatre points de vue principaux :

1° La *synécologie descriptive* étudie la composition floristique des groupements végétaux ;

2° La *synécologie physiologique*, l'écologie au sens restreint, explique les relations de cause à effet entre les groupements et lec facteurs externes (climatiques, édaphiques et biotiques) ;

3° La *synécologie géographique* s'occupe de la répartition des groupements et de leurs différences régionales et altitudinales ;

4° La *synécologie génétique*, enfin, recherche les relations entre les groupements actuels des plantes et leur passé. Elle s'occupe donc de l'évolution des groupements.

La seule base solide de toute étude de cet ordre est la connaissance parfaite de la composition floristique de *l'unité synécologique*. Cette unité a reu son nom. C'est *l'association* (1) (association en anglais, Assoziation ou Bestand en allemand, associazione en italien): *L'association définie « est un groupement végétal de composition floristique déterminée, présentant une physionomie uniforme, croissant dans des conditions stationnelles uniformes »* (Flahault et Schrœter), *et possédant une ou plusieurs espèces caractéristiques.*

Nous avons cru devoir introduire dan cette définition la notion des espèces *caractéristiques* (Charakterflanzen, Leitpflanzen de Gradmann). *Ce sont des espèces localisées exclusivement ou à peu près dans une association donnée* ; elles peuvent être considérées comme l'expression floristique la plus certaine de l'écologie du groupement.

La notion des caractéristiques n'est pas nouvelle : mais elle a été subordonnée, jusqu'à présent, à celle des espèces dominantes. Or, les dominantes sont souvent des ubiquistes, croissant dans des stations très différentes (*Nardus stricta, Poa alpina, Carex curvula* des Alpes et des Pyrénées : *Corylus Avellana, Calluna vulgaris*, etc.) ; dès lors, elles nous instruisent peu sur le caractère écologique.

Il en est autrement lorsqu'il s'agit des espèces caractéristiques. Quelqu'un nous parlera-t-il de l'*Aristida pungens*, de *Cardamine alpina*, de l'*Epipogon aphyllum*, de *Vaccinium Oxycoccus* et de *Carex pauciflora* ; nous savons tout de suite qu'il s'agit d'une association de sable mouvant, des combes à neige, de l'épicea, des tourbières à Sphagnum. Pourquoi ? Parce qu'il a nommé des caractéristiques de ces associations.

---

(1) FLAHAULT et SCHRŒTER. Nomenclature phytogéographique, Rapports et propositions, *IIIe Congrès internat. de Bruxelles*, 1910. — Moss (O.-E.). fundamental units of vegetation. *The new Phyt.* 1910.

Inutile de dire que les caractéristiques peuvent être en même
temps dominantes (*Festuca varia*, *Pinus montana* f. *pumilio*
des Alpes, *Festuca Eskia* des Pyrénées, *Ammophila arenaria*
des dunes littorales, *Salicornia sarmentosa* des prés salés,
etc.).

Chaque station présentant des conditions de vie uniformes
et portant un ou plusieurs espèces caractéristiques, réalise,
suivant nous, une association définie. L'association peut
être une communauté organisée, dont chaque membre tire
profit ; elle ne l'est cependant pas nécessairement. Il nous
est tout à fait impossible de constater jusqu'où s'étend
la concurrence entre les espèces, comme il est impossible de
déterminer la valeur du lien utilitaire, qui réunit les individus.
Les rochers abrupts, considérés souvent comme un milieu
de végétation ouvert, sont parfois tapissés de lichens, d'al-
gues, de mousses (1), formant avec quelques Phanérogames,
vivant dans les touffes des mousses ou dans les fentes des
rochers, une étroite communauté soumise aux lois générales
de l'association. Les études écologiques approfondies de
M. Oettli (2) et M. Hess (3) prouvent suffisamment que la lutte
pour la place se manifeste aussi bien dans les associations
ouvertes (rochers, éboulis) que dans les associations fermées.

L'équilibre entre la concurrence des végétaux d'une part
et les facteurs externes actuels de l'autre marque ce stade, plus
ou moins durable, qui caractérise une association constituée.
La durée de l'état d'équilibre, du développement optimum de
l'association, varie beaucoup. Cette durée est à peu près indé-
finie pour les associations dues au climat (4) (en particulier

---

(1) Cfr. M. SCHADE. Pflanzenœkologische Studien aus den Felswänden der
Sächsischen Schweiz. *Engl. Bot. Iahrb.*, 1912.

(2) OETTLI (Max.). Beiträge z. Oekologie der Felsflora. *Iahrb. St. Gall.
Natf. Ges.*, 1903.

(3) HESS (Eug.). Uber Wuchsformen der alpinen Geröllpflanzen. *Beih. Bot.
Zentralbl.*, 1909.

(4) Cfr. SCHIMPER (W.). *Pflanzengeographie auf physiologischer Grundlage*,
1898, pp. 175-176.

associations des pays arctiques et des hauts sommets [gazon à *Carex curvula*, combes à neige]), elle est éphémère pour certaines associations transitoires (association de *Myricaria* et des saules sur les alluvions de fleuves), et surtout pour des associations culturales et semiculturales (champs de blé, de lin, de chanvre, etc., clairières de bois). La présence des caractéristiques, qui apparaissent en général seulement après l'établissement complet de l'association, et disparaissent les premières, dès que les conditions de vie subissent un changement, nous permet de considérer l'association comme définitivement établie. Nous pouvons citer comme exemple le haut marais avec ses éléments absolument caractéristiques (*Carex pauciflora, Oxycoccus palustris, Andromeda polifolia, Malaxis paludosa, Saxifraga Hirculus*). On trouvera d'autres exemples dans un travail récemment publié par l'un de nous (1).

Citons encore un exemple démonstratif. Les parties élevées de l'Aigoual (Cévennes) sont couvertes de taillis et de futaies de hêtres d'âge très variable. Les jeunes taillis, spontanément reconstitués sur de grandes surfaces, appartiennent bien à l'association du hêtre ; mais l'association ne se révèle avec ses caractères définitifs et ses caractéristiques que dans les anciennes futaies. Dans les jeunes taillis, *Genista purgans, G. pilosa, Calluna, Senecio adonidifolius* et même *Luzula spicata*, éléments héliophiles d'associations diverses, tiennent une grande place. A mesure que les arbres se développent, ces ubiquistes disparaissent jusqu'à manquer absolument dans la futaie, l'association pleinement constituée du hêtre. Dans la futaie seulement se montrent par contre : *Dentaria pinnata et digitata, Asperula odorata, Luzula maxima, Corydalys solida, Allium ursinum*, et d'autres caractéristiques de premier ordre.

Elargissons notre cadre : considérons aussi les associations

(1) BRAUN (Josias). Vegetationsverhältnisse der Schneestufe in den Rätisch Lepontischen Alpen. *Denkschr. Schweiz. Naturf. Ges.*, Bd. 48, 1913.

culturales et semiculturales : nous voyons que, là encore, en dépit de l'homme qui a si bien réussi à jeter le désordre dans la végétation naturelle, les associations définies se dégagent clairement des peuplements quelconques créés par lui. Ce sont encore les espèces caractéristiques qui nous servent de fils conducteurs. Toutes les anciennes formes de cultures, telles que les vignobles, les champs de blé et de lin, la châtaigneraie, les vergers, sont devenues de véritables associations culturales (1). Chacune possède, par conséquent, ses espèces, adaptées exclusivement ou à peu près à l'association donnée. Citons, pour les vignobles du Bas-Languedoc : *Fumaria micrantha* et *parviflora*, *Aristolochia Clematitis*, *Ranunculus parviflorus* et *sardons* comme caractéristiques de premier ou de second ordre. Pour les champs de blé : *Agrostemma*, *Centaurea cyanus*, *Adonis*, diverses espèces, *Veronica acinifolia*, *Ceratocephalus falcatus*, etc. : pour les champs de lin : *Lolium linicola*, *Camelina dentata*, *Silene linicola*. Plus l'introduction d'une culture est récente, plus, en général, les caractéristiques lui font défaut. Les champs de pommes de terre qui existent dans les montagnes de la Suisse depuis peu de siècles n'ont guère que des espèces caractéristiques de second ordre, comme le *Melandryum noctiflorum*. Nos prés, semés dru, constituent souvent un peuplement dense d'une seule espèce (*Arrhenaterum elatius*, *Dactylis glomerata*, *Phleum pratense*, *Lolium italicum*), sans aucune plante caractéristique. Ce sont, dans ce cas, non pas des associations culturales fixées, mais des peuplements purs, purement artificiels (künstliche Reinbestände, Herden), qui, cependant, peuvent devenir des associations définies, si on leur laisse le temps d'évoluer.

Un peuplement est capable de se transformer, même sous nos yeux, en association définie. M. *Warming* (cit. in An-

---

(1) Cfr. aussi ANDERSSON (Gunnar). *Entwicklungsgeschichte der skandinav. Flora*, 1906.

dersson, Gunnar, *l. c.*) en mentionne un exemple. Les forêts de Conifères, plantées il y a 100 à 150 ans en Danemark et en Scanie, ont peu à peu acquis, par immigration du nord, les espèces caractéristiques de l'association naturelle, telles que :

Linnaea borealis.	*Pyrola chlorantha*
Goodyera repens	*Pyrola umbellata*
Pyrola uniflora	*Pyrola media*

Un autre exemple d'un peuplement en train de se constituer en une association est relevé par M. *Negri* (1). La *Robinia pseudacacia*, essence introduite de l'Amérique du Nord, se répand dans la vallée du Pô d'une façon extraordinaire. Plusieurs espèces, également, introduites de l'Amérique : *Phytolacca decandra, Amorpha fruticosa, Oenothera biennis, Erigeron annuus, E. canadensis, Solidago glabra.* recherchent de préférence les peuplements de *Robinia* ; ils apparaissent ainsi comme quelque chose de particulier dans la végétation du pays. M. *Negri* n'hésite pas à appeler ces peuplements « une association acclimatée ».

Il va sans dire qu'une ou plusieurs espèces caractéristiques d'une association en une région donnée peuvent ne pas l'être ailleurs, pour peu que le climat soit différent. Une espèce caractéristique de deuxième ordre, peut devenir de premier ordre lorsqu'elle se rapproche des limites de son aire : *Betula nana*, strictement localisé sur nos hauts marais du Jura et des préalpes, où il trouve sa limite méridionale, croît un peu partout, dans les montagnes de Norvège et de la Laponie, où elle est abondante. Les caractéristiques du Curvuletum, à la limite supérieure dans les Alpes Rético-Lépontiennes : *Leontodon pyrenaicus, Veronica bellidifolia, Avena versicolor*, ne sont plus caractéristiques pour cette

(1) NEGRI. La vegetazione del Bosco Lucedio, 1911.

association, au centre de leur extension altitûdinale (entre 2.200 et 2.600 mètres).

L'étude géographique comparée des associations doit révéler ces faits et nous faire connaître en même temps les espèces absolument caractéristiques, ne dépassant jamais l'aire de l'association.

Il est évident qu'on ne peut distinguer *toutes* les associations possibles par le moyen des seules espèces caractéristiques. Cela n'est pas possible lorsqu'il s'agit : 1° *des groupements passagers*, associations en voie de fixation, ne possédant pas de traits caractéristiques; 2° *des associations mixtes*, formées de la pénétration réciproque de plusieurs associations. Ces associations mixtes couvrent parfois des surfaces considérables (*Hygro-Curvuletum* des Alpes Lépontiennes, *Uebergangsmoor* de la Prusse (1) : la première est une convergence du gazon de *Carex curvula* et de la combe à neige, due au climat pluvieux : la seconde le stade intermédiaire entre le haut et le bas marais (*associations mixtes de chêne vert* et de *chêne pubescent* dans les Cévennes). Les combinaisons sont innombrables, et le phytogéographe peut se contenter de décrire les plus apparentes de ces associations mixtes, celles qui recouvrent une surface plus ou moins étendue. Ne perdons pas de vue que le premier but de l'étude des groupements n'est pas l'inventaire de chaque motte de terre, mais la recherche d'une *unité comparable à l'espèce, pouvant servir de base aux travaux de géographie botanique comparée* et aux investigations synécologiques. L'expérience que nous avons acquise en divers pays nous fait espérer qu'en suivant notre interprétation de l'association, en la perfectionnant au besoin, on arrivera à des résultats positifs comparables. Certes, cette interprétation exige une étude attentive et beaucoup d'expérience ; elle ne se fait pas en parcourant le pays en quelques jours.

(1) ABROMEIT (Joh.). Die Vegetationsverhältnisse von Ostpreussen, etc. *Engl. Bot. Jahrb. Beibl.* z. N° 106, 1912.

L'étude de l'association exige qu'on tienne compte, non seulement des espèces caractéristiques, mais de l'inventaire complet de sa flore, ce qu'a fait remarquer M. *Flahault* dès 1900 (1).

L'inventaire floristique de l'association comprend des espèces *caractéristiques* de premier et de second ordre (Charakterpflanzen), des *constantes* (Konstanten de *Brockmann-Jerosch* (2), apparaissant au moins dans la moitié des relevés d'une même association : des espèces *accessoires*, renfermées encore dans un quart des relevés des espèces plus ou moins accidentelles. Pour désigner la densité de chaque espèce dans l'association, on emploie les termes *solitariae*, *sparsae*, *coposae*, *gregariae* et *sociales*, introduits par *Drude* (3), ou bien les chiffres de 1 à 10. Le chiffre 10 désigne la dominance absolue d'une espèce dans un peuplement fermé.

Les *peuplements purs* ne comprennent qu'une seule espèce : on pourra les appeler *Herden* en allemand. Un grand nombre de végétaux sont capables de former des peuplements purs, ou peu s'en faut (par exemple: *Hippuris, Scirpus lacustris, Allium ursinum, Horminium pyrenaicum, Bellis perennis, Crepis bursifolia*, mousses, lichens, etc.). A l'intérieur d'une même association, nous constatons parfois plusieurs peuplements purs (par exemple dans l'association du hêtre dans les combes à neige, dans le Hochstauden-Läger). Il faut se garder de les confondre avec les associations.

Dans la *méthode* à suivre, pour connaître les associations, pour les individualiser et les délimiter, *deux procédés* se présentent pour le débutant, qui ne connaît pas encore d'espèces caractéristiques.

(1) FLAHAULT (Ch.). Projet de Nomencl. phyogéogr., 1900.
(2) DRUDE (O.). Handbuch der Pflanzengeographie, 1890.
(3) BROCKMANN-JEROSCH (H.). Die Pflanzengesellschaften der Schweizeralpen I. Teil. *Die Flora des Puschlar und ihre Pflanzengesellschaften*, 1907.

Un premier moyen, *floristique*, consiste à délimiter provisoirement des peuplements donnés, grâce à une ou plusieurs espèces dominantes. Rien de plus facile que de distinguer ces peuplements et de les délimiter. On recherche ensuite les espèces caractéristiques de chacun de ces peuplements, qu'il s'agisse d'une forêt d'épicea, d'une prairie à *Nardus stricta*, d'un étang à nénuphars, d'un groupement de *Quercus cocci-fera,* d'un haut marais de *Sphagnum*, d'un rocher à *Rhabdo-weisia*, etc. On reconnaîtra, en suivant cette méthode, qu'un certain nombre de ces peuplements n'ont pas d'espèces caractéristiques : ce ne sont donc pas des associations individualisées, mais souvent des fractions d'autres associations (comme, par exemple, plusieurs peuplements de combes à neige : les peuplements de Pinus halepensis, de P. Pinaster et P. Pinea du Midi de la France, aussi, ne sont peut-être qu'une même association).

La seconde méthode pour arriver à la connaissance des associations est essentiellement *topographique*. Le phyto-géographe choisit des unités topographiques bien déterminées, montrant une végétation du même caractère écologique : les ubacs des rochers, les alluvions d'une rivière, les flaques d'eau stagnante, le bas rivage d'un lac [*Gadeceau* (1)]. des éboulements, etc. Il en fait l'inventaire et détermine les espèces caractéristiques pour chacune de ces stations topographiques. Citons un exemple. En étudiant les éboulis d'un district, on rencontrera d'abord beaucoup d'espèces qui se montrent partout ailleurs (les ubiquistes) : en outre, on constatera des espèces propres à la station étudiée. De ces dernières, un certain nombre habitent exclusivement les éboulis calcaires ; de plus, on s'apercevra que la composition de la flore varie selon l'exposition et la nature des débris, mais aussi que chacune de ces stations possède, en outre, des espèces caractéristiques. De cette manière, on ar-

(1) GADECEAU (E.). Le lac de Grand Lieu, 1909.

rive à établir l'unité inférieure, l'association. Dans les régions arides, semi-arides et froides, où les peuplements d'espèces dominantes font le plus souvent défaut, cette méthode, appliquée avec sagacité, fournira des résultats cadrant bien avec ceux qu'a donnés la méthode floristique. En appliquant *les deux modes* d'investigations pour la même association, le résultat final doit nécessairement être le même.

Un exemple tiré de l'étude détaillée de M. *Guinier* (1), sur le *Roc de Chère*, en Savoie, nous démontre combien concordent les résultats des deux méthodes, pourvu que l'observateur soit précis. La localité, dont M. Guinier nous donne l'inventaire floristique, n'occupe que quelques mètres carrés complètement abrités, bien exposés au soleil, à sol rocheux riche en carbonate de chaux (12 %). L'association végétale y est — nous le comprenons tout de suite — une prairie à *Bromus erectus*, association très répandue dans toute la Suisse. Il est intéressant de la comparer à un exemple de la même association bien développé et situé à plus de 250 kilomètres de là vers l'est, à une prairie de *Bromus erectus* sur une colline morainique, près de Seewis i. O. (Grisons), 750 mètres s. m.

Onze espèces sont communes aux deux listes :

Bromus erectus.	*Helianthemum vulgare.*
Scabiosa columbaria.	*Asperula cynanchica.*
Silene nutans.	*Teucrium chamaedris.*
Dianthus silvestris Wulf.	*Pimpinella saxifraga.*
Sedum album.	*Anthyllis Vulneraria.*
	Thymus serpyllum.

Les *Potentilla verna, Sedum reflexum* et *Hieracium pilosella* du Roc de Chère sont remplacés à Seewis i/O par *Potentilla Gaudini, Sedum sexangulare, Hieracium florentinum*.

_____

(1) GUINIER (Ph.). Le Roc de Chère, 1906, p. 70.

Trois espèces (*Anthericus ramosus*, *Fumana procumbens* et *Arbutus uva ursi*) manquent au relevé de Seewis, où, par contre, on remarque: *Andropogon ischaemum*, *Phleum Boehmeri*, *Kœleria cristata var.*, *Carex nitida*, *Vincetoxicum officinale*, *Globularia Willkommii*, *Teucrium montanum*, *Veronica spicata*, *Euphrasia stricta*, *Lotus corniculatus var. pilosus*.

C'est évidemment la même association, complètement évoluée dans le cas des Grisons, moins évoluée dans l'exemple de la Haute-Savoie.

Quant à la désignation des associations, il convient de s'en tenir aux propositios faites par MM. Flahault et Schrœter (1), en leur donnant un nom employé dans le langage populaire, ou en ajoutant le suffixe -*etum* au nom générique ou spécifique des espèces dominantes. Gardons-nous seulement d'exagérer la valeur des dominantes, dont l'association porte le nom. Elles ne représentent, au fond, qu'un élément de l'association : partout où elles apparaissent en masse, il faut bien s'assurer que les autres éléments s'y trouvent aussi ; à cette condition seule, on peut parler d'association. Le fait que bien des fois, l'espèce dominante est également caractéristique n'y change rien (*Fagus*, *Elyna myosuroides* [*spicata*], *Festuca varia*, etc.). Ces espèces caractéristique peuvent s'entremêler à une autre association dans les zones de contact, là où les conditions sont également favorables à deux associations (associations mixtes, Mischbestände).

M. *Gradmann* (2) et M. *Beck v. Mannagetta* (cfr. *Oest. Bot. Zt.* 52, p. 426) citent plusieurs cas où la même espèce domine dans plusieurs associations. Ajoutons aux exemples, donnés par eux, *Pinus montana*, dominant sur certains hauts marais et formant en même temps des peuplements sur sol calcaire ;

(1) Nomenclature phytogéographique, **Rapports et propositions**, *III' Congrès internat. de Bot. Bruxelles*, 1910.
(2) Uber Begriffsbildung, etc. (*Engl. Bot. Jahrb*, 1909, Beibl. 99.)

Curvuletum (cfr. Braun, *l. c.*) ; *Agrostis tenuis*, dominant dans le pâturage broussailleux (« Buschweide ») et dans des prés un peu humides. Mais l'exemple le plus frappant nous est fourni par le gazon à *Sesleria caerulea*, association des pentes ensoleillées, sèches, d'aspect xérophile, dans les Alpes, formant, au contraire, un groupement hygrophile sur les îles d'Aland, dans le golfe de Botnie. Palmgren (1) y signale la place prépondérante qu'y occupent des *Carex* comme *C. Goodenowii, C. panicea, C. disticha* et *Ranunculus auricomus* et *Galium uliginosum* même en font partie. Voici deux associations complètement différentes, bien que l'espèce dominante soit la même. Nous voyons donc combien il est désirable d'ajouter une courte diagnose mentionnant des espèces caractéristiques dans les cas où la dénomination empruntée à l'espèce dominante ne suffit pas pour exprimer l'association sans équivoque possible.

Jetons encore un coup d'œil sur les *variations éventuelles à l'intérieur même d'une association*. Pour mieux les faire comprendre, nous nous servirons de la *comparaison avec l'unité systématique* (l'espèce), comparaison qui, sans être nouvelle [cfr. Schrœter (2), Pavillard (3)], n'a jamais été soutenue jusque dans le détail.

Chaque association se compose d'individus (Einselbeständen, Einzelassoziationen), de même que l'espèce se compose d'individus nous apparaissant comme identiques. Les deux termes sont donc des abstractions philosophiques, qu'il s'agit de faire entrer dans une sorte de diagnose, en se basant sur les caractères communs reliant les individus du même type. Il y a des associations nettement tranchées (association du hêtre,

---

(1) *Acta Soc. pro Fauna et Flora fennica*, 1912.
(2) Schrœter (C.). u. Kirchner (O.). *Die Veget. d. Bodensee. II. P.*, 1902.
(3) Pavillard (J.). Essai de nomenclature phytogéogr. *Bull. soc. Languedoc. géogr.*, 1912.

de *Salicornia*, de l'*Aristida pungens* ; d'autres se comportent comme des espèces polymorphes et sont difficiles à délimiter. Comme on parle d'espèces transitoires, on pourrait parler d'associations mixtes et d'associations indéfinies. Ces dernières occupent même dans l'Europe moyenne, grâce à l'influence de l'homme, une surface considérable.

Continuant notre parallélisme, nous distinguons, à l'intérieur de l'association :

1. *Races régionales* (Varieties of Associations de Warming, Subtypen de Schrœter, Facies de Brockmann, *l. c.*, et de Braun, *l. c.*). Ces variations floristiques de l'association dépendent du passé historique de la flore. Exemples : l'*Elynetum* des Alpes orientales, des Alpes occidentales et des Pyrénées occupe partout la même station : les endroits exposés au vent, secs et le plus souvent dépourvus de neige en hiver. Les conditions de vie étant partout les mêmes, la physionomie de l'Elynetum pyrénéen et de celui des Alpes est identique. La composition floristique, par contre, tout en montrant beaucoup d'affinités, n'est pas tout à fait la même ; des espèces manquant ici ont été remplacées là par d'autres ayant les mêmes adaptations (cfr. Braun, *l. c.*). Certaines espèces dominantes, comme le hêtre, les *Sphagnum*, le *Carex curvula*, créant, partout où elles se trouvent, des conditions de vie, sinon identiques, du moins très semblables, constituent partout une association floristiquement peu différente. Que l'on observe la forêt de hêtres de la Suisse méridionale, celle des Cévennes ou de l'Apennin du Piémont [cfr. *Gola* (1)] ; le fond de la flore n'a pas changé : à peine y distingue-t-on des *races* différentes, d'après la substitution d'une espèce caractéristique à une autre.

Les associations ouvertes montrent, en général, une varia-

(1) Gola (G.). La vegetazione dell'Apennin piemontese, 1911.

tion plus grande, suivant les diverses régions, et réalisant un plus grand nombre de *races régionales*.

2. *Variations altitudinales.* Les espèces végétales occupant une zone verticale considérable se présentent souvent sous des formes distinctes l'une de l'autre, mais réunies au type par des formes intermédiaires (*Poa annua, Cerastium arvense, Scabiosa columbaria, S. subalpina, S. lucida, Juniperus communis*, var. *intermedia*, var. *montana* [*nana*]). Il en est de même pour certaines associations ayant une répartition altitudinale assez grande. Les variations floristiques, résultant de cette répartition altitudinale, peuvent être appelées *échelons altitudinaux* de l'association (*Höhenglieder der Assoziation*).

Toute association, sous le même climat, a son étage préféré, une ceinture parfaitement favorable à son développement maximum. A mesure qu'on s'éloigne de cette zone privilégiée, l'aspect de l'association et sa composition changent. Il est facile de reconnaître l'étage préféré des associations d'arbres forestiers ; c'est moins facile s'il s'agit d'associations moins apparentes. L'un de nous (B.) a proposé de s'en tenir aux étages principaux de la végétation, en subdivisant l'association en échelons altitudinaux. Ainsi, on aurait à distinguer les échelons montagnards, subalpins et alpins de l'association de *Trichophorum caespitosum*, de *Festuca varia*, des associations de rochers : les échelons montagnards et subalpins de l'association de l'épicea, de *Molinia*, de *Trisetum flavescens*, les échelons subalpins et alpins de *Carex sempervirens*, de *Vaccinium*, etc. Les échelons altitudinaux d'une association ne correspondent cependant pas toujours aux étages de végétation (*Vegetationsstufen*).

L'étude des associations demande qu'on tienne compte des échelons altitudinaux, sans quoi les résultats obtenus dans des districts d'altitude différente ne sont plus comparables. Nous nous bornons ici à renvoyer le lecteur au tableau comparatif de l'association du *Trisetum flavescens* des deux val-

lées voisines de la Bernina et du Puschlav (Grisons), dans
l'ouvrage de M. Rübel (1).

3. Toutes les modifications floristiques de l'association, que
l'on juge bon de distinguer, et qui ne correspondent ni aux
races régionales ni aux échelons altitudinaux, peuvent être
comparés aux *variations biologiques* de l'espèce (Standorts-
variationen). Leur nombre est limité. Ce qui les distingue
d'abord des associations définies, autour desquelles elles
rayonnent, c'est le manque d'espèces caractéristiques. Il s'est
produit un groupement différent des constituants, un changement de la physionomie floristique, mais les espèces caractéristiques sont, en général, celles de l'association mère, et
la physiographie de la station est peu différente.

Retenons, pour ces variations de l'association, le terme
conventionnel de sous-association (Subassoziation, Neben-
bestand). Dans les associations de la *garigue*, il serait facile
d'en distinguer quelques-unes, selon la répartition des espèces
dominantes (2).

Les variations dues au *substratum* (Substratvariation),
qui rentrent dans la même catégorie de variations
biologiques, méritent une mention spéciale. Le *Quer-
cetum Ilicis* (association du chêne vert) se compose d'espèces bien différentes, suivant qu'on l'observe sur le calcaire
ou sur sol siliceux. La différence est grande, les espèces caractéristiques ayant tellement changé que l'on pourrait considérer les peuplements du chêne vert sur sol calcaire et ceux
sur sol siliceux comme deux associations différentes (cf. aussi
M. Flahault). D'autres associations subissent, avec le changement du substratum, des modifications moins profondes. C'est
le cas pour les associations produisant une épaisse couche

(1) RÜBEL (E.). Pflanzengeogr. Monogr. d. Berninagebietes. *Engl. Bot.
Jahrb.*, 1911-1912.

(2) BLANC (L.). La végét. aux environs de Montpellier. *Bull. Soc. bot. Fr.*,
1905.

d'humus : forêt de hêtres, *Curvuletum*, combes à neige. Les espèces caractéristiques y subissent peu de changement ; on est amené à y considérer les variations survenues comme simples variations édaphiques de la même association. Toute sous-association se rattache à l'association mère par les affinités de l'inventaire floristique.

L'étude des groupements végétaux d'un territoire restreint doit nécessairement s'appuyer sur les unités fondamentales. « Les associations sont la base solide des études de géographie botanique. » [Flahault (1)].

Mais il convient aussi de rechercher des *procédés de classement*, permettant de dresser l'inventaire provisoire et d'ébaucher une première synthèse des résultats acquis.

Un des moyens le plus souvent préconisés consiste à grouper les associations d'après leurs affinités physionomiques et écologiques. A cette tendance répondent, par exemple, les notions de « *formation* » et de « *type de végétation* », dans le sens préconisé par les rapporteurs du Congrès de Bruxelles.

« En définitive, il semble que nous pouvons considérer une formation comme une expression actuelle de certaines conditions de vie, *indépendante de la composition floristique* » (p. 6), et « une formation se compose d'associations qui, dans leur *composition floristique, sont différentes*, mais qui correspondent à des conditions stationnelles semblables et revêtent des formes de végétation analogues ».

Dans la pensée de M. Schrœter (*l. c.*, 1902) et de M. Warming (2), cette notion de la formation, qui est basée unique-

(1) FLAHAULT (Ch.). Premier essai de Nomenclature phytogéographique. *Bull. Soc. Languedoc. de géogr.*, 1901.

(2) WARMING (Eug.). Oecologie of Flants, 1910.

ment sur la physionomie et sur l'écologie (1), correspondrait à peu près au genre de la systématique.

Cette dernière manière de voir n'est pas la nôtre. Pour nous, la méthode la plus conforme aux principes posés dans cette étude nous paraît consister à grouper les associations d'après *leurs affinités floristiques* et de raisonner à l'égard *des groupes d'associations* ainsi établis comme à l'égard des associations elles-mêmes. Le terme supérieur à l'association serait alors le groupe d'association, donné *non par le même port*, mais par une *composition floristique semblable, par la présence de caractéristiques communs*.

Quant au terme formation, employé souvent aussi par les auteurs allemands pour exprimer ce que nous désignons ici sous le nom de groupe d'association, il nous paraît désirable de lui garder son sens primitif (voir Grisebach, Drude, dont G. Beck v. Managetta nous semble avor donné le commentaire le plus satisfaisant) (2). La formation serait alors l'expression physionomique et écologique de l'association, comme la forme biologique est l'expression physionomique et écologique de l'espèce (Pavillard, *l. c.*).

Il nous reste à faire quelques remarques au sujet de l'évolution des groupements de plantes, — leur *succession*, pour employer le terme technique.

Il y a une quinzaine d'années qu'on l'étudie méthodiquement, surtout en Amérique et en Angleterre, grâce à l'heureuse initiative de *Fred. E. Clements*. Aujourd'hui, Schimper

(1) Sur cette base, MM. BROCKMANN-JEROSCH et RÜBEL ont élaboré récemment un remarquable système de classement et de dénomination des groupements végétaux. Ils ont su tirer un excellent parti de la notion de la valeur écologique (œkologische Wertigkeit), introduite auparavant par M. Brockmann. (Voir Brockmann-Jerosch H. et Rübel E., Die Einteilung der Pflanzengesellschaften nach œklogisch-physiognomischen Gesichts Nunkten. Leipzig. 1912.)

(2) Uber die Umgrenzung d. Pflanzenformationen, 1902.

n'écrirait plus, comme en 1898 (1) : « En dépit du haut inté-
rêt que présente le développement des formations, on n'a
porté que peu d'attention sur ce sujet. » Pourtant, les observations de ce genre ne sont pas rares dans des œuvres anciennes, mais elles se trouvent disséminées dans des catalogues
de flores, des descriptions de voyage, etc. (2).

En général, on désigne sous le nom de *série* une succession
chronologique de différents groupements de plantes dans une
aire topographiquement déterminée. Elle se compose donc
d'un ou plusieurs *stades* (générations de Kerner, phases, éta-
pes), dont le dernier (climax stage) est plus ou moins stable.

Les opinions sur la délimitation des séries diffèrent beaucoup. L'élément essentiel pour la classification des phénomènes de succession est encore à découvrir : il nous faudrait
une unité fondamentale que l'on pourrait coordonner, par
exemple, avec la notion d'espèce et d'association. Elle correspondrait au milieu typique, autour duquel se groupent d'abord
les variations géographiques (« substitution » [Cowles] de
certains traits floristiques dans différentes régions), ensuite
les variations altitudinales, et enfin celles de différentes conditions édaphiques (sol, exposition, etc.).

La *paléo ttologie* nous révèle que, dans un territoire donné,
différentes végétations se sont succédé d'une époque géo-
logique à l'autre. En Scandinavie, par exemple, la végétation
de *Dryas*, de *Salix polaris*, *retusa*, de plantes aquatiques, etc.,
qui s'était établie après la période glaciaire, a été envahie par
une forêt de *bouleaux*, celle-ci par une forêt de *Pinus sylvestris*, qui, plus tard, a cédé la place à une forêt de *chênes*.
Enfin, dès la période qui confine à la période historique,
Fagus silvatica et *Picea excelsa* ont commencé à s'étendre. Il

(1) *L. c.*, p. 200.
(2) Voir l'aperçu historique dans CLEMENTS (Féd.-E.), The development
and structure of vegetation (*Studies in the veg. of the state*, III). Lincoln,
1904.

paraît généralement difficile de saisir ces successions ; nous les appelons *phylogénétiques* [« régionales » Cowles (1)]; elles nous montreraient, si nous étions assez renseignés, la phylogénèse des associations, effectuée, sans doute, après la phylogénèse des espèces.

Opposons à ces phénomènes *l'ontogenèse des groupements actuels*, qui s'établissent parfois sous les yeux mêmes de l'observateur, et dont il peut saisir avec exactitude la succession.

Quelques exemples : Le long du littoral méditerranéen (près de Montpellier), se trouvent des dunes assez récentes portant une végétation pauvre : *Ammophila arenaria, Vulpia uniglumis, Crucianella marina, Malcolmia maritima, Anthemis maritima* et quelques autres plantes éparses : les sables, filtrés par la pluie, perdent peu à peu leur salure, et (dans la Camargue, par exemple) (2) des arbustes, tels que : *Juniperus phœnicea*, et enfin le Pin d'Alep, s'installent avec un cortège de plantes moins halophiles. Dans les montagnes, les éboulis se fixent en se recouvrant d'un gazon. Sur les alluvions des torrents, s'élèvent des forêts d'*Alnus*, après une phase de *Salix, Hippophaë* et *Myricaria*.

A une altitude qui dépasse 3.000 mètres, il n'y a, dans les Alpes, que peu de groupements méritant de porter le nom de gazon. Une végétation ouverte occupe les rochers, les éboulis, qui ne se couvrent jamais d'un tapis continu. Les séries sont donc réduites au premier stade : il n'y a pas de succession.

En descendant à l'étage du gazon continu, on arrive, par la comparaison attentive des stades successifs plus ou moins avancés, à *reconstruire inductivement la série*. On observe, par exemple, comment des arbrisseaux nains et couchés

(1) COWLES (H.-C.). The causes of vegetative cycles. *Bot. Gaz.*, 1911.
(2) D'après FLAHAULT et COMBRES. Sur la flore de la Camargue. *Bull. Soc. bot. Fr.*, 1894.

(*Dryas*, *Salix retusa*), qui habitent les éboulis, préparent la
station pour des plantes humicoles (*Elyna*, *Carex firma*, etc.).
Celles-ci s'établissent entre les petites branches, envahissent
le petit espalier et finissent par l'anéantir complètement en formant un gazon serré. Ailleurs, des plantes en touffe (*Carex
sempervirens*, *Festuca violacea*, *Sesleria cœrulea*, *Stipa calamagrostis*) débutent sur les pierriers avec leurs gradins ga-
zonnés interrompus. Par suite de l'accumulation d'humus, les
plantes occupant la terre nue entre les touffes succombent
dans la lutte contre les espèces du gazon, en voie de formation.

Mais que l'on se garde d'aller trop loin dans ces interpré-
tations évolutives : très souvent, en effet, l'association demeure
telle qu'elle se présente aujourd'hui, les facteurs topographiques et climatiques locaux empêchant tout développement
ultérieur.

Descendant plus bas encore, dans l'étage des broussailles
(par exemple : *Alnus viridis* ou le Krummholz, formé par
Pinus montana couchée), ou à l'étage subalpin des futaies, la
série s'allonge par l'adjonction de nouveaux stades. Les végétaux ligneux de haute taille établissent progressivement, par
leur propre croissance, des conditions de vie favorables pour
un cortège de plantes d'ombre, qui triomphent définitivement
de la lutte contre leurs prédécesseurs héliophiles.

Après avoir délimité en gros les groupes de séries distinctes se rapportant à chaque étage de végétation, nous nous
empressons d'ajouter qu'il y a des séries ne se soumettant pas
au régime de l'étage de végétation : alors la série ne finit pas
par l'association climatique, au sens de *Schimper*. Les cours
d'eau avec leurs alluvions et leurs bords, les avalanches avec
leurs couloirs et leurs déjections, déterminant, à travers les
étages, des séries autonomes possédant un caractère spécial ;
les marais, les eaux stagnantes auront toujours une végétation différente de celle de l'étage auquel ils appartiennent. Ces
groupes de séries, caractérisées par des traits spéciaux, cor-

respondent en partie aux successions rétrogrades, définies par *Cowles*, comme ayant un stade final édaphique, tandis que ses successions progressives finissent par une association climatique.

A côté de ce groupe de successions, dues aux modifications de la surface terrestre, relevons encore le groupe des séries ontogénétiques provoquées par l'homme qui exerce, comme on sait, une influence profonde sur la végétation.

Ces *successions artificielles* [successions biotiques régressives, *Cowles*] sont généralement privées de régularité et de périodicité, parfois même aussi de stades distincts. La connaissance précise des causes qui déterminent ces successions nous conduit à prendre, pour base de la distinction des séries, la modalité particulière de l'intervention humaine : déboisement, irrigation, pâturage, etc.

Les débutants agiront avec prudence en abordant ces études dans une localité où s'exerce une seule de ces influences élémentaires ; ils pourront alors comparer le résultat cultural avec l'état naturel. On pourrait ainsi considérer d'abord les prés soumis simplement au fauchage ou au pâturage ; ensuite ceux qui subissent, en outre, l'influence de l'irrigation, des amendements, etc.

Les cultures — stade artificiel généralement terminal — ne subsistent que grâce à l'intervention incessante de l'homme. Si l'homme les abandonne, la végétation naturelle, refoulée par les cultures, se met à l'œuvre pour reconstituer le tapis primitif. En général, elle y parvient. Dans les prés des régions forestières, par exemple, qui doivent leur existence au déboisement, au fauchage ou au pâturage annuel, divers arbustes (*Vaccinium*, *Rhododendron* dans les Alpes, *Rubus*, *Ligustrum*, etc., sur le plateau suisse) envahissent le gazon. Les arbres voisins fournissent des graines ; les germes naissent, épargnés par la faux et la dent des animaux, croissent et finissent par reconquérir leur domaine. Nous savons cependant que, vers la limite supérieure de l'étage subalpin des

Alpes, ou vers les zones polaires, la forêt, une fois détruite, peut souvent ne pas retrouver les conditions nécessaires à sa reconstitution.

Dans les lignes précédentes, nous avons voulu exposer méthodiquement les idées fondamentales qui ont inspiré nos travaux phytogéographiques et dirigé les études entreprises par l'un de nous dans le Midi de la France, par l'autre dans les Alpes italiennes orientales. Le contraste frappant de ces deux régions nous a amenés à rechercher ensemble une méthode et des principes pouvant être appliqués partout où il s'agit d'une étude détaillée. Nous serons pleinement satisfaits si les phytogéographes débutants tirent quelque profit de nos efforts.

Institut de Botanique de Montpellier.

Mai 1913.

2

REMARKS ON THE STUDY OF PLANT GROUPS

Josias Braun* and E. Furrer

*This summary was prepared expressly for this Benchmark
volume by Robert P. McIntosh from "Remarques sur
l'étude des groupements de plantes,"* Soc. Languedoc.
Geogr. Bull. **36:**20–41 (1913)

Groups of plants are studied from four points of view—descriptive, physiological, geographic, and genetic or evolutionary. The essential for study of synecology is knowledge of the floristic composition of the *association,* a plant group with a known floristic composition, uniform physiognomy, uniform habitat conditions, and, in addition, one to several *characteristic species* largely restricted to the association. These are better indicators of the association than dominant species, which are often ubiquitous. The association may, or may not, be an organized community of mutually interacting members, and interspecific relations are difficult to determine. The equilibrium and duration of associations vary from almost indefinite in climatic associations to transitory in associations of sand bars along rivers and cultural situations. Associations may develop even in plantations by immigration of natural characteristic species of the association.

Characteristic species of an association in one climatic or altitudinal area may not be characteristic elsewhere, and comparative geographic studies must determine the limits of the area of the association. It is not possible to distinguish all possible associations. The combinations are innumerable and the geographer must describe the associations which cover a more or less large area in the search *for a unit comparable to the species capable of serving as a basis for work in comparative geographical botany* and synecological investigations.

The study of associations requires account of characteristic species, constant species (occurring in 50 percent or more of the stands) and estimates of density of each species. Two methods are available to the beginner: (1) a floristic means, of delimiting stands provisionally by dominant species, and (2) selection of well-defined topographical sites having similar vegetations. The results of the floristic and site or habitat methods must be the same if the researcher is careful. Associations may be named by adding the suffix *-etum* to a generic or specific name of the dominant species, although one should avoid giving too much significance to the dominants.

Every association is composed of individual stands just as a species is

*J. Braun added Blanquet to his name following his marriage.

composed of individuals that seem identical. The two terms are abstractions and must be diagnosed by basing each on the common characters of individuals of the same type. There are clearly distinct associations, but others are difficult to delimit and can be described as mixed or indefinite associations. Within the association we can distinguish:

1. Regional races, dependent on the history of the flora.
2. Altitudinal variations.
3. Biological and substrate variations (subassociations).

The associations are the fundamental units of vegetation and should be grouped according to *their floristic affinities* and classified by common floristic composition, not by physiognomy.

A chronological succession of different plant groupings is called a *series*. It is composed of several stages, the last of which (climax stage) is more or less stable. Delimitations of series differ and the essential element for classification of sequential phenomena is yet to be discovered. Paleontology reveals that different vegetations have followed one another from one geological epoch to the next, but it generally seems difficult to make out these sequences. In addition to these sequences, there are sequences due to changes in the earth's surface (e.g., recent dunes, landslides); and, in some cases, it is possible, by comparing more or less advanced successive stages, to reconstruct the series inductively. In some cases no development occurs, due to topographic or local climatic conditions. Some series correspond to the retrograde sequences defined by Cowles having a terminal edaphic stage, whereas his progressive series end with a climatic association.

In addition to physiographic sequences there are sequences due to the influence of man, deforestation, irrigation, pasturage, etc. When human intervention ends, natural vegetation starts to restore the original ground cover, but in extreme situations (arctic or alpine) conditions may not permit this reconstitution.

3

THE FUNDAMENTAL LAW IN DEVELOPING THE VEGETATION COVER*

L. G. Ramensky

The present report is the first of a series of articles in which the author's general-theoretical and methodological views and conclusions will be set down. They are the first of some 16 years field work by the author and his collaborators devoted chiefly to meadows, to a lesser degree the steppes and forests of Woronesh Province, also the moors and lakes in Nordruss-land and Kamtschatka; these investigations consisted of about 20,000 plant surveys, besides surface measurement, soil analysis "etc."

... The author follows the concept of the coenose introduced by Gams and gives the following definition: "A coenose is an ecologically limited, locally homogeneous plant grouping (the concrete individual coenose) or the totality of plant groups of ecological and floristic delimited similarity (the abstract coenose); the historical and topological resemblance of such groups can be different." He repudiates the phytosociological and plant population viewpoints because of their anthropomorphisms and characterizes the behavior of the coenobiota as decidedly "antisocial.". . .

In considering the moderating law of the coenose one must take into account the production of propagation units (seeds, spores) and their importance for the maintenance of the coenose; in herbaceous coenoses (meadows, steppes and moors) repair appears to follow vegetative processes and the importation of propagation units to play no great part. But this very vegetative propagation often produces a zone of mixing at the limits of two coenoses. In the instance of stenocoenoses, in which the area is segmented into many small lobes or is entirely divided, the influence of these anomalous limits manifests itself mostly over their entire surface; they must not, therefore, be compared directly with macrocoenoses, which homogeneously cover continuously great surfaces (meadows, steppes).

The historical factor is of relevance for a coenose only in the background when an equilibrium is reached between the other factors and the structure of the community, be it a continuous absolute balance (the ideal extreme case) or a dynamic balance which shifts itself slowly in parallel with alterations in external conditions. A diffuse distribution of species is a characteristic of attained equilibrium and at the same time an indication of a high degree of competition among the plants.

The plant cover cannot be understood as a mass phenomenon through the analysis of small parts (neither from uniform coenoses); the ecological moderating law in vegetation composition

*This translation by Edward J. Kormondy is of the abstract by Selma Ruoff of the original Russian aritcle by L. G. Ramensky.

can be determined only through the summarizing of statistical surveys of greater areas and then the obtaining of averages. The surface appearance of a coenose, that is the minimal area in which the correct moderation law makes itself felt numerically, can be very different. In the smallest coenose the significant factor in development of the surface appears to be frequency, whereas abundance is more important in larger coenoses. . . .

On the basis of his numerous plant surveys coupled with different methods, the author comes to the conclusion that the plant cover modifies itself continuously in space. The sharp boundary between coenoses is an individual instance in need of special explanation (influence of culture, discontinuous alteration of other factors, etc.) The rule of continuity in the three dimensional composition of plant cover will be supplemented and clarified by the rule of ecological individuality of plant species; each species reacts to the other

unique factors and occurs as an independent member in the coenose; there are no two groups which end with identical abundance in a coenose. The possible combinations for the association are very great and each one is a group of individual coenoses in which the greatest abundance and the greater annual change is attained.

The kaleidoscopic alteration in the spatial composition of meadows, steppes and low moors (woods and sphagnum bogs appear homogenous) speaks decidedly against a classification of inflexible units. "Groups are not stable, only the rules of plant combinations" and these support research. As a goal of investigation, the author denotes an arrangement of coenoses in ecological sequence with the corresponding abundance curves of their species and according to the coordinates of their factors. The ideal would be the combining of all isolated coenose studies into a uniform coordinated scheme.

4

The individualistic concept of the plant association*

H. A. Gleason

The continued activity of European ecologists, and to a somewhat smaller extent of American ecologists as well, in discussing the fundamental nature, structure, and classification of plant associations, and their apparently chronic inability to come to any general agreement on these matters, make it evident that the last word has not yet been said on the subject. Indeed, the constant disagreement of ecologists, the readiness with which flaws are found by one in the proposals of another, and the wide range of opinions which have been ably presented by careful observers, lead one to the suspicion that possibly many of them are somewhat mistaken in their concepts, or are attacking the problem from the wrong angle.

It is not proposed to cite any of the extensive recent literature on these general subjects, since it is well known to all working ecologists. Neither is it necessary to single out particular contributions for special criticism, nor to point out what may appear to us as errors in methods or conclusions.

It is a fact, as Dr. W. S. Cooper has brought out so clearly in a manuscript which he has allowed me to read, and which will doubtless be in print before this, that the tendency of the human species is to crystallize and to classify his knowledge; to arrange it in pigeon-holes, if I may borrow Dr. Cooper's metaphor. As accumulation of knowledge continues, we eventually find facts that will not fit properly into any established pigeon-hole. This should at once be the sign that possibly our original arrangement of pigeon-holes was insufficient and should lead us to a careful examination of our accumulated data. Then we may conclude that we would better demolish our whole system of arrangement and classification and start anew with hope of better success.

Is it not possible that the study of synecology suffers at the present time from this sort of trouble? Is it not conceivable that, as the study of plant associations has progressed from its originally simple condition into its present highly organized and complex state, we have attempted to arrange all our facts in ac-

---

* Contributions from The New York Botanical Garden, No. 279.

80

cordance with older ideas, and have come as a result into a tangle of conflicting ideas and theories?

No one can doubt for a moment that there is a solid basis of fact on which to build our study of synecology, or that the study is well worth building. It is the duty of the botanist to translate into intelligible words the various phenomena of plant life, and there are few phenomena more apparent than those of their spatial relations. Plant associations exist; we can walk over them, we can measure their extent, we can describe their structure in terms of their component species, we can correlate them with their environment, we can frequently discover their past history and make inferences about their future. For more than a century a general progress in these features of synecology can be traced.

It has been, and still is, the duty of the plant ecologist to furnish clear and accurate descriptions of these plant communities, so that by them the nature of the world's vegetation may be understood. Whether such a description places its emphasis chiefly on the general appearance of the association, on a list of its component species, on its broader successional relations, or on its gross environment, or whether it enters into far greater detail by use of the quadrat method, statistical analysis,[1] or exact environometry, it nevertheless contributes in every case to the advancement of our understanding of each association in detail and of vegetation in all its aspects in general.

It is only natural that we should tend to depart from the various conclusions which we have reached by direct observation or experiment, and to attempt other more general deductions as well. So we invent special terms and methods for indicating the differences between associations and the variation of the plant life within a single community. We draw conclusions for ourselves, and attempt to lay down rules for others as to ways and means of defining single associations, by character species, by statistical studies, by environmental relations, or by successional history. We attempt to classify associations, as individual

[1] Pavillard has cast serious doubt on the efficiency of the statistical method in answering questions of synecology. His argument, based solely on European conditions, needs of course no reply from America, but it may properly be pointed out that the intimate knowledge of vegetational structure obtained in this way may easily lead to a much fuller appreciation of synecological structure, entirely aside from any merits of the actual statistical results.

examples of vegetation, into broader groups, again basing our methods on various observable features and arriving accordingly at various results. We even enter the domain of philosophy, and speculate on the fundamental nature of the association, regard it as the basic unit of vegetation, call it an organism, and compare different areas of the same sort of vegetation to a species.

The numerous conclusions in synecology which depend directly upon observation or experiment are in the vast majority of cases entirely dependable. Ecologists are trained to be accurate in their observations, and it is highly improbable that any have erred purposely in order to substantiate a conclusion not entirely supported by facts. But our various theories on the fundamental nature, definition, and classification of associations extend largely beyond the bounds of experiment and observation and represent merely abstract extrapolations of the ecologist's mind. They are not based on a pure and rigid logic, and suffer regularly from the vagaries and errors of human reason. A geneticist can base a whole system of evolution on his observations of a single species: ecologists are certainly equally gifted with imagination, and their theories are prone to surpass by far the extent warranted by observation.

Let us then throw aside for the moment all our pre-conceived ideas as to the definition, fundamental nature, structure, and classification of plant associations, and examine step by step some of the various facts pertinent to the subject which we actually know. It will not be necessary to illustrate them by reference to definite vegetational conditions, although a few instances will be cited merely to make our meaning clear. Other illustrations will doubtless occur to every reader from his own field experience.

We all readily grant that there are areas of vegetation, having a measurable extent, in each of which there is a high degree of structural uniformity throughout, so that any two small portions of one of them look reasonably alike. Such an area is a plant association, but different ecologists may disagree on a number of matters connected with such an apparently simple condition. More careful examination of one of these areas, especially when conducted by some statistical method, will show that the uniformity is only a matter of degree, and that two sample quadrats with precisely the same structure can scarcely be discovered.

Consequently an area of vegetation which one ecologist regards as a single association may by another be considered as a mosaic or mixture of several, depending on their individual differences in definition. Some of these variations in structure (if one takes the broader view of the association) or smaller associations (if one prefers the narrower view) may be correlated with differences in the environment. For example, the lichens on a tree-trunk enjoy a different environment from the adjacent herbs growing in the forest floor. A prostrate decaying log is covered with herbs which differ from the ground flora in species or in relative numbers of individuals of each species. A shallow depression in the forest, occupied by the same species of trees as the surroundings, may support several species of moisture-loving herbs in the lower stratum of vegetation. In other cases, the variations in vegetational structure may show no relation whatever to the environment, as in the case of a dense patch of some species which spreads by rhizomes and accordingly comes to dominate its own small area. The essential point is that precise structural uniformity[2] of vegetation does not exist, and that we have no general agreement of opinion as to how much variation may be permitted within the scope of a single association.

In our attempts to define the limits of the association, we have but two actually observable features which may be used as a basis, the environment and the vegetation. Logically enough, most ecologists prefer the latter, and have developed a system based on character-species. In northern latitudes, and particularly in glaciated regions, where most of this work has been done, there is a wide diversity in environment and a comparatively limited number of species in the flora. A single association is therefore occupied by few species, with large numbers of individuals of each, and it has not been difficult to select from most associations a set of species which are not only fairly common and abundant, but which are strictly limited to the one association. But in many parts of the tropics, where diversity of environment has been reduced to a minimum by the practical completion of most physiographic processes and by the long-continued cumulation of plant reactions, and where the flora is

---

[2] It has often occurred to the writer that much of the structural variation in an association would disappear if those taxonomic units which have the same vegetational form and behavior could be considered as a single ecological unit.

extraordinarily rich in species, such a procedure is impracticable or even impossible. Where a single hectare may contain a hundred species of trees, not one of which can be found in an adjacent hectare, where a hundred quadrats may never exhibit the same herbaceous species twice, it is obvious that the method of characteristic species is difficult or impracticable.

It is also apparent that different areas of what are generally called the same association do not always have precisely the same environment. A grove of *Pinus Strobus* on soil formed from decomposed rocks in the eastern states, a second on the loose glacial sands of northern Michigan, and a third on the sandstone cliffs of northern Illinois are certainly subject to different environmental conditions of soil. An association of prairie grass in Illinois and another in Nebraska undoubtedly have considerable differences in rainfall and available water. A cypress swamp in Indiana has a different temperature environment from one in Florida.

Two environments which are identical in regard to physiography and climate may be occupied by entirely different associations. It is perfectly possible to duplicate environments in the Andes of southern Chile and in the Cascade Mountains of Oregon, yet the plant life is entirely different. Duplicate environments may be found in the deserts of Australia and of Arizona, and again have an entirely different assemblage of species. Alpine summits have essentially the same environment at equal altitudes and latitudes throughout the world, apart from local variations in the component rock, and again have different floras. It seems apparent, then, that environment can not be used as a means of defining associations with any better success than the vegetation.

At the margin of an association, it comes in contact with another, and there is a transition line or belt between them. In many instances, particularly where there is an abrupt change in the environment, this transition line is very narrow and sharply defined, so that a single step may sometimes be sufficient to take the observer from one into the other. In other places, especially where there is a very gradual transition in the environment, there is a correspondingly wide transition in the vegetation. Examples of the latter condition are easily found in any arid mountain region. The oak forests of the southern Coast Range in California in many places descend upon the grass-covered foothills by a wide transition zone in which the trees become very gradually fewer and farther apart until they ultimately disappear completely. In Utah, it may be miles from the association of desert shrubs on the lower elevations across a mixture of shrubs and juniper before the pure stands of juniper are reached on the higher altitudes. It is obvious, therefore, that it is not always possible to define with accuracy the geographical boundaries of an association and that actual mixtures of associations occur.

Such transition zones, whether broad or narrow, are usually populated by species of the two associations concerned, but instances are not lacking of situations in which a number of species seem to colonize in the transition zone more freely than in either of the contiguous associations. Such is the case along the contact between prairie and forest, where many species of this type occur, probably because their optimum light requirements are better satisfied in the thin shade of the forest border than in the full sun of the prairie or the dense shade of the forest. Measured by component species such a transition zone rises almost to the dignity of an independent association.

Species of plants usually associated by an ecologist with a particular plant community are frequently found within many other types of vegetation. A single boulder, partly exposed above the ground at the foot of the Rocky Mountains in Colorado, in the short-grass prairie association, may be marked by a single plant of the mountain shrub *Cercocarpus*. In northern Michigan, scattered plants of the moisture-loving *Viburnum cassinoides* occur in the xerophytic upland thickets of birch and aspen. Every ecologist has seen these fragmentary associations, or instances of sporadic distribution, but they are generally passed by as negligible exceptions to what is considered a general rule.

There are always variations in vegetational structure from year to year within every plant association. This is exclusive of mere periodic variations from season to season, or aspects, caused by the periodicity of the component species. Slight differences in temperature or rainfall or other environmental factors may cause certain species to increase or decrease conspicuously in number of individuals, or others to vary in their vigor or luxuriance. Coville describes, in this connection, the remarkable variation in size of an *Amaranthus* in the Death Valley, which was three meters high in a year of abundant rainfall, and its progeny only a decimeter high in the following year of drought.

The duration of an association is in general limited. Sooner or later each plant community gives way to a different type of vegetation, constituting the phenomenon known as succession. The existence of an association may be short or long, just as its superficial extent may be great or small. And just as it is often difficult and sometimes impossible to locate satisfactorily the boundaries of an association in space, so is it frequently impossible to distinguish accurately the beginning or the end of an association in time. It is only at the center of the association, both geographical and historical, that its distinctive character is easily recognizable. Fortunately for ecology, it commonly happens that associations of long duration are also wide in extent. But there are others, mostly following fires or other unusual disturbances of the original vegetation, whose existence is so limited, whose disappearance follows so closely on their origin, that they scarcely seem to reach at any time a condition of stable equilibrium, and their treatment in any ecological study is difficult. The short-lived communities bear somewhat the same relation to time-distribution as the fragmentary associations bear to space-distribution. If our ecological terminology were not already nearly saturated, they might be termed ephemeral associations.

Now, when all these features of the plant community are considered, it seems that we are treading upon rather dangerous ground when we define an association as an area of uniform vegetation, or, in fact, when we attempt any definition of it. A community is frequently so heterogeneous as to lead observers to conflicting ideas as to its associational identity, its boundaries may be so poorly marked that they can not be located with any degree of accuracy, its origin and disappearance may be so gradual that its time-boundaries can not be located; small fragments of associations with only a small proportion of their normal components of species are often observed; the duration of a community may be so short that it fails to show a period of equilibrium in its structure.

A great deal has been said of the repetition of associations on different stations over a considerable area. This phenomenon is striking, indeed, and upon it depend our numerous attempts to classify associations into larger groups. In a region of numerous glacial lakes, as in parts of our northeastern states, we find lake after lake surrounded by apparently the same communities, each of them with essentially the same array of species in about the same numerical proportions. If an ecologist had crossed Illinois from east to west prior to civilization, he would have found each stream bordered by the same types of forest, various species of oaks and hickories on the upland, and ash, maple, and sycamore in the alluvial soil nearer the water. But even this idea, if carried too far afield, is found to be far from universal. If our study of glacial lakes is extended to a long series, stretching from Maine past the Great Lakes and far west into Saskatchewan, a very gradual but nevertheless apparent geographical diversity becomes evident, so that the westernmost and easternmost members of the series, while still containing some species in common, are so different floristically that they would scarcely be regarded as members of the same association. If one examines the forests of the alluvial floodplain of the Mississippi River in southeastern Minnesota, that of one mile seems to be precisely like that of the next. As the observer continues his studies farther down stream, additional species very gradually appear, and many of the original ones likewise very gradually disappear. In any short distance these differences are so minute as to be negligible, but they are cumulative and result in an almost complete change in the flora after several hundred miles.

No ecologist would refer the alluvial forests of the upper and lower Mississippi to the same associations, yet there is no place along their whole range where one can logically mark the boundary between them. One association merges gradually into the next without any apparent transition zone. Nor is it necessary to extend our observations over such a wide area to discover this spatial variation in ecological structure. I believe no one has ever doubted that the beech-maple forest of northern Michigan constitutes a single association-type. Yet every detached area of it exhibits easily discoverable floristic peculiarities, and even adjacent square miles of a single area differ notably among themselves, not in the broader features, to be sure, but in the details of floristic composition which a simple statistical analysis brings out. In other words, the local variation in structure of any association merges gradually into the broader geographical variation of the association-type.

This diversity in space is commonly overlooked by ecologists, most of whom of necessity limit their work to a comparatively small area, not extensive enough to indicate that the small observed floristic differences between associations may be of much significance or that this wide geographical variation is actually in operation. Yet it makes difficult the exact definition of any association-type, except as developed in a restricted locality, renders it almost impossible to select for study a typical or average example of a type, and in general introduces complexities into any attempt to classify plant associations.

What have we now as a basis for consideration in our attempts to define and classify associations? In the northeastern states, we can find many sharply marked communities, capable of fairly exact location on a map. But not all of that region can be thus divided into associations, and there are other regions where associations, if they exist at all in the ordinary sense of the word, are so vaguely defined that one does not know where their limits lie and can locate only arbitrary geographic boundaries. We know that associations vary internally from year to year, so that any definition we may make of a particular community, based on the most careful analysis of the vegetation, may be wrong next year. We know that the origin and disappearance of some are rapid, of others slow, but we do not always know whether a particular type of vegetation is really an association in itself or represents merely the slow transition stage between two others. We know that no two areas, supposed to represent the same association-type, are exactly the same, and we do not know which one to accept as typical and which to assume as showing the effects of geographical variation. We find fragmentary associations, and usually have no solid basis for deciding whether they are mere accidental intruders or embryonic stages in a developing association which may become typical after a lapse of years. We find variation of environment within the association, similar associations occupying different environments, and different associations in the same environment. It is small wonder that there is conflict and confusion in the definition and classification of plant communities. Surely our belief in the integrity of the association and the sanctity of the association-concept must be severely shaken. Are we not justified in coming to the general conclusion, far removed from the prevailing opinion, that an association is not an organism, scarcely even a vegetational unit, but merely a *coincidence?*

This question has been raised on what might well be termed negative evidence. It has been shown that the extraordinary variability of the areas termed associations interferes seriously with their description, their delimitation, and their classification. Can we find some more positive evidence to substantiate the same idea? To do this, we must revert to the individualistic concept of the development of plant communities, as suggested by me in an earlier paper.[3]

As a basis for the presentation of the individualistic concept of the plant association, the reader may assume for illustration any plant of his acquaintance, growing in any sort of environment or location. During its life it produces one or more crops of seeds, either unaided or with the assistance of another plant in pollination. These seeds are endowed with some means of migration by which they ultimately come to rest on the ground at a distance from the parent plant. Some seeds are poorly fitted for migration and normally travel but a short distance; others are better adapted and may cover a long distance before coming to rest. All species of plants occasionally profit by accidental means of dispersal, by means of which they traverse

---

[3] I may frankly admit that my earlier ideas of the plant association were by no means similar to the concept here discussed. Ideas are subject to modification and change as additional facts accumulate and the observer's geographical experience is broadened. An inkling of the effect of migration on the plant community appeared as early as 1903 and 1904 (Bull. Illinois State Lab. Nat. Hist. 7: 189.) My field work of 1908 covered a single general type of environment over a wide area, and was responsible for still more of my present opinions (Bull. Illinois State Lab. Nat. Hist. 9: 35–42). Thus we find such statements as the following: "No two areas of vegetation are exactly similar, either in species, the relative numbers of individuals of each, or their spatial arrangement" (l. c. 37), and again: "The more widely the different areas of an association are separated, the greater are the floral discrepancies. . . . Many of these are the results of selective migration from neighboring associations, so that a variation in the general nature of the vegetation of an area affects the specific structure of each association" (l. c. 41). Still further experience led to my summary of vegetational structure in 1917 (Bull. Torrey Club 44: 463–481), and the careful quantitative study of certain associations from 1911 to 1923 produced the unexpected information that the distribution of species and individuals within a community followed the mathematical laws of probability and chance (Ecology 6: 66–74).

distances far in excess of their average journey. Sometimes these longer trips may be of such a nature that the seed is rendered incapable of germination, as in dispersal by currents of salt water, but in many cases they will remain viable. A majority of the seeds reach their final stopping-point not far from the parent, comparatively speaking, and only progressively smaller numbers of them are distributed over a wider circle. The actual number of seeds produced is generally large, or a small number may be compensated by repeated crops in successive years. The actual methods of dispersal are too well known to demand attention at this place.

For the growth of these seeds a certain environment is necessary. They will germinate between folds of paper, if given the proper conditions of light, moisture, oxygen, and heat. They will germinate in the soil if they find a favorable environment, irrespective of its geographical location or the nature of the surrounding vegetation. Herein we find the *crux* of the question. The plant individual shows no physiological response to geographical location or to surrounding vegetation *per se*, but is limited to a particular complex of environmental conditions, which may be correlated with location, or controlled, modified, or supplied by vegetation. If a viable seed migrates to a suitable environment, it germinates. If the environment remains favorable, the young plants will come to maturity, bear seeds in their turn, and serve as further centers of distribution for the species. Seeds which fall in unfavorable environments do not germinate, eventually lose their viability and their history closes.

As a result of this constant seed-migration, every plant association is regularly sowed with seeds of numerous extra-limital species, as well as with seeds of its own normal plant population. The latter will be in the majority, since most seeds fall close to the parent plant. The seeds of extra-limital species will be most numerous near the margin of the association, where they have the advantage of proximity to their parent plants. Smaller numbers of fewer species will be scattered throughout the association, the actual number depending on the distance to be covered, and the species represented depending on their means of migration, including the various accidents of dispersal. This thesis needs no argument in its support. The practical universality of seed dispersal is known to every botanist as a matter of common experience.

An exact physiological analysis of the various species in a single association would certainly show that their optimal environments are not precisely identical, but cover a considerable range. At the same time, the available environment tends to fluctuate from year to year with the annual variations of climate and with the accumulated reactionary effects of the plant population. The average environment may be near the optimum for some species, near the physiological limit of others, and for a third group may occasionally lie completely outside the necessary requirements. In the latter case there will result a group of evanescent species, variable in number and kind, depending on the accidents of dispersal, which may occasionally be found in the association and then be missing for a period of years. This has already been suggested by the writer as a probable explanation of certain phenomena of plant life on mountains, and was also clearly demonstrated by Dodds, Ramaley, and Robbins in their studies of vegetation in Colorado. In the first and second cases, the effect of environmental variation toward or away from the optimum will be reflected in the number of individual plants and their general luxuriance. On the other hand, those species which are limited to a single type of plant association must find in that and in that only the environmental conditions necessary to their life, since they have certainly dispersed their seeds many times into other communities, or else be so far removed from other associations of similar environment that their migration thence is impossible.

Nor are plants in general, apart from these few restricted species, limited to a very narrow range of environmental demands. Probably those species which are parasitic or which require the presence of a certain soil-organism for their successful germination and growth are the most highly restricted, but for the same reason they are generally among the rarest and most localized in their range. Most plants can and do endure a considerable range in their environment.

With the continuance of this dispersal of seeds over a period of years, every plant association tends to contain every species of the vicinity which can grow in the available environment. Once a species is established, even by a single seed-bearing plant, its further spread through the association is hastened, since it no longer needs to depend on a long or accidental migration, and this spread is continued until the species is eventually distributed throughout the area of the association. In general, it may be considered that, other things being equal, those species of wide extent through an association are those of early introduction which have had ample time to complete their spread, while those of localized or sporadic distribution are the recent arrivals which have not yet become completely established.

This individualistic standpoint therefore furnishes us with an explanation of several of the difficulties which confront us in our attempts to diagnose or classify associations. Heterogeneity in the structure of an association may be explained by the accidents of seed dispersal and by the lack of time for complete establishment. Minor differences between neighboring associations of the same general type may be due to irregularities in immigration and minor variations in environment. Geographical variation in the floristics of an association depends not alone on the geographical variation of the environment, but also on differences in the surrounding floras, which furnish the immigrants into the association. Two widely distant but essentially similar environments have different plant associations because of the completely different plant population from which immigrants may be drawn.

But it must be noted that an appreciation of these conditions still leaves us unable to recognize any one example of an association-type as the normal or typical. Every association of the same general type has come into existence and had its structure determined by the same sort of causes; each is independent of the other, except as it has derived immigrants from the other; each is fully entitled to be recognized as an association and there is no more reason for regarding one as typical than another. Neither are we given any method for the classification of associations into any broader groups.

Similar conditions obtain for the development of vegetation in a new habitat. Let us assume a dozen miniature dunes, heaped up behind fragments of driftwood on the shore of Lake Michigan. Seeds are heaped up with the sand by the same propelling power of the wind, but they are never very numerous and usually of various species. Some of them germinate, and the dozen embryonic dunes may thenceforth be held by as many different species of plants. Originally the environment of the dunes was identical and their floristic difference is due solely to the chances of seed dispersal. As soon as the plants have developed, the environment is subject to the modifying action of the plant, and small differences between the different dunes appear. These are so slight that they are evidenced more by the size and shape of the dune than by its flora, but nevertheless they exist. Additional species gradually appear, but that is a slow process, involving not only the chance migration of the seed to the exact spot but also its covering upon arrival. It is not strange that individuals are few and that species vary from one dune to another, and it is not until much later in the history of each dune, when the ground cover has become so dense that it affects conditions of light and soil moisture, and when decaying vegetable matter is adding humus to the sand in appreciable quantities, that a true selective action of the environment becomes possible. After that time permanent differences in the vegetation may appear, but the early stages of dune communities are due to chance alone. Under such circumstances, how can an ecologist select character species or how can he define the boundaries of an association? As a matter of fact, in such a location the association, in the ordinary sense of the term, scarcely exists.

Assume again a series of artificial excavations in an agricultural region, deep enough to catch and retain water for most or all of the summer, but considerably removed from the nearest areas of natural aquatic vegetation. Annually the surrounding fields have been ineffectively planted with seeds of *Typha* and other wind-distributed hydrophytes, and in some of the new pools *Typha* seeds germinate at once. Water-loving birds bring various species to other pools. Various sorts of accidents conspire to the planting of all of them. The result is that certain pools soon have a vegetation of *Typha latifolia*, others of *Typha angustifolia*, others of *Scirpus validus;* plants of *Iris versicolor* appear in one, of *Sagittaria* in another, of *Alisma* in a third, of *Juncus effusus* in a fourth. Only the chances of seed dispersal have determined the allocation of species to different pools, but in the course of three or four years each pool has a different appearance, although the environment, aside from the reaction of the various species, is precisely the same for each. Are we dealing here with several different associations, or with a single association, or with merely embryonic stages of some future association? Under our view, these become merely academic questions, and any answer which may be suggested is equally academic.

But it must again be emphasized that these small areas of vegetation are component parts of the vegetative mantle of the land, and as such are fully worthy of description, of discussion, and of inquiry into the causes which have produced them and into their probable future. It must be emphasized that in citing the foregoing examples, the existence of associations or of successions is not denied, and that the purpose of the two paragraphs is to point out the fact that such communities introduce many difficulties into any attempt to define or classify association-types and successional series.

A plant association therefore, using the term in its ordinarily accepted meaning, represents the result of an environmental sorting of a population, but there are other communities which have existed such a short time that a reasonably large population has not yet been available for sorting.

Let us consider next the relation of migration and environmental selection to succession. We realize that all habitats are marked by continuous environmental fluctuation, accompanied or followed by a resulting vegetational fluctuation, but, in the common usage of the term, this is hardly to be regarded as an example of succession. But if the environmental change proceeds steadily and progressively in one direction, the vegetation ultimately shows a permanent change. Old species find it increasingly difficult or impossible to reproduce, as the environment approaches and finally passes their physiological demands. Some of the migrants find establishment progressively easier, as the environment passes the limit and approaches the optimum of their requirements. These are represented by more and more individuals, until they finally become the most conspicuous element of the association, and we say that a second stage of a successional series has been reached.

It has sometimes been assumed that the various stages in a successional series follow each other in a regular and fixed sequence, but that is frequently not the case. The next vegetation will depend entirely on the nature of the immigration which takes place in the particular period when environmental change reaches the critical stage. Who can predict the future for any one of the little ponds considered above? In one, as the bottom silts up, the chance migration of willow seeds will produce a willow thicket, in a second a thicket of *Cephalanthus* may develop, while a third, which happens to get no shrubby immigrants, may be converted into a miniature meadow of *Calamagrostis canadensis*. A glance at the diagram of observed successions in the Beach Area, Illinois, as published by Gates, will show at once how extraordinarily complicated the matter may become, and how far vegetation may fail to follow simple, pre-supposed successional series.

It is a fact, of course, that adjacent vegetation, because of its mere proximity, has the best chance in migration, and it is equally true that in many cases the tendency is for an environment, during its process of change, to approximate the conditions of adjacent areas. Such an environmental change becomes effective at the margin of an association, and we have as a result the apparent advance of one association upon another, so that their present distribution in space portrays their succession in time. The conspicuousness of this phenomenon has probably been the cause of the undue emphasis laid on the idea of successional series. But even here the individualistic nature of succession is often apparent. Commonly the vegetation of the advancing edge differs from that of the older established portion of the association in the numerical proportion of individuals of the component species due to the sorting of immigrants by an environment which has not yet reached the optimum, and, when the rate of succession is very rapid, the pioneer species are frequently limited to those of the greatest mobility. It also happens that the change in environment may become effective throughout the whole area of the association simultaneously, or may begin somewhere near the center. In such cases the pioneers of the succeeding association are dependent on their high mobility or on accidental dispersal, as well as environmental selection.

It is well known that the duration of the different stages in succession varies greatly. Some are superseded in a very short time, others persist for long or even indefinite periods. This again introduces difficulties into any scheme for defining and classifying associations.

A forest of beech and maple in northern Michigan is lumbered, and as a result of exposure to light and wind most of the usual herbaceous species also die.   Brush fires sweep over the clearing and aid in the destruction of the original vegetation.   Very soon the area grows up to a tangle of other herbaceous and shrubby species, notably *Epilobium angustifolium*, *Rubus strigosus*, and *Sambucus racemosa*.   This persists but a few years before it is overtopped by saplings of the original hardwoods which eventually restore the forest.   Is this early stage of fire-weeds and shrubs a distinct association or merely an embryonic phase of the forest?   Since it has such a short duration, it is frequently regarded as the latter, but since it is caused by an entirely different type of environmental sorting and lacks most of the characteristic species of the forest, it might as well be called distinct.   If it lasted for a long period of years it would certainly be called an association, and if all the forest near enough to provide seeds for immigration were lumbered, that might be the case.   Again we are confronted with a purely arbitrary decision as to the associational identity of the vegetation.

Similarly, in the broad transition zone between the oak-covered mountains and the grass-covered foothills in the Coast Range of California, we are forced to deal arbitrarily in any matter of classification.   Shall we call such a zone a mere transition, describe the forests above and the grasslands below and neglect the transition as a mere mixture?   Or shall we regard it as a successional or time transition, evidencing the advance of the grasslands up the mountain or of the oaks down toward the foothills?   If we choose the latter, we must decide whether the future trend of rainfall is to increase, thereby bringing the oaks to lower elevations, or to decrease, thereby encouraging the grasslands to grow at higher altitudes.   If we adopt the former alternative, we either neglect or do a scientific injustice to a great strip of vegetation, in which numerous species are "associated" just as surely as in any recognized plant association.

The sole conclusion we can draw from all the foregoing considerations is that the vegetation of an area is merely the resultant of two factors, the fluctuating and fortuitous immigration of plants and an equally fluctuating and variable environment.   As a result, there is no inherent reason why any two areas of the earth's surface should bear precisely the same vegetation, nor any reason for adhering to our old ideas of the definiteness and distinctness of plant associations.   As a matter of fact, no two areas of the earth's surface do bear precisely the same vegetation, except as a matter of chance, and that chance may be broken in another year by a continuance of the same variable migration and fluctuating environment which produced it. Again, experience has shown that it is impossible for ecologists to agree on the scope of the plant association or on the method of classifying plant communities. Furthermore, it seems that the vegetation of a region is not capable of complete segregation into definite communities, but that there is a considerable development of vegetational mixtures.

Why then should there be any representation at all of these characteristic areas of relatively similar vegetation which are generally recognized by plant ecologists under the name of associations, the existence of which is indisputable as shown by our field studies in many parts of the world, and whose frequent repetition in similar areas of the same general region has led us to attempt their classification into vegetational groups of superior rank?

It has been shown that vegetation is the resultant of migration and environmental selection. In any general region there is a large flora and it has furnished migrating seeds for all parts of the region alike. Every environment has therefore had, in general, similar material of species for the sorting process. Environments are determined principally by climate and soil, and are altered by climatic changes, physiographic processes, and reaction of the plant population. Essentially the same environments are repeated in the same region, their selective action upon the plant immigrants leads to an essentially similar flora in each, and a similar flora produces similar reactions. These conditions produce the well known phenomena of plant associations of recognizable extent and their repetition with great fidelity in many areas of the same region, but they also produce the variable vegetation of our sand dunes and small pools, the fragmentary associations of areas of small size, and the broad transition zones where different types of vegetation are mixed. Climatic changes are always slow, physiographic processes frequently reach stages where further change is greatly retarded, and the accumulated effects of plant reaction often reach a condition beyond which they have relatively little effect on plant life. All of these conspire to give to certain areas a comparatively uniform en vironment for a considerable period of time, during which continued migration of plants leads to a smoothing out of original vegetational differences and to the establishment of a relatively uniform and static vegetational structure. But other physiographic processes are rapid and soon develop an entirely different environment, and some plant reactions are rapid in their operation and profound in their effects. These lead to the short duration of some plant communities, to the development, through the prevention of complete migration by lack of sufficient time, of associations of few species and of different species in the same environment, and to mixtures of vegetation which seem to baffle all attempts to resolve them into distinct associations.

Under the usual concept, the plant association is an area of vegetation in which spatial extent, describable structure, and distinctness from other areas are the essential features. Under extensions of this concept it has been regarded as a unit of vegetation, signifying or implying that vegetation in general is composed of a multiplicity of such units, as an individual representation of a general group, bearing a general similarity to the relation of an individual to a species, or even as an organism, which is merely a more striking manner of expressing its unit nature and uniformity of structure. In every case spatial extent is an indispensable part of the definition. Under the individualistic concept, the fundamental idea is neither extent, unit character, permanence, nor definiteness of structure. It is rather the visible expression, through the juxtaposition of individuals, of the same or different species and either with or without mutual influence, of the result of causes in continuous operation. These primary causes, migration and environmental selection, operate independently on each area, no matter how small, and have no relation to the process on any other area. Nor are they related to the vegetation of any other area, except as the latter may serve as a source of migrants or control the environment of the former. The effect of these primary causes is therefore not to produce large areas of similar vegetation, but to determine the plant life on every minimum area. The recurrence of a similar juxtaposition over tracts of measurable extent, producing an association in the ordinary use of the term, is due to a similarity in the contributing causes over the whole area involved.

Where one or both of the primary causes changes abruptly, sharply delimited areas of vegetation ensue. Since such a condition is of common occurrence, the distinctness of associations is in many regions obvious, and has led first to the recognition of communities and later to their common acceptance as vegetational units. Where the variation of the causes is gradual, the apparent distinctness of associations is lost. The continuation in time of these primary causes unchanged produces associational stability, and the alteration of either or both leads to succession. If the nature and sequence of these changes are identical for all the associations of one general type (although they need not be synchronous), similar successions ensue, producing successional series. Climax vegetation represents a stage at which effective changes have ceased, although their resumption at any future time may again initiate a new series of successions.

In conclusion, it may be said that every species of plant is a law unto itself, the distribution of which in space depends upon its individual peculiarities of migration and environmental requirements. Its disseminules migrate everywhere, and grow wherever they find favorable conditions. The species disappears from areas where the environment is no longer endurable. It grows in company with any other species of similar environmental requirements, irrespective of their normal associational affiliations. The behavior of the plant offers in itself no reason at all for the segregation of definite communities. Plant associations, the most conspicuous illustration of the space relation of plants, depend solely on the coincidence of environmental selection and migration over an area of recognizable extent and usually for a time of considerable duration. A rigid definition of the scope or extent of the association is impossible, and a logical classification of associations into larger groups, or into successional series, has not yet been achieved.

The writer expresses his thanks to Dr. W. S. Cooper, Dr. Frank C. Gates, Major Barrington Moore, Mr. Norman Taylor, and Dr. A. G. Vestal for kindly criticism and suggestion during the preparation of this paper.

CLASSIFICATION AND NOMENCLATURE OF VEGETATION.[1]

BY

G. EINAR DU RIETZ.

The present confusion in the terminology of vegetational units is a fact well known to all ecologists. What a real chaos still prevails in this terminology is clearly shown by Tab. 1, which presents an attempt to parallelize some of the most prominent terminological systems of present-day ecology with each other and with the new compromise-system recently proposed by myself in ABDERHALDEN's Handbuch der biologischen Arbeitsmethoden (Abt. XI, Teil 5). From that table it is easily seen, that even such fundamental terms as »association» and »formation» are still used for quite different concepts by different authors. As this confusion appears to be a serious obstacle to progress and international collaboration in present plant ecology, the attaining of a better agreement in this respect certainly ought to form an important part of the program of any international botanical congress.

In order to simplify the problem and to facilitate the discussion only c l i m a x   u n i t s have been included in the table. Only of the   p h y t o c o e n o s e s, or »complete plant-communities», all units of higher and lower rank have been included. To the one-layered elementary communities, or   s y n u s i a e, into which those phytocoenoses may be dissolved, we shall return later on.

---

[1] Lecture presented at the Fifth International Botanical Congress (Cambridge 1930), as an introduction to the general discussion on »The Classification and Nomenclature of Vegetation» in the Section for Phytogeography and Ecology (Aug. 19 th, 1930).

The most elementary units in the series of phytocoenoses, or the s o c i a t i o n s, have until now been studied nearly only by the Scandinavian School of Ecologists (= Phytosociologists). Until 1928 they were called »associations», but in order to facilitate an international agreement, Scandinavian ecologists have now agreed to accept this term in its Middle-European sense, following Rübel's proposition to apply the new term »sociation» to the earlier Scandinavian »associations» (or »micro-associations»).

Even c o n s o c i a t i o n s are rarely recognized in the present ecology of Middle Europe, except in forests, where they are often given the rank of associations, or even a still higher rank. In Clements' system my consociations correspond not only to the »consociations» (with o n e dominant in the highest layer) but also to the »groupings» (with several dominants in it).

The term a s s o c i a t i o n is now used in the same sense — at least theoretically — all over the continent of Europe, and also by many ecologists in America, New Zealand, etc. If Clements and Tansley would sacrifice their use of this term as Scandinavian ecologists have sacrificed theirs, international agreement in the use of this important term could now be easily attained. That the unit recognized as »association» in European ecology is lacking in Clements' system, probably depends only upon the lack of a suitable term. Certainly also Clements must sometimes feel the need of dividing his »Petran montane forest» — or »*Pinus-Pseudotsuga*-association» — into two subordinate units, or one *Pinus ponderosa - Pinus flexilis* - association and one *Pseudotsuga - Abies concolor - Picea pungens* - association, each of them consisting of several consociations.

As to the term f e d e r a t i o n, it is only an international translation, equally usable in all languages, of Braun-Blanquet's »Verband». As no other term is in use for the unit concerned, there ought to be good hope of getting it generally accepted.

The term s u b f o r m a t i o n is used here for any distinct geographical facies of a formation, analogously with the modern tendencies in the use of the term »subspecies». While my subformations are quite identical with the »associations» of Clements and Tansley, they only rarely coincide with Braun-Blanquet's »Associations-Ordnungen», owing to the rather different principles applied by Braun-Blanquet and myself to the delimitation of the units concerned.

Tab. 1. — Comparison of the plant-sociological terminology proposed by G. E. Du

Examples	G. E. Du Rietz 1928—1930	H. Gams	E. Rübel	H. L. Shantz	G. E. Nichols
The *Cetraria islandica* - layer of the alpine *Vaccinium uliginosum - Cetraria islandica* - heath of Middle Europe	Socion (Synusia)	Synusia	Synusia	Society	Society
The alpine *Vaccinium uliginosum - Cetraria islandica* - heath of Middle Europe	Sociation		Sociation	—	—
The alpine *Vaccinium uliginosum* - heath of Middle Europe (with variable bottom-layer)	Consociation		Consociation	—	Consociation
The alpine *Empetrum - Vaccinium uliginosum* - heath of Middle Europe (Braun-Blanquet's *Empetreto-Vaccinietum*)	Association	Phytocoenoses	Phytocoenoses	Association	Association	Association
The alpine *Loiseleuria - Empetrum - Vaccinium uliginosum* - heath of Middle Europe (Braun-Blanquet's *Loiseleurieto-Vaccinion*)	Federation			(Associations-) Verband	—	—
The Middle-European ericaceous heath (Braun-Blanquet's *Rhodoreto-Vaccinietalia* in somewhat widened sense)	Subformation			Formation	—	—
The boreal ericaceous heath	Formation			—	Formation	—
The boreal, tropical-subalpine, and austral ericaceous heaths, scrubs, and dwarf-forests	Panformation			—	(World) formation group	—
The physiognomic group of lichenous dwarf-shrub heaths	Isocoenoses	Isocoenoses		—	—	—
The physiognomic group of all dwarf-shrub heaths				Formation-group	—	Association-type
The deflation-complex of the low-alpine morain-ridges	Mosaic-complex	Phytocoenose-complexes		—	—	—	Association-complex (physiographic formation)
The *Loiseleuria - Empetrum - Vaccinium uliginosum* - region of the low-alpine belt of the Alps	Vegetation-region		Gebiet	Vegetations-stufe	Formation	Climatic formation

102

RIETZ 1930 with some terminological systems of present plant-sociological literature.

L. Cockayne	F. E. Clements	A. G. Tansley	J. Braun-Blanquet	O. Drude	A. K. Cajander	G. E. Du Rietz 1917—1927
Colony	Society	Society	—	—	Bestand	Bestand(1918) Boden-schichttypus
—	—	—	Part of a Facies	Elementar-associationen mit floristischer Facies	Facies	Association
Subassocia-tion	Consociation	Consociation	—		Association	
Association	—	—	Association	Association	Association-classes	Association-groups
Groups of associations	—	—	(Associa-tions-) Verband	—		
	Association	Association	(Associa-tions-) Ordnung	Haupt-association (earlier formation)		
(Minor) Formation	Formation (Climax)	—	(Associa-tions-) Klasse	Associations-gruppe		
(Major) Formation	Panclimax	—	—	Vegetations-typus		
—	—	—	—	—	—	Formation
—	—	—	Homologous Formations	Formation (earlier Formations-klasse)	—	Formation-group
—	—	(Physio-graphic) Formation	Association-complex	Komplex von Elementar-associationen	Siedlungs-komplex	Association-complex
—	Formation with its seral (ontogenetic) stages	(Climatic) Formation	Klimax-gebiet	Vegetations-region	Formation	Vegetation-region

Also my present f o r m a t i o n s closely correspond to those of
CLEMENTS, as far as climax units are concerned. Practically all
formations recognized by CLEMENTS in North American vegetation
are accepted also by myself. This use of the term »formation»
appears now to be accepted by nearly all extra-European and many
European ecologists, and only in this sense there appears to be
any hope ot getting it generally accepted.

As to my p a n f o r m a t i o n, it coincides with CLEMENTS' »pan-
climax» (not with his »eoclimax» as erroneously stated in my
earlier tables). Certainly CLEMENTS will not object to using it as
a synonym to that term, as he himself uses »formation» and
»climax» alternatively.

The i s o c o e n o s e s, or physiognomic groups of phytocoenoses,
are losing much of their earlier importance in present-day ecology,
and certainly they cannot compete for the term »formation» with
our formations. Many modern ecologists, like CLEMENTS, TANSLEY,
and PAVILLARD, do not even consider them worth recognizing.

To the phytocoenose-complexes we shall return later on.

As already mentioned, the matter was somewhat simplified in
the table by restricting the examples of phytocoenoses to climax
units. For s e r a l units it may prove convenient to substitute
the terms »sociation», »consociation» etc. with »socies», »consocies» ... etc., following the method proposed by CLEMENTS. Owing
to the great difficulties in deciding whether a certain phytocoenose
should be considered a climax unit or a seral unit, and to the
great divergences in the opinion of present ecologists as to this
question, I propose to restrict the terms »socies», »consocies» etc.
to o b v i o u s l y u n s t a b l e phytocoenoses (mostly representing
products of human influence upon the more stable primeval
vegetation).

While the lowest units of phytocoenoses, the sociations, are
founded upon a relative homogeneity in a l l layers, most consociations are homogeneous only in one layer, and the higher units in
no layer at all, being founded mainly upon the s o c i o l o g i c a l
a f f i n i t y of the dominants of the layer arbitrarily chosen as
base for the classification (mostly the highest one of the layers
represented). Thus most phytocoenoses of higher rank than sociations are natural units only in o n e layer, each of the other
layers consisting of alternating synusiae with very little relationship
to each other. A good example of this is afforded by the Scandina-

vian facies of the arctic subformation of the boreal ericaceous heath formation, containing 79 sociations known at present (Tab. 2). If only the field-layer is taken into consideration, these sociations may be naturally grouped into 16 consociations, forming 7 associations and 2 federations. But as even these consociations consist of sociations with the most different bottom-layers, some authors have preferred to found the phytocoenoses superior to the sociations not upon the field-layer but upon the bottom-layer. In this way we get 18 consociations with a homogeneous bottom-layer but a very variable field-layer, as is shown by the table. And these bottom-layer consociations may be grouped to 11 associations, 4 federations, and 3 formations, all of them with a very variable field-layer. It is impossible to say that the consociations, associations, etc. founded upon the field-layer are more natural than those founded upon the bottom-layer, or vice versa. — The same conflict between the layers is met with in most other formations, as the grouping of the sociations to phytocoenoses of higher rank mostly includes an obvious violence done to the natural units of the layers not chosen as base for the classification. The only way out of this conflict appears to be that proposed by Gams in 1918, *i. e.* the method of dividing each sociation into its elementary one-layered units, or s y n u s i a e, and grouping the synusiae of each layer independently of those of the other layers to s y n u s i a e of higher rank. The practical application of this method is also seen in the table. It includes the establishment of a whole series of higher units of synusiae corresponding to that of the phytocoenoses. The sociations are thus dissolved into their elementary s o c i o n s, and the socions of each layer are grouped independently of each other into c o n s o c i o n s, a s s o c i o n s, f e d e r i o n s, and f o r m i o n s, as shown in the table. In Tab. 3 these higher units of synusiae are parallelized with the corresponding phytocoenoses and with the units of idiobiological taxonomy. — If this synusiological method is generally applied, the p h y t o c o e n o s e s superior to sociations lose, of course, much of their importance for the classification of vegetation.

P h y t o c o e n o s e - c o m p l e x e s are vegetational units consisting of phytocoenoses with little or no relationship to each other but more or less regularly alternating. They are of several kinds. Good examples for m o s a i c - c o m p l e x e s are furnished by the »association-complexes» described by Osvald and myself from

Tab. 2. — Phytocoenoses and synusiae of the alpine ericaceous heath in Scandinavia. (16 dwarf-shrub consocions, combined with 8 lichen-consocions and 9 moss-consocions to 79 sociations known at present.)

Field-layer synusiae / Boreal ericaceous heath form consoc.	«Cesiolichen» - consoc.	Alectoria ochroleuca - consoc.	Cetraria nivalis - consoc.	Cladonia alpestris - consoc.	Clad. rangiferina-silvatica - consoc.
Calluna vulgaris - consoc.					
Phyllodoce coerulea - consoc.					
Vaccinium myrtillus - consoc.				+	+
Cassiope tetragona - consoc.					
Betula nana - consoc.		+	+	+	+
Empetrum - Betula nana - consoc.				+	+
Empetrum nigrum - consoc.		+	+	+	+
Empetrum - Vacc. ulig. - consoc.		+			
Vaccinium uliginosum - consoc.	+		+		
Vaccinium vitis idaea - consoc.			+		
Arctostaphilos alpina - consoc.		+	+		
Arctostaphilos uva ursi - consoc.		+			
Rhododendron lapponicum - consoc.	+				
Diapensia lapponica - consoc.	+				
Diapensia - Loiseleuria - consoc.	+				
Loiseleuria procumbens - consoc.	+	+	+		

Lichen form: Cetraria - Alectoria - feder. — Alect. - Cetr. niv. - Cesiolich. - ass. ; Cladina - ass.

Bottom-layer synusiae																			
Boreal heath-	*Cladina-Stereocaulon-*	*St. pasch.-* ass.	*Stereocaulon paschale* - consoc.	(+)								+	+	+	+		+	+	
		Cetr. isl.- Delis. - ass.	*Cetraria islandica* - consoc.												+				
			Cetraria Delisei - consoc.	+							+				+				
			No bottom-layer (»naked»)	+		+		+	+	+	+	+	+	+	+	+	+	+	+
Boreal heath-moss form.	*Campt. - Rhytid.* - feder.	*Rhytid.* - ass.	*Rhytidium rugosum* - consoc.										+						
		Campt. nit.- Sphaeroceph.- ass.	*Camptothecium nitens* - consoc.				+						+			+			
			Sphaerocephalus turgidus - consoc.				+						+			+			
	Hylocomium - Dicranum - Racomitrium - feder.	*Hyl.*- ass.	*Hylocomium parietinum - proliferum* - consoc.									+	+	+	+	+	+	+	
		Bleph. - D. fusc.- ass.	*Blepharozia ciliaris* - consoc.								+	+	+	+	+				
			Dicranum fuscescens - consoc.										+	+	+		+	+	
		Rac.- ass.	*Racomitrium hypnoides* - consoc.			+					+		+						
		D. elong.- ass.	*Dicranum elongatum* - consoc.										+						
Bor. *Sphagn* - form.	*S. fusc.*- feder.	*S. fusc.*- ass.	*Sphagnum fuscum* - consoc.										+						

Tab. 3. — System of plant-sociological units and their analogies
with the units of idiobiology.

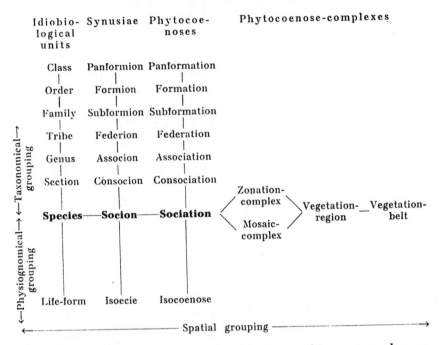

Scandinavian bogs, maritime rocks etc. Z o n a t i o n - c o m p l e x e s
are found on the shore of every lake or river, and have been
described by many authors. Phytocoenose-complexes of greater extension and mostly of less regularity are the v e g e t a t i o n -
r e g i o n s, in most cases practically identical with CLEMENTS'» climaxes» or »formations» (if the seral stages of CLEMENTS' are included) and with the »Klimaxgebiete» of BRAUN-BLANQUET. Contrary to the m o n o c l i m a x - t h e o r y developed by CLEMENTS'
and BRAUN-BLANQUET, I find it necessary, however, to maintain
the p o l y c l i m a x - t h e o r y developed by DOMIN, GAMS, GLEASON,
NICHOLS, NORDHAGEN, SCHARFETTER, TANSLEY, myself, etc. (comp.
DU RIETZ, Vegetationsforschung p. 346), *i. e.* to admit that several climax-phytocoenoses — in many cases with very little relationship to each other — may occur side by side in the same
vegetation-region, but in edaphically different habitats (»edaphic
climaxes» of NICHOLS, TANSLEY, etc.). My own field-experience
certainly does not support the theory of the power of a uniform
climate to transform a l l these edaphically different habitats into

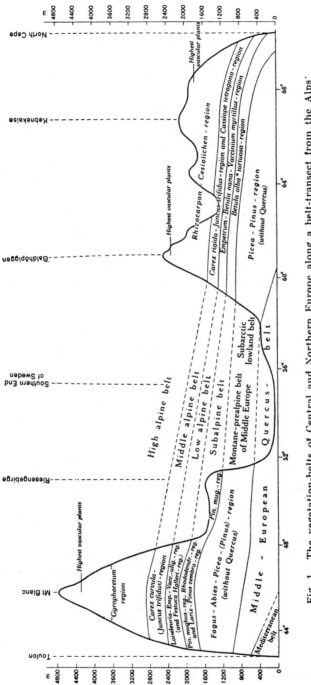

Fig. 1. The vegetation-belts of Central and Northern Europe along a belt-transect from the Alps'
to the northern end of Scandinavia.

one comparatively uniform »climax-habitat» (»climax-soil» etc.), as postulated by the monoclimax-theory.

Vegetation-regions homologous in their altitudinal position constitute together one v e g e t a t i o n - b e l t (Vegetationsstufe, étage de vegetation). More local altitudinal divisions within a vegatation-belt are called v e g e t a t i o n - h o r i z o n s.

The practical application of the concepts »vegetation-region» and »vegetation-belt» is demonstrated in Fig. 1, showing the vegetation-belts and the main vegetation-regions of Middle Europe and Scandinavia along a belt-transect from the Alps to the Arctic Sea.[1] The contour-line of the profile is drawn after the highest elevations within the belt concerned. In many of the regions concerned several local horizons may be developed. Most of those regions would certainly coincide with the »climaxes» of CLEMENTS and the »Klimaxgebiete» of BRAUN-BLANQUET. Nevertheless, according to my opinion, each of them contains not one but many climax-phytocoenoses, even of highest rank. In the names of the vegetation-regions only some of the dominating climax-phytocoenoses, or »regionale Hauptphytocoenosen», have been included. But in the *Quercus*-region of South Sweden, for instance, certainly both *Quercus*-forests and *Picea-Pinus*-forests, among others, must be admitted as climax-phytocoenoses, the former on better and the latter on poorer soils. In the case of the alpine vegetation-regions of the Alps and of Scandinavia, there is a complete agreement among Scandinavian ecologists that the dominating dwarf-shrub-heath and grass-heath of the acid and moderately snow-covered ground in the low-alpine and middle-alpine belts respectively, are by no means the only climax-phytocoenoses of these regions as assumed by BRAUN-BLANQUET. Parallel with them there also occur not only subneutrophilous (pH 5,5—7) and neutrophilous climax-phytocoenoses (the latter only in the Alps) of *Dryas*-heath, *Elyna*-heath, various meadows etc., but also other acidophilous climax-phytocoenoses both chionophobous (on wind-exposed ridges) and chionophilous (or snow-loving, i. e. »Schneeboden» communities), consisting of grass-, herb-, dwarf-*Salix*-, moss-, and lichen-communities very different from the dominating subchionophobous heath. Usually, since the same main ve-

---

[1] For further details comp. Du RIETZ, Vegetationsforschung auf soziationsanalytischer Grundlage (ABDERHALDEN's Handbuch der biologischen Arbeitsmethoden, Abt. XI, Teil 5, 1930), p. 353—358.

getation-regions are admitted by monoclimaxists and polyclimaxists, this theoretical divergence appears to be of rather small practical importance.

For the comparison of the vegetation-belts in different parts of the world it is most important to use the terms »alpine», »subalpine» etc. only for really homologous belts in different mountain-districts. In present literature chaos prevails also in this respect, especially in the use of the term »subalpine belt». In Middle-European literature, much of the p r e a l p i n e forest-belt of the profile is often called »subalpine belt». Also most of the »subal-

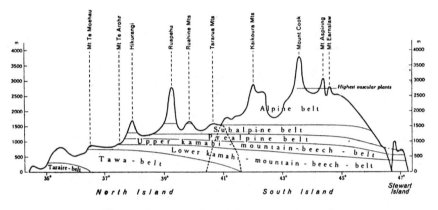

Fig. 2. The vegetation-belts of New Zealand.

pine belt» of North American ecologists I would rather call p r e a l-p i n e, as the so-called »subalpine» *Picea - Abies* - forest of North America is clearly homologous to the prealpine forest-belt of Europe. The real subalpine belt of North American mountains appears to be formed by *Alnus* - scrub in Alaska, by *Pinus aristata* - Krummholz in Colorado, and possibly by a *Larix Lyallii - Pinus albicaulis* - region in the northern Rocky Mountains, analogous to the subalpine *Larix - Pinus cembra* - region of Middle Europe.

Another example of vegetation-belts is given in Fig. 2, showing an analogous sketch-profile of the vegetation-belts of New Zealand (founded upon a synthesis of my own observations during half a year of field-work in 1926—1927 with the literature concerned, especially with the fundamental works of L. Cockayne). Owing to the sloping of most of the belts from north to south, no belt can be called montane over the whole country (just as in Europe),

111

Tab. 4. — The vegetation-regions of New Zealand.

		Mixed forest facies (maritime)	Beech-forest facies (maritime→subcontinental)	Grassland facies (continental)
	Alpine belt		Danthonia crassiuscula (short grass) - region (up to about 2 000 m), etc.	
	Subalpine belt		Danthonia Raoulii (tall tussock) - subalpine scrub - region	
Mountain-beech belt	Prealpine belt	Libocedrus Bidwillii - Dacrydium biforme - region	Prealpine mountain-beech region (Prealp. Nothofagus cliffortioides-Menziesii-reg.)	Festuca novae-zelandiae - Poa cæspitosa - region
Mountain-beech belt	Upper kamahi-mountain-beech - belt	Kaikawaka-kamahi-region (Libocedrus Bidwillii - Weinmannia racemosa - reg.)	Middle mountain-beech region (Middle Nothofagus cliffortioides-Menziesii-reg.)	Festuca novae-zelandiae - Poa cæspitosa - region
Mountain-beech belt	Lower kamahi-mountain-beech - belt	Rimu-kamahi-region (Dacrydium cupressinum - Weinmannia racemosa - reg.)	Lower mountain-beech region (Lower Nothofagus cliffortioides-Menziesii-reg.)	Festuca novae-zelandiae - Poa cæspitosa - region
Beilschmiedia-belt	Tawa-belt	(Beilschmiedia) tawa - region		
Beilschmiedia-belt	Taraire-belt	(Beilschmiedia) taraire - region		

while no such difficulties are met with in the prealpine, subalpine, and alpine belts. In Tab. 4 the division of those vegetation-belts into vegetation-regions is shown. Most of these vegetation-regions would certainly be accepted by CLEMENTS as good »climaxes» and by BRAUN-BLANQUET as good »Klimaxgebiete» — though in some cases two or three of my regions would rather form one »climax» together (*e. g.*, the three mountain-beech regions, which, however, are entirely different in the composition of the lower layers). Also in those vegetation-regions, however, it is evident that many of them contain not one, but several climax-phytocoenoses even of highest rank. In the *Beilschmiedia tawa* - region, for instance, the dominating climax-phytocoenose of the richer and moister soils is a mixed rain-forest with *Beilschmiedia tawa* as the main dominant and a lighter upper tree-layer of *Podocarpus, Dacrydium,* or *Metrosideros robusta*, while the dominating climax-phytocoenose of the dryer and poorer soils is an entirely different *Nothofagus Solandri-truncata* - forest. And in the subalpine belt both the *Danthonia Raoulii* - tussock and the subalpine scrub must certainly be recognized as climax-phytocoenoses, as well as the various dwarf-shrub heath phytocoenoses of wind-exposed ridges etc. In the alpine belt there is — just as in Europe — a whole series of very different climax-phytocoenoses from the extremely chionophobous cushion-plant communities on wind-exposed ridges to the various extremely chionophilous »Schneeboden»-communities (the *Danthonia crassiuscula* - consociation belongs to the intermediate group). — But also in this case the attitude taken by the investigator towards the monoclimax- and polyclimax-theories, probably is of very little practical importance for the actual delimitation of the vegetation-regions.

It appears to me that we have now reached such a general agreement regarding the delimitation of vegetational units of various kinds, that we should not allow this enjoyable fact to be obscured by divergences in the theoretical interpretation of the structure and future of those units — monoclimax- and polyclimax-theories — nor by divergences of purely terminological character. Especially the divergences last mentioned should certainly not be allowed any longer to hamper international collaboration in ecology. I firmly believe that the time is ripe now to take up the idea again of formally regulating this terminology by international agreement. And I herewith propose that an international committee should be appointed to investigate the possibility of attaining general agreement at least in some of the more important concepts and terms in Plant Sociology, and if sufficient agreement can be attained, should make proposals for an international terminology to be presented to the next international botanical congress.[1]

Plant-biological Institution of Upsala University, August, 1930.

---

[1] This proposition was accepted by the Congress, and the author was appointed to act as a recorder in the committee elected by the Congress. The author will be very indebted to his colleagues in and outside this committee for any constructive criticism of the proposals made in this paper and in ABDERHALDEN's Handbuch (comp. above p. 499). Such criticism will be of great value for the organization of the work of the committee.

6

Reprinted from *Am. Midl. Nat.* **21**(1):147–158, 176–181 (1939)

THE CLIMAX AND ITS COMPLEXITIES

Stanley A. Cain

I. Introduction

No biologist doubts the existence of individual socially integrated communities—the stands, the association-individuals, the biocoenoses—in their concrete expression at a certain place and at a certain time. The facts concerning the individual community are definite; but descriptions of communities are seldom complete owing to the complexity of community composition and structure, and to the narrow point of view of many investigators.

The association in the abstract, or community of any rank, is another problem. The confusion, misunderstanding, and diverging principles found among students of communities are nearly all related, not to the association-individual but to the association as an abstraction. Science requires the formation of concepts and scientific thought attempts clarification of these. Ideas of the association in the abstract are obtained by the selection of the common features of numerous association-individuals and synthesizing them into a "type." This does not mean that the association is a "pigeon-hole." In dynamic ecology the association is very large, but it is a concrete entity.

Because of the broad, more or less obvious features held in common, the numerous differences among stands frequently are neglected or minimized. The varying degree to which investigators recognize differences constitutes one of the main causes of divergence in community interpretation and classification. The other principal cause of divergence seems to be the varying emphasis placed on aspects of structure and processes in vegetation. Floristic, physiognomic, edaphic, and successional ecology all have produced systems of classification. This paper aims to describe the complexity of vegetation and thereby to suggest the necessity for comprehensive investigations; and to show, in part at least, wherein various interpretations coincide or diverge from one another.

Rübel says, "The fundamental need of the human mind is for order, for grouping, so to be able to apprehend the infinity of things existing side by side. . . . We need a system of the plant communities of the world. . . . A classification may not be on a single factor, nor on a few factors. . . . All is relative." Schmid (1936) recently emphasized the highly complicated structure of the biocoenosis which can only be understood by an organized method combining the different approaches: the edaphic-physiognomic, the floristic-statistic, the biocoenologic, and the historic.

II. What is a Plant Community?

Community is a term of convenience which usually is employed to designate sociological units to every degree from the simplest one-layered aggregation to the most complex phytocoenosis. There are, however, certain aggregations of organisms which do not exhibit the characteristics generally attributed to a true community.

Alechin (196) says, "A plant community is a complex of plants with mutual adaptations. . . . It is obvious that the community is not made up of plants of one and the same life-form. . . . It is not correct that a community is a group of plants showing the same adaptations. . . . When a pure community occurs, it may be because no ecological equivalent species is in the same region. . . . The fundamental characteristics of a plant community are: layering in space, layering in time [aspection], variability, stability." Paczoski (1917, quoted by Alechin, 1926) says, "Under community must be placed only such a plant complex as consists of elements of unlike value lawfully bound together, which build stable combinations (in variable equilibrium) and which can come to exist only in the course of a long time." According to Alechin (1926), Boris Keller states that with respect to the ecological and social qualities of communities, they must be divided into more or less marked plant groups. He calls them "Genossenschaften" (brotherhoods?) and claims that they have a certain independence and a uniform inner social structure. His "Genossenschaften" are in part phases or aspects, and in part layers; only the concept is ecological whereas the "Schicht" (society) can embrace two or more alternating "Genossenschaften." Alechin considers Keller's (1907) subdivisions of the phytocoenosis identical with the synusiae of Gams (1918). It would seem likely that they coincide with the unions of Lippmaa (1938). See also DuRietz (1936).

Lüdi (1928) believes that constance and fidelity are the most important community characteristics. (Constance is a purely structural unit, whereas fidelity, especially on the basis of a group of species, has diagnostic-indicative value.) At the same time a great importance must be attached to the genetic rôle of the species. The "aufbauenden" species are usually the dominants, but dominance is not a primary criterion for the demarcation of a community. Rübel (1936)', also speaking of the association as the fundamental sociological unit, states that the species known as constants and characteristics are the main fetures of a community; they form the normal characteristic combination of species. Wangerin (1925) on the other hand claims that the basis of the community lies in the floristic assemblage but that its ecology is also important and it is correlated, in most cases, with the floristic assemblage. He considers the serviceability of purely statistical methods limited and sometimes misleading, and accordingly he leans toward the definition of the association given by Flahault and Schröter in 1910.

Tansley (1935) apparently postulates sociological integration as a characteristic of a community, for in criticizing the terms "organism" and "complex organism" as applied to a community (association), and in defending his term "quasi-organism" he says,

It is precisely this mass action, together with the actions due to the close and often delicate interlocking of the functions of the constituent organisms, which gives coherence to the aggregation, forces us to call it a unit, justifies us in considering it as an organic entity, and makes it reasonable to speak of the development of that entity.

In describing the fundamental unity of the phytocoenosis, as opposed to the synusia, Gleason (1936) points out certain characteristics of communities. He finds that since its first recognition the community has always been considered a geographic unit. "It occupies space and it has boundaries. Moreover, it exhibits uniformity of structure within the area. . . . Without uniformity both extent and boundary lose all significance. . . . There must be a boundary in time as well as in space; there must be a beginning as well as an end." Gleason states that the cause of uniformity is two-fold. "The first [cause] is the existence over a considerable area of an environment which is itself uniform. . . . The second is the existence throughout the area of a uniform type of physiological interference." By physiological interference Gleason refers to the "intersecting spheres of influence" of the plants of the community. ' 'In the plant community the essential nature, the fundamental cause, the real *ens* of its existence, is non-areal, while its manifestation is distinctly areal, characterized by uniformity, extent, and boundary."

Braun-Blanquet (1932) writes: "The community life of plants rests upon relations of dependency and commensalism; its universal and ever present expression is competition. Wherever the struggle for a place in which to germinate and to grow, or to obtain light or food, can be demonstrated, there is competition, there is relationship." Commensal organisms are those which enter into competition separately, whereas their common relation consists in the fact that they utilize simultaneously the various life conditions of a given habitat. Completely open stands of plants hardly can be considered true communities because there is no competition among them, no physiological interference. It must be remembered, however, that a community can be open above ground and yet be sociologically closed on account of root competition. Apparently Clements (1928) also excludes from true communities all aggregations which are strictly open: "Competition occurs whenever two or more plants make demands in excess of the supply. It is a universal characteristic of all plant communities, and is absent only in the initial stages of succession, when the pioneers are still isolated."

So far little has been said about animal members of communities, but it is probable that every community is a biocoenosis. In fact, the terms community and biocoenosis are actually synonymous, but both are so general that they have lost nearly all force. Communities may range all the way from the *biome* (of formational rank) to those of the smallest *biotope* (the life area of smallest space). As an example of investigations of the latter type of community (excluding microscopic communities), see Heinis (1937) who studied quantitatively the microbiocoenoses of *Sphagnum* polsters.

The following quotation from Alechin (1926) aptly concludes this section concerning the "requirements" for the recognition of a community.

We place in a plant community a complex of plant species which has a definite structure and is composed of ecologically and phenologically different elements ("Schichtung" in space and time), which moreover—in spite of its mobility—forms a persevering system depicting botanically the physico-geogpraphic relations and the history of the region.

Plant groupings
   I. Open
  II. Closed
     1. Settlements ("Siedlungen," without social integration)
     2. Pure stands
       a. Temporary
       b. Enduring
     3. Communities

   a. Dearranged equilibrium
   b. Flowing equilibrium
   c. Stable equilibrium
     a) Unsaturated (without the full complement of societies)
     b) Saturated.

Obviously sharp delimitations of the above cannot be drawn.

The nature of the plant community has always interested plant sociologists. Detailed discussions of this and related problems may be found in the following papers: Alechin (1925, 1926), Braun-Blanquet (1921, 1932), Cajander (1922), Carpenter (1936), Clements (1928), Däniker (1928, 1928a, 1936), Gams (1918), Gleason (1936), Katz (1930), Kylin (1926), Lüdi (1928), Phillips (1931, 1934), Scharfetter (1924), Shelford (1932), Sukatschew (1929), Wangerin (1922, 1925).

III. A Brief Conspectus of Clements' Concepts and Terminology of the Climax

The following statements derived almost entirely from a recent essay by Clements (1936) form the basis for subsequent discussions. Irrespective of how widely his concepts and terminology are accepted, he has evolved a philosophy, a group of principles, and a terminology which meet *nearly* every exigency.

The *climax* constitutes the major unit of vegetation and it is regarded as synonymous with *formation* and *biome;* it is considered a more or less permanent and final stage of a *complex organism* under the paramount control of climate. The visible unity of the climax is due to the life-form of the *dominants,* which is a concrete expression of the climate. Inasmuch as the climax constitutes the major unit of vegetation (or of the biosphere), it is often continental in extent. This being the case, the climax can not be expected to exhibit uniformity in biotic composition and sociological structure, on account of the gradual but marked shift in rainfall and temperature from one boundary of the climax to the other. What unity the climax has is dependent on the life-form of the dominants and on certain characteristic *genera* and families of plants and animals. Considerable differences in rank or territory occur among the dominants of each formation. *Perdominants* are those which, because of their wide range (not necessarily as dominants), link the associations of a climax. Within an association, the abundant and controlling species of a characteristic life-form are termed *dominants* and *subdominants* are the controlling species of a different life-form which belong to the inferior synusiae.

Although *development* is a primary characteristic of vegetation, *stabilization* is the universal tendency of all vegetation under the ruling climate and climaxes are characterized by a high degree of stability if reckoned in thousands or even millions of years. It is a fundamental part of Clements' concept that, given time and freedom, a climax vegetation of the same general type will be produced and stabilized, irrespective of earlier site differences. *Change* is constantly at work, but in primeval vegetation it is within the fabric of the climax, and not destructive of it. *Aspection, annuation,* and natural *coaction* leave no permanent impress on the climax, but *successional changes* are an intrinsic part of the stabilizing process.

The hierarchy of climax units recognized by Clements consists of the following categories:

```
Climax (formation, biome)
   Association ⎫
      Faciation ⎬_____Consociation
      Lociation ⎭
   Society
      Sociation
      Lamiation
   Clan
```

The *association* of Clements represents the primary division of the climax and differs entirely from what may be called the international interpretation as approved by recent Botanical Congresses (Du Rietz, 1929, 1930, 1930b, 1936). As the unity of the formation rests upon the wide distribution of perdominants, so the association is marked by one or more dominants peculiar to it, and by differences in rank and grouping of dominants held in common.

The *faciation* is the concrete subdivision of the association; the entire area of the latter being made up of the various faciations corresponding to particular regional climates within the general climate of the formation. Faciations appear to be geographical varieties of the association, or subassociations, related to temperature and rainfall variations in different portions of the area of the association.

The *lociation* is a subdivision of the faciation characterized by more or less local differences in the abundance and grouping of two or more dominants of the faciation. Lociations may depend in some part upon edaphic variations, within the area of a faciation, such as variations in soil, slope, exposure, altitude. Lociations occur frequently as *alternes* with each other and with proclimaxes.

The *consociation* is constituted by a single dominant. Since the actual area of the association is divided into faciations, and the faciations into lociations, the consociation is apparently a phytocoenosis with only one dominant. Mono-dominant lociations, faciations, and associations, i.e., consociations of all ranks can easily be conceived. The term consociation is therefore not important since one might well speak of 1-dominant, 2-dominant, 3-dominant, and poly-dominant locations, faciations, and associations, rather than to single out the 1-dominant phytocoenoses as consociations.

The term *society* is employed as a general term for all communities of subdominants, that is, for all synusiae or life-form groups below the dominant synusia. Societies partake of both space and time differentiation.

The *sociation* is an aspect society related to the life history of the subdominants and the seasonal march of the environment factors.

The *lamiation* is a layer society based on life-form and environmental differences within the phytocoenosis. The term lamiation appears less suitable than synusia because the latter includes all subdominant life-form groups, whether in horizontal layers or not (corticolous, epixylic, epilithic societies), as well as the dominant community. It is probably over-emphasis of the dominant layer which causes American ecologists to separate it so definitely from subdominant layers. The term *union* is now proposed for these "einschichtige Pflanzenvereine" (Du Rietz, 1936).

The *clan* is a small community of subordinate importance but commonly of distinctive character. It is marked by high density and efficiency in competition of the dominant plants resulting from vegetative reproduction by rhizomes, stolons, corms, bulbs, tubers, etc.

It is essentially a patch or a consociation.* Although clan is usually applied to members of the field layer, the dynamics of clan formation may occur also in bryophytic, frutescent, and arborescent societies. Consequently, the idea of consociation can be applied to all other life-form subdivisions of the phytocoenosis. Thus the consociation apparently is not a fundamental unit of the climax.

IV. The Complexity of the Climax

A. THE STRUCTURE OF THE CLIMAX

1. *Chorological complexity of the climax*

The earth's surface is covered by relatively few world formations (Rübel, 1930, 1936) such as rain forest, summergreen forest, grassland, etc., each covering thousands of square miles and possessing a fundamental life-form unity, — the climaxes (formations, biomes) of Clements. Speaking of North America, Clements (1936) says:

> The visible unity of the climax is due primarily to the life-form of the dominants, which is the concrete expression of the climate. In prairie and steppe this is the grass form, with which must be reckoned the sedges, especially in the tundra. The shrub characterizes the three scrub climaxes of North America, namely, desert, sagebrush, and chaparral, while the tree appears in three subforms, coniferous, deciduous, and broad-leaved evergreen.

Weaver and Clements (1938) recognize only 12 extra-tropical climaxes for North America.

Within a climax formation, however, there are many different communities variously treated and designated by different authors. For example, Rübel (1932) lists for Switzerland 169 communities (133 associations and 36 subassociations) grouped into 54 orders and 31 classes, whereas Conard (1935) enumerates about 70 associations for Central Long Island. On the other hand Weaver and Clements recognize only 29 associations for all of extra-tropical North America. Undoubtedly numerous different stands of vegetation exist within a climax; consequently the difference of interpretation

---

* In correspondence Clements states, "It seems best to restrict consociation to single extensive dominants, such as *Picea engelmanni* in the Front Range of Colorado, *Pinus pondorosa* in the Southwest, *Picea alba* in the western part of the boreal forest, etc. Naturally, in associations of several or many dominants, they are lacking or relatively rare."

must be largely due to the differences between static and dynamic treatments.

The beech forests of Europe probably have been studied more intensely by a group of plant sociologists than any other single complex type. Careful reading of "Die Buchenwälder Europas" (Rübel, 1932) reveals two important points: 1) the fundamental variability in composition and structure of the beech forests of Europe and 2) that despite an organized plan of investiagtion, differences of method hinder greatly the comparison of results and descriptions.

Within the climax, Clements recognizes a small number of associations which are divided chorologically into faciations and lociations, occupying progressively smaller areas. In original vegetation the associations showed some interdigitation according to the influence of topography on climate; the faciations represented geographical subassociations and zones, and the lociations undoubtedly appeared as belts and alternes within the faciations. The patchwork nature of vegetation is further complicated by the fact that not all vegetation is in the climax state since in all physiographically young regions and in disturbed areas occur numerous seral stages, slowly on their way to stabilization.

Moreover, relatively stable communities neither actively developing nor true climaxes, represent the proclimaxes of Clements, namely, subclimaxes, and disclimaxes. See also Tüxen (1937) and Ciferri (1936). Seral communities, climatic climaxes (lociations and faciations), and the various deflected (Godwin, 1929) and edaphic climaxes of Nichols, Tansley, et al., all have been regarded as associations. Whatever their rank, they compose a tremendously complicated chorological pattern of vegetation.

An association-individual, regardless of rank, is concrete, but the association itself is a concept to the extent that different stands of an association lack identity. The farther apart the stands the more different they become in composition and structure so that the extremes hardly represent the same "association," except as broadly conceived by Clements.

2. Chronological Complexity of the Climax

Vegetation is complicated not only in space but also in time. Cooper (1926) emphasizes the fundamental nature of change in vegetation and Clements (1936) says, "No one realizes more clearly than the devotee of succession that change is constantly and universally at work, but in the absence of civilized men this is within the fabric of the climax, and not destructive to it. . . . It can still be confidently affirmed that stabilization is the universal tendency of all vegetation under the ruling climate." See also, Phillips (1934).

Vegetational development has long been recognized as consisting of series of communities (associes) which can be grouped into xeroseres and hydroseres, according to the relative dryness or moistness of the primary condition, leading toward a climax under conditions of relative mesophytism for the given climate. Polunin (1936) wanted a term to describe certain conditions found in the Arctic; "In considering dynamic ecology of many regions, at least in the Arctic, it seems necessary to distinguish a mesosere, or succession starting with the colonization of suitable bare areas, neither by hydrophytes nor by extreme xerophytes but, instead, by some of the more mesophytic plants existing in the region." Moreover, a condition of relative mesophytism must not be always the end of the successional tendency. Bourne (1934) says, "Confusion has followed from the assumption that the climax must be relatively mesophytic, no allowance being made for degradation of the habitat. . . . Clements has failed to realize that a mature soil in many climates is relatively degraded. . . . He has clearly had no experience of either podsol or laterite soils." At any rate, autogenic and allogenic successions (Tansley, 1935) add tremendously to the complexity of the vegetational pattern.

To these successions must be added the secular changes of the clisere which lead, especially in ecotonal regions, to an abundance of relict preclimax or postclimax communities. Clements (1916) defines a clisere as a series of climax formations or zones which follow each other in a particular climatic region subsequent to a distinct change of climate.

Still another group of successions are those of the *serule*, including a great variety of short, relatively swift successions starting on such bare substrata as bark, rotting wood, rocks, cadavers, etc. Oliver (1930) describes such successions among corticolous New Zealand epiphytes, whereas Cain and Penfound (1938) and Cain and Sharp (1938) recently called attention to such successions among bryophytes.

Other chronological complications in vegetation, such as those introduced by aspection and annuation, are discussed in connection with synusial structure.

Although American and British workers generally are concerned with successional phenomena certain papers in the field less familiar to Americans can be recommended: Braun-Blanquet and Jenny (1926), Ciferri (1936), Furrer (1922), Lüdi (1920, 1921, 1930).

3. *Internal Structural Complexity of Communities*

a. *The synusial structure of vegetation.* — Alechin (1926) emphasizes that the division of a phytocoenosis into layers is its first noticeable characteristic. A plant community is a complex of plants with mutual adaptations, and the layers indicate a form of this adaptation. Salisbury (1931) says that "The restriction of the concept of physiognomy to that of the dominants of a community is really to characterize the whole by a part that commonly represents a single stratum. The physiognomy of the dominants is often, in fact, that of a partial habitat." The recognition of the importance of layers in the structure of vegetation goes back, at least, to Hult (1881) whereas a system of organization of plant and animal life on a basis of life-form groups (synusiae) was proposed by Gams (1918). Obviously much of the complexity of communities is due to the fact that each biocoenosis usually is composed of many synusiae. Consequently, many systems of community classification ranging all the way from extremes in which only the dominant layer in considered to systems based primarily on the field layer (Cajander and the Finnish School) have been proposed. Other examples are the unistratal system of Lippmaa and the complicated twin associations of the Russians.

Plant associations show stratification above ground as well as in the soil. Therefore some investigators have arranged the communities of a region on a basis of sociological progression (Braun-Blanquet, 1932; Conard, 1935) proceeding from the simplest one-layered to the most complex many-layered associations. Mats of algae, moss, and lichens on soil, rock, and bark usually are considered one-layered (without rooting), but even these communities are likely biocoenoses of two or more synusiae. Heinis (1937), for example, studying quantitatively the microbiocoenoses of alpine plant polsters, did not find within the *Sphnagum* polsters an accumulation of dead organic remains but a complicated web of life based on food chains containing holophytes, herbivores, and carnivores, dominated by bacteria, diatomes, protozoa, and nematodes. In fact, in the detritus of these polsters he found 56 species including Flagellatae, Diatomaceae, algae, Ciliata, Rhizopoda, Rotaria, and Nematoda.

Forest phytocoenoses present the most complicated situations. Alechin (1926) states that temperate oak forests may have as many as seven layers, three of woody plants, three of herbaceous plants, and one of mosses. If the communities of all biotopes are included, the number of synusiae in a phytocoenosis becomes considerably greater. According to Lippmaa (1935), the Ulmus Acer-Tilia association consists of five layers with six vertically distributed corticolous communities of bryophytes and lichens. The number of synusial communities would be even greater in forests with special biotopes such as erratic boulders, rotting logs, etc. The synusial complexity of many tropical rain forests is scarcely greater than that of temperate deciduous forests of mesophytic situations.

Subterranean root-layers also may be numerous, as shown by the extensive work of Weaver. Woodhead (1916) describes four root-layers in the forests of Huddersfield.

According to Alechin (1926), layering seems to be the result of long continued competition and at the same time of processes of adaptation. Layering results in the production of maximum plant mass on a definite surface and in the possibility of coexistence of plants of radically different ecological types. Competition is considerably more severe within a synusia than among members of different synusiae; the latter, consequently, are much less likely to destroy one another, but are able to reach a dynamic equilibrium. Such social integration apparently requires much time and as a result saturation, that is, a maximum number of societies, usually is found only in climax vegetation.

The development of synusiae in vegetation is really a four-dimensional matter. Alternating stands occur with changes in temperature and moisture from north to south and from east to west, as the case may be, in each layer of vegetation. The third dimension, of course, is layering and the fourth is time, as aspection and annuation.

Aspection refers to seasonal phases and every type of vegetation shows such rhythms. Some such phases are well marked by flowering periods and others by a radical shift in life-form. This time-"Schichtung" allows the development of a large number of ecologically different plants on the same area. Alechin (1926) recognizes as many as 11 phases in the steppe near Kursk.

Annuation results from seasonal differences — a dry year following a wet one, weather conditions correlated with sun-spot cycles, etc. These influences are especially evident in grassland and desert where certain "ingredients" may fail completely for a series of years. According to Paczoski (1917; quoted by Alechin, 1926) this makes for "antagonistic plant groups" which flourish under different optima, occurring in time at the same place. In the steppe, legumes and grasses constitute such antagonistic groups. Iljinsky (1921; quoted by Alechin, 1926) found in a moist year on the steppe that grasses formed 66 per cent and legumes 2.3 per cent of the dry weight; whereas in a dry year the grasses formed 50 per cent and the legumes 18 per cent. A description of vegetation without consideration of time variation is clearly an approximation.

Certain plants find their required habitats only in association with other plants. This is essentially true of the hemicryptophytic-cryptophytic union of north temperate deciduous forests (Lippmaa, 1938). Certain inferior synusiae occur in many places associated with quite different superior synusiae. Recognition of these facts led Lippmaa to his emphasis on the unistratal community now called a union. Keller (1927), for example, in describing the vegetation of the plains of Russia, says: "The most important type of virgin spruce forest is the *Piceetum hylocomiosum*, the soil of which is characterized by a thick unbroken moss covering consisting of the following species: *Hylocomium proliferum* (= *H. splendens*), *H. triquetrum* (= *Rhytidiadelphus triquetrus*), *Hypnum Schreberi*, *H. cristacastrensis*, *Dicranum undulatum*. These mosses grow very well together and form a remarkably independent community, which may exist in association with various conifer trees": *Picea excelsa*, *Pinus sibirica*, *Abies sibirica*, *Larix sibirica*, *Pinus silvestris*. Essentially the same soil layer is characteristic of other conifer forests such as the *Picea rubens-Abies Fraseri* forests of the Southern Appalachians (Cain and Sharp, 1938). Cain (1934) pointed out that the small tree synusia of *Cornus florida*, *Cercis canadensis*, *Ostrya virginiana*, and *Carpinus caroliniana* occurs with a wide variety of forest types and associated herbaceous communities. It must be remembered, however, that the dynamic bonds are essentially vertical (Gleason, 1936) and that the lateral relations among plants of a synusia are relatively weak.

In conclusion, the development of synusiae is typical of all vegetation. The synusiae increase the complexity of vegetation but at the same time give it order. They are the practical units for investigation, because of their relative structural and ecological homogeneity. No study of a plant community can be considered complete unless it takes into account the synusiae.

b. *Quantitative relations within the community.* — The richness of its flora is certainly a characteristic of the community. Lippmaa (1932) says that, given time for stabilization, a certain number of species (variable within limits) is characteristic for a certain community. He cites the following: *Festuca ovina* association, 40 species; *Meum-Anemone* meadow, 52 species; *Tricho-*

phorum meadow, 14 species, for minimal areas of 50, 8, and 2 sq. m., respectively. Cain (1935) found *Piceetum rubentis* to consist of 9 tree species and *Abietum Fraseri* of 5. A stand of virgin hardwood forest, dominated by *Liriodendron tulipifera-Quercus alba-Acer saccharum-Nyssa sylvatica* consists of 32 species of trees, 9 shrubs and 105 herbs (Cain, 1934). Several of Conard's (1935) seaside associations, such as *Spartinetum patentis*, are composed of only 4 to 6 species each. At the other extreme stand certain tropical rain forests, each with a very large flora. Richards (1936) reports in the climax montane rain forest, on a plot of less than 4 acres, a total of 130 species of trees. Brown found 290 species of trees on 2.5 acres in the Dipterocarpous forest of the Philippines. In every case, however, the species number must be taken only in a general way.

Mere lists of species cannot give more than a very poor idea of the complexity of a piece of vegetation. Every description of a plant community should consider the density and manner of distribution of the plants of each species. Density, frequently estimated under the name of abundance, should be distinguished clearly from dominance, the area or space covered by the plants. The homogeneity of the community depends on the regularity with which plants of each species reoccur. Usually it is expressed by the term frequence (local constance) and constance (occurrence in quadrats from several stands of the same type). Homogeneity tends to develop because of the random nature of the distribution of propagules and of biotopes within a community. On the other hand, heterogeneity tends to develop as a result of the sociability, or gregariousness, of plants, especially those with vegetative methods of reproduction, with the resultant mosaic nature of vegetation. A discussion of these concepts is found in Braun-Blanquet (1932), in a résumé by Cain (1932), and Alechin (1932) gives a good review of the methods of the Russian phytosociologists.

Often minimal area is regarded as a characteristic of the community. Cain (1938) has recently considered minimal area, that area necessary for the development of the characteristic combination of species. In some bryophytic communities, this area may be as small as 0.1 sq. m. or less. For associations dominated by trees, it may be as large as a few thousand sq. m. Stands smaller than the minimal area are likely to be fragments of the association, without the characteristic combination of species (constant-dominants and species of high fidelity).

Salisbury (1931) believes that "the concept of minimal area cannot be given a very precise meaning and, though perhaps affording a convenient method of indicating differences in special cases, it is doubtful if it affords any additional information to that furnished by other data." In many tropical forests with a very rich flora and in certain African scrub communities described by Adamson (1931), the species-area curve used to indicate minimal area will not flatten out on account of the lack of species dominance. In these communities there is usually physiognomic dominance (Cain, 1932) and sometimes generic or family dominance. Richards (1936) describes the dominance (60 per cent) by Dipterocarpaceae in the rain forest but with no one

125

species attaining anything approaching dominance. In such cases, minimal area does not exist as the least area on which a characteristic combination of species can develop, or else it is exceedingly large.

The numerous aspects of community structure offer different criteria for the recognition and delimitation of communities. The available systems range from those emphasizing physiognomy to those based on ecology or floristic assemblages. Either the dominants are regarded most important or the key species are those of high constance or of high frequence. Braun-Blanquet bases his associations largely on species of high fidelity occurring primarily within the association type. However, in certain regions, such as the Sylene of Norway (Nordhagen, 1927), characteristic species with marked exclusiveness are almost entirely absent and constant-dominants must be relied upon.

Constancy and fidelity are both locally, but not generally valid. The composition and quantitative relations within an association change with the faciations. No association, except in the narrowest sense, has the same constant species throughout its entire range. Likewise, fidelity of a species to a community must be a rather local phenomenon since the same species will live under different conditions in different parts of its range and hence be associated with different communities.

[*Editor's Note:* Material has been omitted at this point.]

REFERENCES

No effort has been made to give a complete bibliography, nor to infer that the papers cited here are always the best or earliest examples of a given type of work or point of view.

ADAMSON, R. S. 1931—The plant communities of Table Mountain. II. Life-form dominance and succession. Jour. Ecol. **19**:304-320.

ALECHIN, W. W. 1925—Ist die Pflanzengesellschaft eine Abstraktion oder eine Realität? Englers Bot. Jahrb. **60**,(Beibl. 135):17-25.

————1926—Was ist eine Pflanzengesellschaft? Ihr Wesen und ihr Wert als Ausdruck des sozialen Lebens der Pflanzen. Fedde Repert. spec. novar. reg. veg. Beih. **37**:1-50. (Transl. from the Russian by Selma Ruoff.)

————1932—Die ältere russische Steppenforschung mit besonderer Berücksichtigung der quantitativen Methoden. Beitr. Biol. Pflanz. (Cohn's Beträge) **20**:25-58.

APPALACHIAN FOREST EXPERIMENT STATION. 1935—Forest improvement measures for the Southern Appalachians. U.S.D.A., Tech. Bull. **476**:1-45.

BERRY, E. W. 1930—Revision of the Lower Eocene Wilcox flora of the Southeastern States. U. S. Geol. Surv., Prof. Paper **156**.

BOURNE, R. 1934—Some ecological conceptions. Empire Forestry Journ. **13**:15-30. (See also, review by Barrington Moore, Jour. Forestry **32**:894-5. 1934.)

BRAUN, E. L. 1916—The physiographic ecology of the Cincinnati region. Ohio Biol. Surv. Bull. **7**.

———1935—The vegetation of Pine Mountain, Kentucky. Amer. Midl. Nat. **16**:517-565.

———1935b—The undifferentiated deciduous forest climax and the association-segregate. Ecology **16**:514-519.

BRAUN-BLANQUET, J. 1921—Prinzipien einer Systematik der Pflanzengesellschaften auf floristischer Grundlage. Jahrb. d. St. Gallischen naturw. Gesell. **57**:305-351.

———1932—Plant sociology: The study of plant communities. (Transl. by Fuller and Conard) McGraw-Hill, N. Y.

BRAUN-BLANQUET, J. AND H. JENNY. 1926—Vegetationsentwicklung und Bodenbildung in der alpinen Stufe der Zentralalpen. Neue Denkschr. Schweiz. Naturf. Ges. **63**:175-349.

CAIN, STANLEY A. 1930—An ecological study of the health balds of the Great Smoky Mountains. Butler Univ. Bot. Studies **1**:177-208.

———1931—Ecological studies of the vegetation of the Great Smoky Mountains. I. Soil reaction and plant distribution. Bot. Gaz. **91**:22-41.

———1932—Concerning certain phytosociological concepts. Ecol. Monogr. **2**:475-508.

———1934—Studies on virgin hardwood forest. II. A comparison of quadrat sizes in a quantitative phytosociological study of Nash's Woods, Posey County, Indiana. Amer. Midl. Nat. **15**:529-566.

———1935—Ecological studies of the vegetation of the Great Smoky Mountains. II. The quadrat method applied to sampling spruce and fir forest types. Amer. Midl. Nat. **16**:566-584.

———1936—The composition and structure of an oak woods, Cold Spring Harbor, Long Island, with special attention to sampling methods. Amer. Midl. Nat. **17**:725-740.

———1936b—Synusiae as a basis in plant sociological field work. Amer. Midl. Nat. **17**:665-672.

———1938—The species-area curve. Amer. Midl. Nat. **19**:573-581.

CAIN, STANLEY A. AND WM. T. PENFOUND. 1938—Aceretum rubri: the red maple swamp forest of Central Long Island. Amer. Midl. Nat. **19**:390-416.

CAIN, STANLEY A. AND A. J. SHARP. 1938—Bryophytic unions of certain forest types of the Great Smoky Mountains. Amer. Midl. Nat. **20**:249-301.

CAJANDER, A. K. 1922—Zur Begriffsbestimmung im Gebiet der Pflanzentopographie. Acta forest. fennica **20**.

CARPENTER, J. R. 1936—Concepts and criteria for the recognition of communities. Journ. Ecol. 24: 285-289.

———1938—An ecological glossary. Univ. Okla. Press. Norman.

CIFERRI, R. 1936—Studio geobotanico dell'Isola Hispaniola (Antille). Atti d'Inst. Bot. d'Univ. Pavia. Ser. 4, **8**:1-336.

CLEMENTS, F. E. 1928—Plant succession and indicators. H. W. Wilson, N. Y.

CLEMENTS, F. E. ET AL. 1932—Climate and climaxes. Carnegie Inst. Wash., Yearbook **31**:214-217.

———1934—The relict method in dynamic ecology. Journ. Ecol. **22**:39-68.

———1936—Nature and structure of the climax. Journ. Ecol. **24**:252-284.

CONARD, H. S. 1935—The plant associations of Central Long Island. Amer. Midl. Nat. **16**:433-516.

COOPER, W. S. 1926—The fundamentals of vegetational change. Ecology **7**:391-413.

COWLES, H. C. 1899—The ecological relations of the vegetation of the sand dunes of Lake Michigan. Bot. Gaz. **27**:95-116, 167-202, 281-308, 361-391.

———1901—The physiographic ecology of Chicago and vicinity; a study of the origin, development, and classification of plant societies. Bot. Gaz. **31**:73-108, 145-182.

DÄNIKER, A. U. 1928—Ein ökologisches Prinzip zur Einteilung der Pflanzengesellschaften. Beiblatt z. Vierteljahrsschr. d. Nat. Gesell. Zürich. Festschrift Hans Schinz: 405-423.

———1928a—Die Grundlagen zur ökologischen Untersuchung der Pflanzengesellschaften. Vierteljahrsschr. d. Naturf. Ges. in Zürich **73**:

———1936—Die Struktur der Pflanzengesellschaft. Ber. Schweiz. Bot. Ges. **46**:576-593, Festband Rübel.

DAVIS, T. A. W. AND R. W. RICHARDS. 1934—The vegetation of Moraballi Creek, British Guiana: an ecological study of a limited area of tropical rain forest. II. Journ. Ecol. 22:160-133.

DAVY, J. BURTT. 1935—A sketch of the vegetation and flora of Tropical Africa. Empire Forestry Journ. **14**:191-201

DEL VILLAR, E. H. 1929—Geobotánica. Barcelona.

DU RIETZ, G. E. 1924—Studien über die Vegetation der Alpen, mit derjenigen Skandinaviens verglichen. Veröffentlichungen d. geobotan. Inst. Rübel in Zürich **1**:31-138.

———1929—Fundamental units cf vegetation. Proc. Intern. Congress Plant Sci., Ithaca 1926, **1**:623-7.

———1930—Classification and nomenclature of vegetation. Svensk Botan. Tidskrift **24**(4):489-503.

———1930b—Classification and nomenclature of vegetation. Rept. Proc. (Sect. Abstr.) Fifth Intern. Botan. Congress, Cambridge: 72-77.

———1936—Classification and nomenclature of vegetation units 1930-1935. Svensk Botan. Tidskrift **30**(3):580-589.

FURRER, E. 1922—Begriff und System der Pflanzensukzession. Vierteljahrsschr. Naturf. Ges. Zürich **67**:132-156.

GAMS, H. 1918—Prinzipienfragen der Vegetationsforschung. Vierteljahrsschr. d. Naturf. Ges. Zürich **63**:293-493.

GLEASON, H. A. 1923—The vegetational history of the middle west. Ann. Soc. Amer. Geogr. **12**:39-85.

———1936—Is the synusia an association? Ecology **17**:444-451.

GLINKA, K. 1927—The great soil groups of the world and their development. Edwards Bros., Ann Arbor.

GODWIN, H. 1929—The subclimax and deflected succession. Journ. Ecol. **17**:144-7.

GORDON, R. B. 1932—The primary forest types of the East-Central States. Abstr. Doctor's Disser. 8, Ohio State Univ.

GRADMANN, R. 1909—Über Begriffsbildung in der Lehre von den Pflanzenformationen. Englers Bot. Jahrb. **43**(Beibl. 99):91-103.

GRISEBACH, A. 1872—Die Vegetation der Erde naach ihrer klimatischen Anordnung. Leipzig.

HAWLEY, R. C. ET AL. 1932—Forest cover types of the Eastern United States: Report of committee on forest types. Journ. Forestry **30**:451-498.

HEINIS, FRITZ. 1937—Beiträge zur Mikrobiocoenose in alpinen Pflanzenpolstern. Bericht über d. Geobotan. Forschungsinst. Rübel in Zürich **1936**:61-76.

HULT, R. 1881—Försök till analytisk behandling af Växtformationerna. Meddel soc. pro fauna et flora fennica 8.

JENNY, H. 1929—Climate and climatic soil types of Europe and United States. Soil Res. **1**:139-189.

KATZ, N. 1929—Die Zwillingsassoziationen und die homologen Reihen in der Phytosoziologie. Ber. deutsch. botan. Ges. **47**:154-164.

————1930—Die Haüptgesetzmässigkeit der Pflanzengesellschaften und der Begriff der Assoziationen. Bull. de la Soc. d. Natural. d. Moscou N. S. **39**(1-2): 147-176.

KELLER, B. A. 1927—Distribution of vegetation on the plains of European Russia. Journ. Ecol. **15**:189-233.

KNOWLTON, F. H. 1927—Plants of the Past. Princeton Univ. Press.

KYLIN, H. 1926—Über Begriffsbildung und Statistik in der Pflanzensoziologie. Bot. Notiser **1926**:81-180.

LIPPMAA, T. 1932—Pflanzensoziologische Betrachtungen. Fedde Repert. spec. nov. reg. veg. Beih. **66**:88-95.

————1933—Taimeühingute Uurimise Metoodika Ja Eesti Taimeühingute Klassifi-katsiooni Pohi jooni. (Refeiat, Grundzüge der Pflanzensoziologischen Methodik nebst einer Klassifikation der Pflanzenassociationen Estlands). Acta Inst. Horti Botan. Univ. Tartu. **3**(4):1-169.

————1934—La méthode des associations unistrates et la systeme écologique des asso-ciations. Acta Inst. Horti Botan. Univ. Tartu. **4**(1-2):1-6.

————1935—Une analyse des forêts de l'île estonienne d'Abruka (Abro) sur la bade des associations unistrates. Acta inst. Horti Botan. Univ. Tartu. **4**(1-2):1-97.

————1938—Areal und Altersbestimmung einer Union (Galeobdolon-Asperula-Asarum-Union) sowie das Problem der Charakterarten und der Konstanten. Acta Inst. Horti Botan. Univ. Tartu **6**(2):1-152.

LIVINGSTON, B. E. AND F. SHREVE. 1921—The distribution of vegetation in the United States as related to climatic conditions. Carnegie Inst. Wash. Publ. **284**.

LÜDI, W. 1920—Die Sukzession der Pflanzenvereine. Mitteil. d. Naturf. Ges. in Bern a. d. Jahr 1919.

————1921—Die Pflanzengesellschaften des Lauterbrunnentales und ihre Sukzession. Beitr. z. geobot. Landesaufnahme **9**. Zürich.

————1928—Der Assoziationsbegriff in der Pflanzensoziologie. Bibliotheca Botanica **96**:1-93.

————1930—Die Methoden der Sukzessionsforschung in der Pflanzensoziologie. Handb. Biol. Arbetismeth. Abderhalden. XI. **5**:528-728.

McLUCKIE, J. AND A. H. K. PETRIE. 1926—An ecological study of the flora of Mount Wilson. Proc. Linn. Soc. N. S. W. **51**:94-113.

LUNDEGARDH, H. 1931—Environment and plant developent ("Klima und Boden in ihrer Wirkung auf das Pflanzenleben," Transl. E. Ashby). London.

MERRIAM, C. H. 1898—Life zones and crop zones of the United States. U.S.D.A., Bull. **10**.

MICHELMORE, A. P. G. 1934—Vegetation succession and regional surveys with special reference to tropical Africa. Journ. Ecol. **22**:313-7.

MILNE, G. ET AL. 1936—A provisional soil map of East Africa with explanatory memoir. Amani, T. T., E. Afric. Agric. Res. Sta.

MILNE, G. 1937—Notes on soil conditions and two East African vegetation types. Joourn. Ecol. **25**:254-8.

MOSS, C. E. 1910—The fundamental units of vegetation. New Phytol. **9**:18-53.

NICHOLS, G. E. 1917—The interpretation and application of certain terms and con-cepts in the ecological classification of plant communities. Plant World **20**:305-319, 341-353.

————1923—A working basis for the ecological classification of plant communities. Ecology **4**:11-23, 154-180.

NORDHAGEN, R. 1927—Die Vegetation und Flora des Sylenegebietes. Oslo.

OLIVER, W. R. B. 1930—New Zealand epiphytes. Jour. Ecol. **28**:1-50.

PALLMANN, H. AND P. HAFFTER. 1933—Pflanzensoziologische und bodenkundliche Untersuchungen im Oberengadin mit besonderer Berücksichtigung der Zwerg-

129

strauchgesellschaften der Ordnung Rhodoreto-Vaccinietalia. Ber. Schweiz. Botan. Ges. **42**:357-466.

PARKER, DOROTHY. 1938—Plant succession at Long Pond, Long Island. In manuscript.

PASTAK, ELSA. 1935—Harilaiu Taimkate (English summary, The vegetation of the Peninsula of Hariland, Estonie). Acta Inst. Horti Botan. Univ. Tartu. **5**(1-2):1-44.

PETRIE, A. H. K., P. H. JARRETT, AND R. T. PATTON. 1929—The vegetation of the Black Spur regiono: A study in the ecology of some Australian mountain Eucalyptus forests. I. Jour. Ecol. **17**:221-48.

PHILLIPS, J. 1931—The biotic community. Journ. Ecol. **19**:1-24.

————1934—Succession, development, the climax, and the complex organism: An analysis of concepts. Journ. Ecol. **22**:554-571; **23**:210-246, 488-508.

POLUNIN, NICHOLAS. 1936—Plant succession in Norwegian Lapland. Journ. Ecol. **24**:372-391.

RICHARDS, P. W. 1936—Ecological observations on the rain forest of Mount Dulit, Sarawak. I. Journ. Ecol. **24**:1-37.

RÜBEL, E. 1930—Pflanzengesellschaften der Erde. Hans Huber. Bern/Berlin.

————1932—Versuch einer Übersicht über die Pflanzengesellschaften der Schweiz. Bericht über d. geobotan. Forschungsinst. Rübel in Zürich: 19-30.

————1932—Die Buchenwälder Europas. Veröffentl. Geobotan. Inst. Rübel in Zürich. **8**:1-502.

————1936—Plant communities of the world. Pp. 263-290 of "Essays in Geobotany in Honor of William Albert Setchell." Univ. Calif. Press.

SALISBURY, E. J. 1924—The change in habitat of certain plants. Veröffentlichungen d. geobotan. Inst. Rübel in Zürich. **1**:285-8.

————1931—The standardization of descriptions of plant communities. Journ. Ecol. **19**:177-189.

SAMPSON, H. C. 1930—The mixed mesophytic forest community of northeastern Ohio. Ohio Journ. Sci. **30**:358-367.

SCHARFETTER, R. 1924—Die Grenzen der Pflanzenvereins. Festschrift für Robert Sieger: 54-69. Wien.

SCHMID, E. 1922—Biozönologie und Soziologie. Natur. Wochensch. N. F. **21**:518-523.

————1936—Was ist eine Pflanzengesellschaft? Schweiz. Botan. Gesell. Berichte **46**:565-575. Festband Rübel.

SIRGO, V. 1935—Emajoe Alamjooksul Peipsiäärsel Medalikul Asuvaist Taimeühinguist. (English summary, Plant unions of the swamps at the mouth of the river Emajogi.) Acta Inst. Horti Botan. Univ. Tartu. **4**(3-4):1-64.

SUKACHEV, V. N. 1928—1928—Principles of classification of the spruce communities of European Russia. Journ. Ecol. **16**:1-18.

SUKATSCHEW, W. 1929—Über einige Grundbegriffe in der Phytosoziologie. Berricht. d. deutsch Botan. Gesell. **47**:296-312.

SZYMKIEWICZ, D. 1923-1927—Études climatologiques. I-XIII. Acta Soc. Bot. Pol. 1-4.

TANSLEY, A. G. 1935—The use and abuse of vegetational concepts and terms. Ecology **16**:284-307.

THORNTHWAITE, C. W. 1931—The climates of North America. Geogr. Rev. **21**:633-655.

TRANSEAU, E. N. 1927—Vegetation types and insect devastations. Ecology **8**:285-288.

TÜXEN, R. and H. DIEMONT. 1937—Klimaxgruppe und Klimaxschwarm, ein Beitrag zur Klimaxtheorie. Jahresber. Naturhist. Ges. Hannover **88/89**:73-87.

VIERHAPPER, F. 1919—Über echten und falschen Vikarismus. Oest. Bot. Zeitschr. **68**:1-22.

WANGERIN, W. 1922—Die Grundfragen der Pflanzensoziologie. Die Naturwissenschaften 10:574-582.

——1925—Beiträge zur pflanzensoziologischen Begriffsbildung und Terminologie. I. Die Assoziation. Fedde Repert. spec. novar. reg. veg. Beihefte 36:3-59.

WEAVER, J. E. AND F. E. CLEMENTS. 1938—Plant Ecology. McGraw-Hill, N. Y.

WOODHEAD, E. W. 1906—Ecology of woodland plants in the neighborhood of Huddersfield. Journ. Linn. Soc. 37.

YAPP, R. H. 1922—The concept of habitat. Journ. Ecol. 10:1-23.

——1925—The interrelations of plants in vegetation, and the concept of association. Veröff. Geobotan. Inst. Rübel in Zürich 3:684-706. Festschrift Carl Schröter.

Part II

SELECTED VEGETATION STUDIES

Editor's Comments
on Papers 7 Through 11

Phytosociology developed as an extremely polymorphic discipline following its initial coalescence. By 1935 it had some basic concepts and was exploring new and more elaborate quantitative methods of analyzing vegetation. Diverse ideas concerning the nature and definition of the plant association persisted with little convergence (Gleason 1936). Braun-Blanquet (1932) criticized the deductive classification of Clements based on climax concept; and the major American phytosociologists prior to 1950 did not adopt the methods of the European schools (Egler 1951), although a well-known symposium (Just 1939) did include papers representing the Braun-Blanquet approach and the stratal approach of the Finnish school as well as Gleason's individualistic concept and a review of Clements's climax concept. At that symposium, Stanley Cain phrased the perennial question, "How many species must remain to preserve the identity of the association?", and commented, "It is doubtful that a satisfactory answer will ever be found." Paul Sears (1956) attributed the divergence among plant ecologists to an "Ecology of Ecologists" (cf. Whittaker 1962, p.72). Various ecologists from Conard (1939) to Becking (1957) attempted to introduce Braun-Blanquet's methods to North American phytosociologists without notable success, and one major American textbook on plant communities did not even cite Braun-Blanquet directly. The major effort in phytosociology prior to midcentury continued to be description of plant communities and grouping these into classes according to the tenets of one or another school, primarily Clements in North America and Braun-Blanquet in continental Europe.

The second group of papers is not designed as a chronological segment but attempts to illustrate some characteristic approaches to phytosociology. It is not possible to represent all of the schools or traditions much less the changes that could be illustrated in some of these through several decades. The major Russian tradition is represented by Sukachev (Paper 7), who developed an idea of the plant community *(phytocoenose)* as any plot of vegetation with uniform composition and uniform synusial structure and considered the association as a discrete unit of vegetation. Sukachev also developed the concept of a series of plant communities related to an environmental gradient, or multiple gradients (compound ecological series), recognized by differences in layer societies, or synusia, in the understory vegetation. These communities, he noted, correspond with the ecological responses of the species to the environmental gradient and also to the successional ("genetical") sequence. Contrary to Clements's dictum of progressive succession, Sukachev noted that bog could replace forest—a retrogressive change in phytosociological parlance. The development of Russian phytosociology is traced by Aleksandrova (1973).

The ecological series or series of communities along a gradient of soil or climatic change was widely used in classification of vegetation types. The recognition of species responding differentially to environmental gradients was explicit in the individualistic dissent of Ramensky in Russia and Gleason in the United States and continues today in the gradient analysis techniques discussed in Part III. However, in the view of Sukachev, and most other phytosociologists, the series or sequence was of communities of distinctive and integrated species compositions, not of the plants species acting independently. Russian phytosociology incorporated the concept of *zonal* vegetation of upland well-drained habitats which have stable or climax communities as against *intrazonal* vegetation of poorly drained or extreme sites. This idea is analogous to the emendation of Clements's climax concept which recognized climax as the vegetation developing on intermediate or mesic sites. This concept is seen also in Daubenmire (Paper 9).

The concept of the climax association, recognized by its dominant species, is illustrated in the study of the Abietum Magnificae by Oosting and Billings (Paper 8). They took a somewhat hybird approach, also recognizing several unions or synusia comprising the association and using the -*etum* ending characteristic of the Braun-Blanquet school. This study illustrates the increasing emphasis on quantitative sampling approaches of phytosociology of the 1930s and 1940s (McIntosh 1974), the use of presence, or constance, as synthetic characters of the association, and the analytic characters density, frequency, and dominance (measured as basal area). Oosting and Billings used the phytograph, a diagramatic representation of these characteristics of the performance of species on four axes, in an effort to visualize the importance value of species (McCormick and Harcombe 1968).

Daubenmire (Paper 9 and 1966, 1976) illustrates the view of many phytosociologists in arguing that there can be no science of vegetation without a system of classification. This reflects the emphasis of some phytosociologists on management of vegetation and productivity, which requires mapping, necessitating some sort of classification. In a long series of studies of the vegetation of the northwestern United States, Daubenmire developed the concept of *habitat type,* defining it as "The collective area which one association occupies or will come to occupy." Daubenmire, in a manner similar to Sukachev, recognized the association as a homogeneous community defined by a combination of tree and understory dominants (unions). The habitat types are indicated by the vegetation similarly to the site qualities of the Finnish school (Cajander 1926, Heimberger 1934). Recognizing that most areas have been disturbed, Daubenmire calls for identification of areas having the potential of supporting a particular relatively stable climax plant community by extrapolating from undisturbed sites. He delimits vegetation *zones* based on the recognition of one association as the climatic climax on normal sites. Climatic climax habitats are those with fairly deep, loamy soils with adequate nutrients and moisture. Specific associations may occur in one zone as zonal climax vegetation of normal sites and also in other zones as *azonal* vegetation where they are controlled by local site (e.g., edaphic) conditions rather than by macroclimate.

The archtypal physiognomic approach to vegetation study persists in modern phytosociology (Fosberg 1961) and is represented here by Beard's classification of tropical American vegetation types (Paper 10). Beard's approach relates vegetation physiognomy to sequences along environmental gradients (cf. Sukachev's compositional series) recognizing, however, that vegetation is a continuum (Beard 1973). Physiognomic approaches, particularly to tropical vegetation, have been widely used, and various diagrammatic and symbolic representations have been developed (Davies and Richards 1933, Dansereau 1951).

Tüxen is one of the primary figures of the Braun-Blanquet school (Braun-Blanquet 1969). The paper by Tüxen (Paper 11) illustrates the classification of a series of relevés by tabular comparison and rearrangement to form a synthesis table that shows the differences among the sample stands. The ubiquitous species of high constancy (Kennart) are those which characterize the samples. The table is then differentiated, on the basis of species of intermediate constancy, to determine so-called differential species (Trennart) that occur together in several relevés. If two or more such species groups can be identified that are associated with each other, but not with members of the other groups, the table identifies the differential species on the vertical axis and the communities on the horizontal. Other species occuring with low constancy are termed accidentals (Begleiter). Tüxen is also the leader of recent efforts to develop an international classification and nomenclature of vegetation units based on the principles of the Braun-Blanquet school.

Daubenmire (1966) described phytosociology as "a spectrum of concepts, terms and methods so broad as to discourage the novice and confuse even the specialist at times." This section has displayed some of that spectrum and noted some of the convergence which can be seen between diverse regional and linguistic traditions in phytosociology. It is still possible for the novice to be confused and specialists to misunderstand, and thus misrepresent the positions of various phytosociologists.

The relatively recent rise of the *ecosystem*, a term coined in 1935 by the British phytosociologist Sir Arthur Tansley, is noted in Daubenmire (Paper 9) and Sukachev (1960) in relation to some of the more traditional phytosociological uses.

REFERENCES

Alexsandrova, V. D. 1973. Russian Approaches to Classification of Vegetation. In *Ordination and Classification of Vegetation,* ed. R. H. Whittaker, pp. 493–527. The Hague, The Netherlands: W. Junk Publ.

Beard, J. S. 1973. The Physiognomic Approach. In *Ordination and Classification of Vegetation,* ed. R. H. Whittaker, pp. 355–433. The Hague, The Netherlands: W. Junk Publ.

Becking, R. 1957. The Zurich-Montpellier School of Phytosociology. *Bot. Rev.* **23**:411–488.

Braun-Blanquet, J. 1928. *Pflanzenziologie: Grundzuge der Vegetationskunde.* Berlin: Springer Verlag. Second edition 1951, Vienna: Springer Verlag, 631 pp.; third edition 1964, Vienna: Springer Verlag. First edition translated by G. D. Fuller and H. S. Conrad, *Plant Sociology,* New York: McGraw-Hill, 1932, 438 pp.

Braun-Blanquet, J. 1969. Reinhold Tüxen, Meister—Pflanzensoziologe. *Vegetatio* **16**:1–25.

Cajander, A. K. 1926. The Theory of Forest Types. *Acta for Fenn.* **29**:1–108.

Conard, H. S. 1939. Plant Associations on Land. *Am. Midl. Nat.* **21**:1–26.

Dansereau, P. 1951. Description and Recording of Vegetation upon a Structural Basis. *Ecology* **32**:172–229.

Daubenmire, R. 1966. Vegetation: Identification of Typal Communities. *Science* **151**:291–298.

Daubenmire, R. 1976. The Use of Vegetation in Assessing the Productivity of Forest Lands. *Bot. Rev.* **42**:115–143.

Davies, T. A. W., and P. W. Richards, 1933. The Vegetation of Moraballi Creek, British Guiana. *J. Ecol.* **21**:350–384.

Egler, F. E. 1951. A Commentary on American Plant Ecology Based on the Textbooks of 1947–1949. *Ecology* **32**:673–695

Fosberg, F. R. 1961. A Classification of Vegetation for General Purposes. *Trop. Ecol.* **2**:1–28.

Gleason, H. A. 1936. Twenty-five Years of Ecology 1910–1935. *Mem. Brooklyn Bot. Gard* **4**:41–49.

Heimberger, C. C. 1934. Forest-type Studies in the Adirondack Region. *Mem. Cornell Univ. (N.Y.) Agric. Exp. Stn.* **165**:1–22.

Just, T. 1939. Plant and Animal Communities. *Am. Midl. Nat.* **21**:1–255.

McCormick, J., and P. A. Harcombe. 1968. Phytograph: Useful Tool or Decorative Doodle? *Ecology* **49**:13–20.

McIntosh, R. P. 1974. Plant Ecology 1947–1972. *Ann. Mo. Bot. Gard.* **61**:132–165.

Sears, P. B. 1956. Some Notes on the Ecology of Ecologists. *Sci. Monthly* **83**:22–27.

Sukachev, V. N. 1960. The Correlation Between the Concept "Forest Ecosystem" and Forest Biogeocoenose and Their Importance for the Classification of Forests. *Silva Fenn.* **103**:94–97.

Whittaker, R. H. 1962. Classification of Natural Communities. *Bot. Rev.* **28**:1–239.

7

Reprinted from *J. Ecol.* **16**(1).1–18 (1928)

PRINCIPLES OF CLASSIFICATION OF THE SPRUCE COMMUNITIES OF EUROPEAN RUSSIA

By V. N. SUKACHEV[1].

(With one Figure in the Text.)

The spruce is the most characteristic tree of the forests of the northern half of European Russia; therefore the forest or taiga zone is called the spruce zone.

Within the limits of European Russia the northern limit of the area of this tree passes through the northern part of the Kola Peninsula, not reaching, by the way, the limit of tree vegetation. In the Kola Peninsula *typical Picea excelsa* Linn. (European spruce) already does not occur. The forms approach *Picea obovata* Ldb. (Siberian spruce), and to the east of the White Sea the northern limit of spruce is represented by this latter species alone.

The southern limit of the spruce coincides in European Russia exactly enough with the northern limit of the chernozëm. Beginning from the central part of the province Volhynia, the limit of the spruce runs north of the province Kiev, through the province Bryansk, the western part of the province Tula and the southern part of the province Nizhni-Novgorod, forming a large tongue along the River Sura. Then the limit again rises north of the Volga, which it crosses near Kazan and runs towards the Kama, along which it takes a north-eastern direction, crossing the Kama in the southern part of the province Perm and proceeding eastward as far as the Ural, along which it descends to $53\frac{1}{2}°$ of N. lat. But here it passes over into *P. obovata* Ldb.

Thus the spruce occupies one-half the area of European Russia but its distribution is not uniform. Proceeding southward, it becomes thinner and its dominance is shared by pine and oak. In the more northerly provinces spruce is the most widely spread species. East of the Ural Mountains it is not easy to distinguish the areas of European and Siberian spruce, because over a large area these two species present a series of intermediate forms. If, then, we take the two together, it can be said that in the northern provinces, including those of the north-west, the spruce occupies on the average 44 per cent. of the forest area. But a considerable part of the area occupied by deciduous trees, partly by the pine, and also the part now under tillage, were, before the introduction of culture by man, doubtless covered by the spruce; and we may say that in the primitive covering of the whole north of Russia the spruce forests prevailed (dominated) and occupied the greater part of the territory.

[1] The author's name (with the rest of the Russian names in this paper) is spelt in accordance with the recommendations of the British Academy's Committee on the transliteration of Russian names into English. The author, however, desires it to be understood that he is better known outside his own country by the German form of his name—"Sukatschew."—EDIT., *Journal of Ecology*.

In the last few decades many botanists, and foresters in particular, have dealt with the investigation of spruce forests[1]. But the descriptions of the different types of spruce community have been so brief that it is impossible to compare them accurately or to give them a satisfactory classification. Below an attempt is made to arrive at the basis of such a classification.

In order to elucidate the laws of distribution of the spruce communities, it is necessary first to characterise, if only briefly, the geo-morphological features of the regions in which they occur.

The greater part of the spruce region of European Russia is, on the whole, a plain, except towards the Ural, the Timan range and the peninsula of Kola, and most of it consists of sandy and loamy glacial and fresh water deposits. Only in the region of the northern marine transgression do we find marine deposits. In a few places rocks, including limestones (which are sometimes dolomitised), reach the surface. Omitting the rock outcrops, we have, for the spruce region of European Russia, particularly the northern part, the following geo-morphological schema. In a sufficiently developed river valley beyond the low banks which are subject to inundation, and occupied by meadows, comes a terrace, partly sandy, which is followed by the true bank. The areas between the rivers nearest to the river valleys are usually drained by them and also by their tributary gorges, the valleys of brooks, etc.; at a distance from the river valleys we meet, on the contrary, with less drained and often more level areas, which are thus more liable to become moory. The good forests, therefore, not infrequently border the rivers in broader or narrower bands, while further away from the rivers an inferior forest is sometimes succeeded by extensive high-moors. This fundamental regularity in the distribution of forests is typical in the northern part of the spruce region not only of European Russia but also of West Siberia.

Hence we are able to trace the following five typical habitats of spruce forest: (1) more or less nutritive clayey loam or sandy loam soils, well drained in places with sufficiently pronounced relief, in the north for the most part adjoining the rivers, (2) similar soils but becoming already moory, with worse drainage, less developed relief, in the north for the most part situated farther from the rivers, (3) soils still more moory, without differentiated relief, level, situated still farther from the rivers. These three types of habitat form, as it were, a connected series. Besides these there are (4) the bottoms of narrow valleys where there is excessive moisture, but the water is for the most part in motion, and (5) places carrying (for the north) exceptionally rich soils, most often connected with neighbouring limestone rocks.

According to these five fundamental habitats of the spruce we can divide the spruce forest into five types which differ in the phytosociological structure of their communities, as well as in the size of the trees. A

---

[1] These papers having been all published only in Russian, I did not think it requisite to annex any list of them.

conspicuous character is the presence and composition of the lower strata of the community.

Therefore the following scheme of classification for spruce associations may be given:

1. Relief adequately developed; site well drained, soils more or less nutritive, loams, clays or sandy loams, not moory—**Piceeta hylocomiosa.**

2. Relief less developed, sites feebly drained, soils the same, but already somewhat moory—**Piceeta polytrichosa.**

3. Relief undeveloped, surface flat, site not drained, soils moory—**Piceeta sphagnosa.**

4. Bottom of depressions with moory soils, but running water—**Piceeta herbosa.**

5. Sites with nutritive well-drained soils, for the most part in the neighbourhood of limestone deposits—**Piceeta fruticosa.**

Each of these types is composed of a series of plant communities. Owing to insufficient investigation full species lists of these communities cannot yet be given.

1. PICEETA HYLOCOMIOSA.

These are characterised by the first layer consisting of spruce only (in the north-east not seldom with admixture of fir), sometimes also with soft wood deciduous trees (aspen and birch), absence of a special second layer, absence of any marked shrub stratum, a grass stratum which is neither dense nor abundant but consists of characteristic species, and a strong continuous stratum of mosses (*Pleurozium schreberi, Hylocomium proliferum, Dicranum undulatum* and others). On the whole, the Piceeta hylocomiosa occupy the poorer and fresh or moist soils. Within these limits there is, however, a certain variety of soils. In some cases they may be simply sand, in others loam, in others, again, a drier sandy loam. The spruce, being a plant of marked sociological dominance, grows successfully on these various soils, and levels the other conditions to such a degree, that it creates on all the soils communities of very similar character.

The second tree stratum is usually wanting. The shrub stratum is either completely absent or represented by rare shrubs of *Sorbus aucuparia*, and, less frequently, *Lonicera xylosteum, Rhamnus frangula, Daphne mezereum* and, in the north, *Rosa acicularis*. In places which have preserved their primitive character to a high degree, these shrubs are feebly developed, owing to the deep overshadowing of the soil. *Lonicera xylosteum* and *Daphne mezereum* stand the overshadowing better than the other shrubs.

The herb stratum is usually not continuous and is poor in species, developing in patches consisting sometimes of one species, sometimes of a mixture of several. The following is a list of plants most characteristic of the herb stratum of this group of communities. In this list the degree of the *exclusiveness* or fidelity (*Fidélité, Gesellschaftstreue*) of each species is added, according to the system of J. Braun-Blanquet and J. Pavillard, where the number 5 indicates the strictest connection of the given species to this group of associations: a similar estimation is given to the plants according to the *constancy* or persistence with which they are found in the Piceeta hylocomiosa. These notes are given on the basis of the conception of these plants drawn from the study of the literature on the spruce forests of the north and personal acquaintance with these forests.

	Exclu-siveness (Fidelity)	Constancy (Per-sistence)		Exclu-siveness (Fidelity)	Constancy (Per-sistence)
Phegopteris polypodioides	5	3	Hepatica triloba	4	3
Ph. dryopteris	4	4	Anemone nemorosa	4	5
Polystichum spinulosum	3	4	Oxalis acetosella	4	5
Lycopodium annotinum	3	3	Rubus saxatilis	3	3
L. selago	5	1	Fragaria vesca	2	5
Poa nemoralis	2	3	Circaea alpina	5	4
Aira flexuosa	3	3	Pirola rotundifolia	3	4
Millium effusum	3	3	P. secunda	4	5
Luzula pilosa	3	5	P. media	3	3
Maianthemum bifolium	4	5	P. minor	3	2
Paris quadrifolia	3	4	P. uniflora	5	2
Convallaria majalis	3	3	Monotropa hypopitys	3	2
Epipogon aphyllus	5	1	Vaccinium vitis-idaea	2	5
Neottia nidus avis	3	1	V. myrtillus	3	5
Platanthera bifolia	3	3	Trientalis europaea	4	5
Gymnadenia cucullata	4	1	Melampyrum nemorosum	3	3
Goodyera repens	5	4	M. pratense	4	4
Corallorhiza innata	3	3	Veronica officinalis	3	3
Calypso bulbosa	5	1	Galium triflorum	5	2
Asarum europaeum	3	3	Linnaea borealis	5	4
Stellaria frieseana	4	3	Solidago virgaurea	3	5
Actaea spicata	3	3	Lactuca muralis	5	2

If the herb stratum of the Piceeta hylocomiosa has no considerable density, the moss stratum, on the contrary is usually quite continuous, consisting chiefly of *Hylocomium proliferum, Rhytidiadelphus triquetrus, Pleurozium schreberi, Ptilium crista-castrensis, Dicranum undulatum, D. scoparium*, more rarely *Polytrichum commune, P. juniperinum, Rhodobryum roseum* and some species of *Sphagnum*.

This stratum usually covers the soil with a continuous soft thick carpet. Under the layer of living moss there is not seldom a rather thick dead covering (up to 5–8 cm.), consisting partly of the remains of Phanerogams, chiefly of dead débris of moss. In some cases it has the character of peat.

From all this we see that the Piceeta of this type is very consistent in its structure. In this case the spruce may be called, in the full meaning of the word, the *builder* (*édificateur*) of the association. It greatly influences the conditions of the habitat, the soil as well as the atmosphere under its canopy, and through them the composition of flora and the ecology of the other members of the adult association. Notwithstanding a certain general monotonousness of this type there may still be distinguished (from the composition of the herb and moss strata on the one hand, the character of the growth of the spruce on the other, and partly from the participation, though a feeble one, of other species in the uppermost tree stratum) separate communities closely connected with the soil conditions, among which the first place must be assigned to the conditions of moistening and aeration of the soil connected with it, or more exactly, the conditions of penetration into the soil of oxygen, and the second place to the abundance or poverty of the soil in nutritive substances.

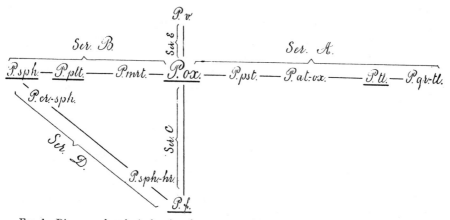

Fig. 1. Diagram of ecological series of spruce associations (Piceeta) of European Russia.

P.ox. = Piceetum oxalidosum	P.f. = P. fontinale
P.mrt. = P. myrtillosum	P.sph.-hr. = P. sphagnoso-herbosum
P.plt. = P. polytrichosum	P.pst. = P. polystichosum
P.sph. = P. sphagnosum	P.at.-ox. = P. atragenoso-oxalidosum
P.cr.-sph. = P. caricoso-sphagnosum	P.tl. = P. tiliosum
P.v. = P. vacciniosum	P.qr.-tl. = P. quercoso-tiliosum

Therefore the different communities belonging to this type may be characterised on the one hand by the process of growth and the volume at a certain age of the tree stratum and on the other by the character of the herb and moss strata, but only within the limits of a certain natural region. The communities characterised, for instance, by the prevalence of *Oxalis acetosella* will, under different physical conditions, differ in growth and volume as well as partly by a different secondary vegetation. In this case we may speak of the presence of definite geographical vicarious subassociations.

The chief associations belonging to this group are the following:

(a) *Piceetum oxalidosum.* Characteristic is the considerable participation in the herb stratum of *Oxalis acetosella, Maianthemum bifolium, Circaea alpina,* etc. It usually occupies the best, well-drained soils in the region covered with spruce. The tree stratum is high, well formed, and clear of branches to a great height. The second tree stratum and the shrub stratum are usually wanting, but sometimes there occur infrequent bushes of *Salix caprea, Lonicera xylosteum, Daphne mezereum.* Of mosses *Rhytiadelphus triquetrus* (*Hylocomium triquetrum*) is often dominant, but *Pleurozium* (*Hypnum*) *schreberi* is also abundant; *Hylocomium proliferum* (*H. splendens*), *Rhodobryum roseum* and others are less frequent; here belongs the *Oxalis-Myrtillus* type of Cajander and other Finnish botanists.

(b) *Piceetum myrtillosum* is characterised by the prevalence in the field stratum of *Vaccinium myrtillus* and the absence or feeble distribution of *Oxalis acetosella* and other indicators of better soils. It occupies somewhat inferior and moister soils than the *P. oxalidosum*. The tree stratum is of lower growth and yields a smaller volume of timber. The second layer is absent. There is no shrub stratum, or only here and there *Sorbus aucuparia, Rhamnus rangula* and *Lonicera xylosteum*. The most frequent moss is *Hylocomium proliferum* (*H. splendens*), less frequent are *Pleurozium* (*Hypnum*) *schreberi, Ptilium crista-castrensis, Dicranum undulatum, D. scoparium, Polytrichum juniperinum* and *P. commune*. Here belongs the *Myrtillus* type of Cajander.

In its typical form this community is widely distributed in the more level areas of north and north-west of Russia.

(c) *Piceetum atragenoso-oxalidosum.* This is doubtless a distinct community represented by the type of the spruce-larch forest in the Orlovsk forest in the district Veliki-Ustyug, where above the spruce layer there towers a thin stand of immense larch trees. In the thin shrub stratum are *Tilia cordata, Lonicera xylosteum, Sorbus aucuparia* entwined by the creeper *Atragene sibirica*. In the herb stratum, together with indifferent spruce indicators, we have the indicators of rich soils: *Asarum europaeum, Oxalis acetosella* and others. In this case we have, properly speaking, one of the stages of the replacement of the larch by the spruce. The limits of the area of this community are unknown.

(d) *Piceetum polystichosum* is another distinct spruce community with a dense fern stratum. This occurs on good soils and is distinguished by good growth of the spruce. *Polystichum spinulosum* is especially characteristic. Up to the present time this association has been described only by Damberg for the forest district of Tikhvin in the district of Cherepovets.

(e) *Piceetum empetroso-vacciniosum.* An association often developed in the far north of the forest region occupying cool and even somewhat dry loamy and sandy loam soils. The first layer consists of spruce with an admixture of birch. The trees are low, of slow growth, hung with lichens; and there are great quantities of dead wood. The shrub stratum is thin, consisting of *Sorbus aucuparia* and low juniper. The herb stratum is composed of *Vaccinium vitis idaea, V. myrtillus, Empetrum nigrum, Aira flexuosa, Lycopodium complanatum,* and in some places *Cornus suecica*. The moss stratum is continuous, but not too thick, consisting of *Pleurozium schreberi, Ptilium crista-castrensis* with admixture of *Polytrichum commune* and lichens (*Cladonia* and others).

This community is essentially vicarious in the north with the two first described and is distinguished phytosociologically by a somewhat lesser influence of the spruce on the rest of the vegetation, since tree growth is here strongly depressed by the general unfavourable climatic conditions.

2. Piceeta polytrichosa.

This type is characterised by the first layer consisting of spruce more or less intermingled with birch. The second layer is entirely wanting. The shrub stratum is even less developed than in the preceding type. The growth too is much worse: the trees do not attain a large size, and are often hung with *Usnea barbata*. The herb stratum is considerably poorer in species and number of individuals. On the other hand, there is a very characteristic continuous carpet of *Polytrichum commune* into which the foot sinks. The communities belonging to this type occupy, as a rule, level, rather low lying, areas with podsol soil super-saturated with water.

Compared with the Piceeta hylocomiosa, the spruce as the creator and the builder of the community is somewhat weaker, since the massive stratum of *Polytrichum commune* in some degree reduces its rôle, taking part of it upon itself. Therefore, though the influence of the spruce on the environment is still great even here, it nevertheless does not attain its full force, and the inferior development of the tree canopy allows of the development of less shade-resistant forms, while the massive stratum of *Polytrichum* has an unfavourable influence on the natural regeneration of the spruce and on most of the representatives of the herb stratum peculiar to Piceeta hylocomiosa. The compact mass of *Polytrichum commune*, decomposing with difficulty, offers an important obstacle to the development of the superficial root system of the spruce and of horizontally growing rhizomes. If the preceding type of spruce forest was characterised by an unequal distribution of the herb stratum, growing in patches, and consisting of one or another species of herb, here, on the contrary, we have a more equal distribution of these plants, not showing a tendency to form large patches.

Particularly typical of the herb stratum is the considerable share taken by *Equisetum silvaticum*, which is a sign of the poor quality of the soil and of the excessive moisture. Further there is a luxuriant growth of *Vaccinium myrtillus*, while *Vaccinium vitis idaea*, though it does occur, obviously takes a second place. Not infrequent are *Orchis maculata, Coralliorrhiza innata, Lycopodium annotinum, Melampyrum pratense* and *M. silvaticum, Linnaea borealis* (particularly on tussocks and elevations around stumps), sometimes the rare *Ledum palustre* and even *Rubus chamaemorus*.

Besides *Polytrichum commune* the moss carpet contains a larger or smaller amount of *Pleurozium schreberi, Hylocomium proliferum, Dicranum scoparium*, and more rarely an admixture of some species of *Sphagnum*.

The Piceeta of this type are rather monotonous and do not present such a variety of communities as the Piceeta hylocomiosa; they are widely distributed over the whole north of Russia. The two following may be distinguished:

(*a*) *Piceetum polytrichosum* to which applies everything said of the entire group. Here must also be referred the type "paksu-sammal-typpi" (thick moss type, Dickmosstyp, HMT of the Finns).

(b) *Piceetum empetroso-polytrichosum.* The forests of this community occupy large stretches in the far north of the forest region, particularly in the Pechora district. They occupy moist podsol soils. The uppermost tree layer consists of spruce with an admixture of birch; it is low, thin, of very bad growth. The second tree layer is wanting. The shrub stratum, owing to the first layer being thin, is more developed but nevertheless thin, consisting of *Salix bicolor*, *S. nigricans*, *S. caprea*, *Sorbus aucuparia*, *Juniperus communis* and *Rosa acicularis*. The herb stratum is more varied, consisting of a mixture of the representatives of the preceding community and *Empetrum nigrum*, *Ledum palustre*, *Rubus arcticus*, *R. humulifolius*, *Carex globularis* and, to some extent, of the representatives of the community Piceetum myrtillosum. The moss stratum is composed of *Polytrichum commune*, with *Sphagnum*, *Ptilium crista-castrensis*, *Cetraria islandica* and some other lichens.

In comparison with the preceding one this community is distinguished by the circumstance that both the "builders"—the spruce and *Polytrichum*— are strongly depressed by the unfavourable general climatic conditions and therefore their influence on the other elements is considerably less.

This community essentially succeeds the preceding one geographically, but is rather sharply separated from it.

3. PICEETA SPHAGNOSA.

The general features of the communities of this group are the very depressed growth of the spruce, which yields almost no building material, poor development of the tree canopy, the admixture, sometimes considerable, of birch, and sometimes of pine and black alder, a considerable participation in the moss stratum of *Sphagnum*, which sometimes supplants all other mosses. Soil very moist, often with a considerable peat horizon.

If in the preceding type the rôle of the spruce as "builder" of the association is disputed by *Polytrichum*, here the competitor, and a successful one, is *Sphagnum*. The spruce has already no considerable influence on the environment, it is rather itself under the control of the environment, which is in a considerable degree a creation of the sphagnum carpet. Therefore in the herb stratum the participation of the usual representatives of the spruce forest is reduced to a minimum; there are, on the contrary, predominating bog plants.

In dependence on the degree of bogginess, whether the water stagnates or is running, the thickness of the peat layer, the degree to which the water is mineralised, there arise different communities distinguished by the character of the growth of the spruce, as well as by the admixture of other species.

To this group there can be referred a considerable number of communities, which are however up to the present very inadequately described. As instances there may be mentioned two as examples:

(a) *Piceetum sphagnosum.* This community usually represents a stage of the conversion of Piceetum polytrichosum into bog. It occupies depressed, for the most part level stretches, with stagnating or slowly running water and a more or less considerable layer of peat. As the process of conversion into bog proceeds and the sphagnum peat becomes thicker the spruce begins to be supplanted by the pine. The growth of the spruce is usually very depressed and the spruce itself is shortlived. The admixture of birch is usually inconsiderable. The second tree layer and the shrub layer are wanting; in the region farther to the north the dwarf birch (*Betula nana*) is frequent. In the herb stratum along with the species peculiar to Piceeta polytrichosa, such as *Equisetum silvaticum, Vaccinium myrtillus, V. vitis-idaea* (on the tussocks), *Orchis maculata*, there are already met with representatives of the sphagnum bog, such as *Ledum palustre, Vaccinium uliginosum, Rubus chamaemorus*, sometimes *Comarum palustre, Menyanthes trifoliata*. There are also not a few small sedges.

The moss stratum consists for the most part only of *Sphagnum*, sometimes with an admixture, chiefly on tussocks, of *Polytrichum* or *Pleurozium schreberi*.

In its typical form this association seems to be distributed chiefly in the north-west of Russia and partly in the north, without spreading to the northeast.

(b) *Piceetum caricoso-sphagnosum*. Boggy Piceetum with varied herb stratum in which, for the most part, species of *Carex* predominate. This association occupies low, level, basin-like sites, with a surface covered with tussocks, and holes filled with water between the tussocks. Water running slowly or stagnating. The tree stratum consisting of thin, low spruce, intermingled with birch and often pine. Shrub stratum thin; *Rhamnus frangula, Alnus incana, Ribes nigrum* occur the most frequently; in some places *Lonicera xylosteum* and others. Herb stratum varied, consisting of a mixture of the representatives of spruce forest on tussocks, and different bog plants, sedges, grasses, with sometimes *Phragmites communis*, often *Athyrium filix femina* and others. Moss stratum usually not continuous; on tussocks consisting of sphagnum, *Polytrichum commune*, and less frequently *Pleurozium schreberi*. This community has a rather wide distribution in the north, and is mostly called "sogra."

4. PICEETA HERBOSA.

Piceeta with dense, high herb stratum. The characteristic feature of this piceetum is its connection with the bottoms of the valleys of small rivers and brooks, forming narrow bands with nutritive, moist, even wet soil, flooded by spring freshets, but always with running water. The tree stratum is not dense and consists of spruce and birch. The shrub stratum is fairly well developed. The herb stratum is dense, high and luxuriant; the moss stratum on the contrary is feebly developed.

The representatives of this group of communities are:

(a) *Piceetum fontinale*. This community has a considerable distribution in the valleys of small rivers, brooks and springs, where between the trees, in the open or covered by densely growing grass, more seldom by moss carpets, murmurs the cold running water. It is characterised by the following features (including the features of the whole group). The spruce reaches a great size, grows strongly, but not with well-formed trunks. The second tree stratum is wanting. The shrub stratum is developed rather strongly and consists of *Ribes nigrum, R. rubrum, Sorbus aucuparia, Prunus padus, Juniperus communis, Rosa acicularis, Lonicera coerulea, Tilia cordata, Viburnum opulus, Ulmus pedunculata,* and in the north-eastern part of the region also fir.

The herb stratum is vigorous, dense, high and rich in species. The following are characteristic: *Aconitum excelsum, Ulmaria palustris, Cirsium oleraceum, Angelica silvestris, Aegopodium podagraria, Comarum palustre, Calla palustris* and other species. On tussocks: *Vaccinum vitis-idaea, V. myrtillus.* The degree of development of the moss carpet varies considerably but it is as a rule insignificant. On tussocks there are found the usual mosses of spruce forests as well as *Polytrichum commune* and (rarely) *Sphagnum.*

(*b*) *Piceetum sphagnoso-herbosum.* This community is characterised by a much worse growth of the spruce, which seldom reaches normal size. The shrub stratum (mainly *Alnus incana*) is poor in species, and also the herb stratum which has fewer tall herbs (*Cirsium oleraceum* and *Aegopodium podagraria* entirely disappear). A great rôle is played by ferns and by the representatives of the usual spruce forest species. The moss stratum is thicker and a much greater part is played in it by *Sphagnum.* The appearance of this community is caused by diminished flow of the water, less enrichment of the soil by spring freshets and by somewhat poorer soils in general.

5. Piceeta fruticosa.

This group is characterised by a good growth of the spruce and the participation of broad-leaved trees. These either enter into the uppermost layer or take part in the second layer only, or in the shrub stratum. Very characteristic is the dense and varied shrub stratum. The herb stratum is moderately developed, often with an admixture of species characteristic of broad-leaved forest. The moss carpet is mostly not thick and not continuous. The soils of these communities are fresh or even a little dry, and always fairly rich in nutritive salts.

These communities are characteristic of the southern part of the spruce region, where they are distributed rather widely.

The degree of participation in the community of other species of trees, as well as the development and composition of the shrub, herb and moss strata and the growth of the spruce itself depends on the greater or lesser fertility of the soil and its humidity. This forms the basis of distinction between the associations belonging to this group. Full lists, however, must wait for the future.

Instances are the following:

(*a*) *Piceetum tiliosum*. A piceetum with a *Tilia* stratum. In its region of distribution the lime, as a shade-resisting plant, sometimes occurs as an isolated shrub in the shrub stratum in Piceeta hylocomiosa. On better soils, however, *Tilia* forms a dense shrub stratum, into which enter a considerable number of other species of shrubs. In such cases the spruce usually makes good growth, but there is always some admixture of aspen and birch, also fir (in its region of distribution) as well as isolated pines. The second tree layer does not attain any perceptible development. In the shrub stratum besides *Tilia cordata*, there are *Euonymus verrucosus, Rhamnus frangula, Lonicera xylosteum, Corylus avellana, Daphne mezereum, Viburnum opulus* and other shrubs, the composition varying somewhat according to the region.

The herb stratum is of average or considerable density. Along with the usual representatives of spruce forest such as *Oxalis acetosella, Maianthemum bifolium, Linnaea borealis, Pirola rotundifolia, P. secunda, Trientalis europaea, Vaccinium myrtillus, Convallaria majalis, Goodyera repens*, we find *Asarum europaeum, Pulmonaria officinalis, Mercurialis perennis, Asperula odorata* and other species characteristic of broad-leaved forest. The moss stratum is feebly developed or altogether wanting; it consists mostly of *Rhytidiadelphus triquetrus, Dicranum undulatum, Pleurozium schreberi*.

Even within the limits of Central Russia and the southern parts of northern and north-eastern Russia this association varies considerably, and when studied in detail several subordinate communities will probably have to be distinguished. Of these attention may be called to the type described by Finnish botanists for southern Finland: "Käen-kaali-lehdet" (*Oxalis-Maianthemum* type, OMaT).

(*b*) *Piceetum quercoso-tiliosum*. Piceetum with oak and other broad-leaved species. This is connected with a further amelioration of the soil and at the same time a complication of the community. Usually it develops on the richest soils, often with underlying limestone and marl.

The spruce makes very rapid growth, attaining great height. Together with it in the same layer there are found oak, and often lime, maple, ash, elm; also pine, birch, aspen. In the region of distribution of fir the shrub stratum is rich, sometimes even richer than in the preceding communities, consisting of the same forms. Especially luxuriant development is often shown by *Euonymus verrucosus*.

In the herb stratum, which in its composition is near to that of the preceding group though usually thinner, there already predominate forms peculiar to broad-leaved forests supplanting the representatives of spruce forests. The moss stratum is either entirely wanting or feebly developed; it consists chiefly of *Rhodobryum roseum*.

This community is distinguished by considerable variability and after more detailed investigation will certainly have to be divided into a series of distinct communities, among which there will be found such as succeed one another geographically. It is developed chiefly in the southern part of the forest region, nearer to the forest-steppe, seldom penetrating into the depth of the forest region in places where limestone exists close beneath the soil.

Being variable, as already mentioned, this community in some cases approaches the broad-leaved forest, where the oak and concomitant species begin to prevail; in other cases it approaches the preceding communities. The interference of man or injury to the spruce caused by bark-beetles is ready to alter the interrelation of the species of the coniferous and deciduous trees in this community, giving prevalence to the latter.

Considering the different communities formed by the spruce, we see that the Piceeta hylocomiosa represent communities, the interrelation of whose elements is particularly perfect. This type is the most persistent, and must be regarded as the most fundamental of the spruce communities, for not only has the dominant tree a great influence on the rest of the vegetation of the community which is very closely adapted to this influence, but the dominant itself is closely adapted to the conditions so brought about: for instance, the carpet of characteristic mosses is specially suited to the regeneration of the spruce from seed. The closeness of these interrelations of different parts of the community is the criterion of high social integration, and probably of long duration.

In this respect the first place among all the communities belonging to this group must be assigned to the Piceetum oxalidosum and the P. myrtillosum. These two, which are, on the whole, near each other morphologically, appear to be also genetically close. The development of phytosociological integration has proceeded in both during thousands of years. But nevertheless the P. myrtillosum is the starting point of a series of communities, which departing from P. oxalidosum and passing through P. polytrichosum, leads to the extermination of spruce forest, a series connected with the process of conversion of forest into bog, and terminating, so far as Piceeta are maintained, with the P. sphagnosum, which is, in its turn, followed by the succession to Pinetum, and the ultimate replacement of forest by bog.

This process, the succession accompanying the increasing bogginess of the soil, is observed in spruce forests where the soil is not enriched by springs or the freshets of small rivers and brooks which so abundantly water our northern forests. The most complex of these wet forests and at the same time those in which the growth of the spruce is best is the Piceetum fontinale (Russian *log*). The further development of this association has two possible courses. As the river develops its valley and the bed becomes deeper, the soil becomes better drained, and in the course of time the community may pass over to one of the group of Piceeta hylocomiosa. In the second case, where the drainage is bad and peaty deposits continue to accumulate, the feeding of the upper layers of soil by springs and floods decreases every year. This causes an impoverishment of the herb stratum, an increased development of the mosses and a worse growth of the spruce, which result in the replacing of Piceetum fontinale by P. sphagnoso-herbosum (*sogra*), whose natural course of development again leads in the course of time to sphagnum bog with stunted crooked pine. Thus in this second case we have a series of successions analogous with that mentioned above (P. oxalidosum to P. sphagnosum). The two series in the end lead to the same thing—the extermination of the spruce forests and their replacement by moss bogs. In both cases the process, according to the concomitant external conditions, may proceed sometimes more rapidly, sometimes more slowly, lingering or stopping at certain stages.

Some of the other communities of the first four groups clearly enter into certain of the successions described. Piceetum alnoso-caricosum, for example, being allied to P. fontinale also changes into P. sphagnoso-herbosum and later into moss bog. But just as in the second series described the communities generally succeed each other less rapidly in the progress to bog, so P. alnoso-caricosum too may for a considerable time maintain itself on an area.

Let us turn to the consideration of the interrelations in the group P. fruticosa. Since the two communities which have been distinguished as belonging to it are closely linked, the limit between them is an artificial one, and since the group is characterised by the intrusion of broad-leaved species, the elucidation of the interrelations between the two communities as well as those between them and other Piceeta is equivalent to the elucidation of the interrelations between the spruce and these broad-leaved species, among which the oak occupies the fundamental position. Therefore it is a question of the interrelations between the spruce and the oak. On this question there are extant a series of works by Russian investigators, botanists as well as specialists in forestry, among which those by S. T. Korzhinski, G. F. Morosov and A. I. Gordyagin are the most important.

Comparing all that has been written on this subject, it must be recognised that the oak and its concomitants are supplanted by the spruce, and that the group of Piceeta fruticosa is but a stage in this process. The process is however a slow one, in which the victory of the spruce over the oak is brought about not only by the shade resistance of Picea but by other peculiarities, among which is its influence on the soil. Thus Piceetum querceto-tiliosum may remain without sharp change during several generations and even sometimes owing to fortuitous causes, e.g. an attack of bark-beetles on the spruce, may undergo temporary alterations in the opposite direction, i.e. towards the predominance of the oak and its concomitants. Nevertheless the final result will be the supplanting of the broad-leaved species by the spruce, together with the replacement of grey forest-soils and soils resembling the black earth, characteristic of the oak, by podsol-soils.

This process of the supplanting of deciduous summer forest by spruce forest connected with the gradual degradation of the soils is generally marked by a dropping out of the oak and its concomitants, the first to disappear being the ash, followed by the oak and the maple, the last being the lime. *Tilia cordata*, being the most shade resistant and the least exacting as to soil (satisfied even with the podsol-soil) holds on the longest, assuming in the end the form of a shrub stratum in the community Piceetum tiliosum. This must be considered as the last stage in the supplanting of broad-leaved by spruce forests; and the lime, as G. F. Morosov put it, "is the witness, as it were, of those times when in a certain region there lived an oak forest." Logically we must admit that a further step in this process, in connection with a further impoverishment of the soil, will be the dropping out of the lime also, together with the rest of the shrub stratum of the oak forest, the impoverishment of the herb flora and the development of a continuous moss stratum, i.e. the formation of P. oxalidosum, which, as it were, rounds off the whole succession.

From the preceding characterisation of spruce forest it is evident that the fundamental type, in which the most important features of spruce forest are expressed most completely, is the group of Piceeta hylocomiosa. The other groups include either communities formed under conditions of excessive moisture, not characteristic of the usual life conditions of the spruce, or under the influence of conditions nearer to those of broad-leaved forest, and consequently also not characteristic of the spruce. We see, as it were, three fundamental ecological series of communities[1], originating from the group P. hylocomiosa. One of them is connected with the beginning of excessive water supply together with stagnation of the water, i.e. a worse supply of oxygen to the roots and deterioration in the supply of mineral food—this is the series P. hylocomiosa—P. polytrichosa—P. sphagnosa. The second series, too, is connected with excessive moisture, but the water is in motion and the roots, consequently, are sufficiently supplied with oxygen, while the mineral food supply is not always improved—this is the series P. hylocomiosa—P. herbosa. The third series is connected with an improvement of the mineral food supply without a change in the water régime of the habitat or (and this is the commoner case) with a certain increase of dryness and consequently without deterioration and even with improvement of the supply of the roots with oxygen—this is the series P. hylocomiosa—P. fruticosa.

In each of these groups there is one principal community in which the characters of the group are represented most typically. Next to this come the communities connected with conditions usually already changing in the direction of an approach to the conditions of another group. This leads to an approach of the structure of the community of the given group to the structure of another group. Here within the limits of each group series of two categories are suggested:

---

[1] The idea of distinguishing ecological series of communities for the purpose of studying the vegetation was first advanced (in Russia) by B. A. Keller, and later by V. V. Alekhin, B. N. Gorodkov, and some other Russian authors.

(1) *ecological-edaphic series* connected with a change of the nature of the soil within the limits of a definite region.

(2) *ecological-geographic*, or more exactly *ecological-climatic* series.

Each series will represent vicarious associations.

It has been already mentioned that at the present time it is difficult to give even an approximately complete list of these series and of the principal communities composing them, still less of the subordinate communities, owing to the fact that the spruce forests have been little studied from the phytosociological point of view. We have been able to give only instances illustrating general propositions. But these fundamental ideas must serve as a guide in the further investigation of the spruce forests. Then the establishment of new spruce associations, even if their number be considerable, or the subdivision of the old ones, notwithstanding inevitable subjectiveness in their extent and distinction, will not produce confusion, and the different grades of community distinguished will form a definite system.

On further consideration of the examples of communities given for each group, it may be seen that in the fundamental group Piceeta hylocomiosa, the principal community is P. oxalidosum. This realises most completely the type of spruce forest in general. Here the rôle of the spruce as the "builder" of the community finds its strongest expression. Within the limits of this group we see the series moving through P. myrtillosum in the direction of somewhat greater moisture and decrease of oxygen supply, i.e. in the direction of P. polytrichosa. Another series leads, on the contrary, in the direction of an amelioration of the mineral régime of the soil; here belong P. polytrichosum and P. atragenoso-oxalidosum, though neither of these communities have been sufficiently investigated. This series leads to the group P. fruticosa. Finally P. empetroso-vacciniosum concludes the ecological-climatic series, where it is vicarious climatically with P. oxalidosum and P. myrtillosum.

In the second group the principal community is P. polytrichosum, with which P. empetroso-polytrichosum forms an ecological-climatic series. Further investigation will show whether within the limits of this group there may be distinguished ecological-edaphic vicarious communities or whether P. polytrichosum is so persistent and sharply delineated that there are no others.

In the third group the principal community is P. sphagnosum, with P. caricoso-sphagnosum on the side leading to P. herbosa, thus forming an ecological-edaphic series determined by ameliorated oxygen supply of the roots resulting from a less complete stagnation of the water.

In the fourth group, where the principal community is doubtless P. fontinale, P. sphagnoso-herbosum on the contrary finds itself in the series of deteriorating oxygen supply of the roots, i.e. diminished flow of the water. Thus we have within the limits of the two last groups the following series: P. sphagnosum—P. caricoso-sphagnosum—P. sphagnoso-herbosum—P. fontinale, connected with an increase of the water flow, where the extreme members are phytosociologically very sharply distinguished, while the two intermediate ones form connecting links between them. To separate these two groups by tracing the limit between the two inner members would, of course, be somewhat arbitrary.

Finally the principal community of the fifth group is P. tiliosum, and P. quercoso-tiliosum has its place in the series determined by an amelioration of the mineral régime and further leading to broad-leaved forest. These ecological series of Piceeta, leaving out of consideration the ecological climatic vicarious associations and limiting outselves to the ecological-edaphic series, may be represented as in the diagram on p. 5[1].

On closer examination of these series we see that they correspond not only with the ecological but also the genetical connections of the communities. The process of succession takes its course along these very series. So the series, let us call it the series A, connected with alteration of the nutritive mineral content of the soil and the absence of excessive moisture is in its essence a succession from oak forest communities to typical Piceeta, being the result of the impoverishment of the soil due to the influence of the forest communities themselves and characterised by the clearing from the Piceeta of the alien elements of broad-leaved forest. This series leads to the formation of the most characteristic association of Piceeta—P. oxalidosum.

The series B, leading in the direction of increased moisture and its stagnation, is the succession of Piceeta to bog. If the impulse to its appearance is usually given by external factors, its whole further course is connected with a change of environment produced by the plant communities themselves. Thus their own action on the environment is the chief cause of the succession of communities in this series.

The median series C, on the contrary, has its origin in communities dependent on excessive soil moisture and running water; it is a succession whose fundamental cause is the change in the external factors of existence, the influence of the plant communities on the environment taking no considerable part in the process. In this series, developing parallel to the development of the valley of the rivulet or brook and the deepening of its bed, and hence to the draining of the bottom of the valley, we have a succession to the community P. fontinale from communities of the group P. hylocomiosa and ultimately from P. oxalidosum.

---

[1] It seems possible to establish one more series of communities, again beginning with P. oxalidosum and dependent on increasing depletion of the soil in nutritive material, but without any increase in its moisture. To this series may be referred the spruce forest with a dominant herb stratum composed of *Vaccinium vitis-idaea* (Piceetum vacciniosum), which the Finnish authors usually identify as "Vaccinium-type" (VT). According to the statement of A. P. Shennikov, the well-known investigator of Northern Russia, this community proves to be fairly common in some localities of the province of Archangel. Nevertheless there exist as yet no descriptions of it in the Russian literature; I have therefore had to withhold any detailed characteristics. This community undoubtedly belongs in the group of P. hylocomiosa, differing from P. oxalidosum in greater dryness of the soil. A further study of this series (E) is very desirable.

If the succession from P. fontinale is determined, not by a gradual drying of the soil, but by impeded flow of water and gradual accumulation of peat deposits, we shall have the series D, i.e. the succession from P. fontinale to P. sphagnosa and then to pure bog.

The geographically vicarious communities which form ecological-climatic series are here left without discussion because they are almost uninvestigated. But they also must represent genetical series in a certain sense, i.e. communities succeeding one another when the climate changes in the course of time. If we wanted to present in one diagram these ecological-climatic series as well as the ecological-edaphic ones, the most convenient way would be to trace them upwards and downwards from the latter communities represented on a plane, i.e. to plot our diagram in three dimensions.

If we take into consideration that the series of Piceeta under discussion represent also successive changes of the structure (morphology) of the communities and the interaction of the members of which they are composed, we shall be fully justified in calling them also phytosociological series.

The farther in each series a community is from the P. oxalidosum the weaker is the rôle of the spruce as the "builder" of the association, and the more prominent on the contrary are the importance of the other elements of the community, which in their nature are alien to the type community P. oxalidosum. In the series A such a rôle is played by the broad-leaved species, in the series B by Sphagnum, and in the series C by the herb stratum.

With an improvement in the quality of the soil, according to which the ecological series was established, we shall obtain communities where the spruce is already absent, being supplanted by another "builder" (édificateur of Braun-Blanquet). Thus in the series A such a community will be the broad-leaved forest with the oak and its concomitant species, in the series B Sphagnum bog with gnarled and stunted pine, an association in which the "builder" is Sphagnum, and in the series C the grassy bog.

Although, chiefly from the morphological point of view, we have established five groups of Piceeta, it would perhaps be more correct from the phyto-sociological and genetic standpoints to establish only four, considering P. polytrichosa as transitory associations between P. hylocomiosa and P. sphagnosa and including them in the fundamental group of P. hylocomiosa, as the nearest links to P. myrtillosum.

It was pointed out above that the P. herbosa may, under certain conditions, be genetically succeeded by communities of the group P. sphagnosa. It is, however, theoretically quite imaginable that P. herbosa would pass over into P. fruticosa, viz. when in the P. herbosa there is an abundance of running water combined to very rich soil, and when after the natural draining of the land owing to the development of the river valley the soil remained as rich as it was before. But such successions are not as yet described, and connecting links between these groups have not been observed.

Thus we see that the establishment of ecological series of spruce communities gives us a clear idea of the phytosociological and genetical interrelations between these communities.

Further, the same series allow us to foresee, within certain limits, the character of new communities, not yet known or described, but which may be detected in the future. Thus communities connecting P. oxalidosum and P. fontinale are not yet known, but no doubt they must exist since there are thinkable transitional natural conditions between those peculiar to these communities in their typical forms. Since we are already acquainted with the extreme members of this series we may in some degree imagine the composition and structure of intermediate links yet to be discovered. The same applies also to the other series, where we are far from knowing all the members.

Hence, it seems to me, it can be seen that the method of ecological series may have a much more extensive significance than that of elucidating the connection of the vegetation series with environment. In its further development it promises to be of assistance in the construction of such a system of communities, as will offer not only a harmonious picture of the phytosociological and genetical interrelations between them, but will allow us to foretell the existence and character of communities not yet described.

Forestry Institute
  Leningrad

8

Reprinted from *Ecol. Monogr.* **13**:259, 261–274 (1943), with permission of the publisher, Duke University Press, Durham, North Carolina

THE RED FIR FOREST OF THE SIERRA NEVADA: ABIETUM MAGNIFICAE[1]

H. J. Oosting

Duke University, Durham, N. C.

AND

W. D. Billings

University of Nevada, Reno, Nevada

[1]This study was made possible by a grant-in-aid from the Research Council of Duke University.

INTRODUCTION

Occupying a prominent place in the subalpine forests of the southern part of-the Cascade-Sierra Nevada mountain chain is the red fir, *Abies magnifica*.[2] This species, the largest of all American true firs, reaches its northern distributional limit at latitude 43° 35′ N in the Cascade Mountains of southern Oregon. Extending southward into northern California, its range is divided by the northern part of the Sacramento Valley. A relatively short western arm extends down the Coast Range at high altitudes as far as Lake County, California, while the much longer eastern extension follows the Sierra Nevada to the head of Poso Creek, latitude 35° 40′ N, its southern limit (Sudworth 1908). The species does not occur east of the Sierra Nevada, its nearest approach to the aridity of the Great Basin being on the southeastern slopes of Mt. Rose in the Carson Range, a Sierran spur in southwestern Washoe County, Nevada.

The red fir is pre-eminently a tree of high altitudes, being almost confined to the so-called subalpine zone. It ranges altitudinally from about 5,000 to 8,000 feet in northern California and Oregon and from 7,000 to over 9,000 feet in the southern Sierra. In general, this species like many others descends lower on the moist western slopes of the Sierra than on the drier eastern slopes and spurs.

At the lower elevations of its growth in the Sierra Nevada, the red fir occurs with *Pinus jeffreyi*, *Pinus ponderosa*, and *Abies concolor*. It is only a straggler when associated with these species typical of the altitudinally lower yellow pine forest. At its upper limits, the fir is associated principally with *Pinus contorta*, *Pinus albicaulis*, *Pinus monticola*, and *Tsuga mertensiana*, although it scarcely tolerates the rocky soil and wind-buffeting which these species endure at high elevations.

Between the extremes of its altitudinal distribution, on relatively deep well-drained soils, the red fir forms extensive, almost pure stands. These old virgin forests of massive trees are to be found in many parts of the Sierra Nevada. The present relative abundance of undisturbed stands of the species is due in large part to the comparative inaccessibility of the type to lumbering. The wood of the red fir, though brittle, is the heaviest of any species of Abies. For some uses it can replace that of *Pinus ponderosa*, the principal timber species of the lower slopes. As the old growth pine disappears through cutting, it is quite probable that the reasonably accessible old fir in time will be completely utilized. This has already occurred in some places, notably in the Carson Range of western Nevada. On this spur of the Sierra, the large pine timber was lumbered during the latter half of the nineteenth century to supply the mine timbers and lumber for Virginia City, Nevada. As a result, when the virgin pine had disappeared from the eastern slopes of this range, it became profitable to cut the red fir forest as high as 9,500 feet. Thus most of the old red fir on this eastern outpost of the type has already disappeared. The slopes are now covered with an extremely sparse second growth of red fir and white, lodgepole, and whitebark pines. Natural stocking, even on areas from 50 to 70 years old, has been very poor and the spaces between the trees are barren or support a patchy growth of shrubs.

Probably most of the undisturbed mature red fir type will disappear eventually through cutting or fire and will be replaced slowly by the scrubby second growth. Even under full protection, from 400 to 500 years would be needed for the forest to mature to its present condition considering the growth rate observed from stumps on the cutover land.

Should the type eventually disappear, a detailed sociological record of the original virgin condition would be particularly valuable since no quantitative ecological analyses of the community have ever been reported. To obtain such a record the following investigations were carried on during the summer of 1941. The viewpoint is essentially geobotanical with emphasis on the plant population pattern and its general relation to climate and geological history.

LITERATURE

Muir (1911), with his incomparable descriptions, has drawn an excellent picture of the Sierran forests as they appeared in the early days. His observations recorded during the summer of 1869 describe magnificent red fir forests in the Yosemite region. His notes on the composition, accompanying biota, and environment of this forest type provide a valuable record of its primeval condition. Fortunately, in most places, the forest appears much the same today, the result of inaccessibility and governmental protection.

Pratt (1907), in a short paper, gave condensed but accurate silvical information regarding the species as he found it in the Tahoe forest reservation. Red fir was then scarcely used commercially, but Pratt predicted its use in the future and showed how and why silvicultural practices should differ from those applied to white fir.

Sudworth (1908), while not primarily concerned with forest types, in describing the habitat of *Abies magnifica* briefly summarized some of the characteristics of this fir community. His detailed data on the distribution of the species are especially valuable.

[2] Synonymy for arborescent species is that of Harlow and Harrar (1941); for shrubby and herbaceous species, Jepson (1925).

158

Smiley (1915) described the red fir association as a characteristic feature of the Canadian zone in the Lake Tahoe region. His observation of the optimum development of the forest on the deep soil of glacial moraines coincides with that of Muir and that of the present authors. Smiley concludes, "The fir forest is an exclusive association, few of the other Canadian species entering into it, doubtless excluded by the insufficient light for seedlings."

Clements (1920) considered *Abies magnifica* as a dominant species in the southern extension of the Sierran subalpine forest, which he designated as the Pinus-Tsuga association. Although an *Abies magnifica* consociation was mentioned by Clements his data on the type are scanty and many of the shrubs and herbs listed are confined to the pine types. He stated that most of the societies have been derived from the montane forest or subalpine meadows, perhaps a good indication of their successional status in the wooded part of the subalpine zone.

METHODS

The region selected for study lies in the northern Sierra Nevada and southern Cascade ranges, from the southern rim of Yosemite Valley to Lassen Peak, approximately 220 miles airline to the northwest along the mountain axis. Within these extremes, 16 stations in the red fir type were chosen for study. Fourteen stations were south of the Feather River in the Sierra Nevada but two were located on the slopes of Lassen Peak at the southern end of the Cascade Range. The stations were as evenly distributed between the extremes as possible. Figure 1A shows the location of the stations in relation to the principal geographic features of the region. Pertinent data concerning the stations are listed in Table 1.

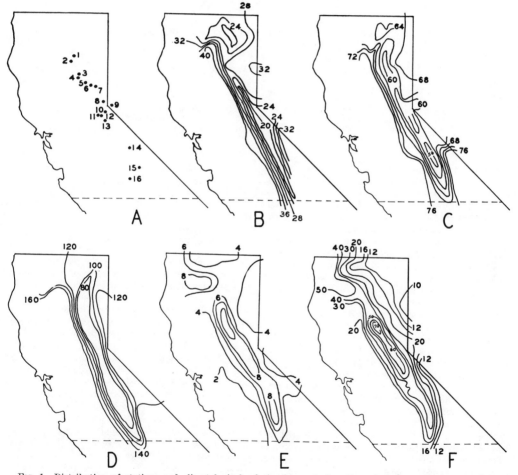

FIG. 1. Distribution of stations and climatological relationships. A, Location of stations; B, Average January temperatures; C, Average July temperatures; D, Average number of days without killing frost; E, Average warm season precipitation (inches) April-September, inclusive; F, Average annual precipitation (inches). B, C, D, E, F adapted from Sprague (1941).

TABLE 1. General data concerning stations.

Station number	Name	Location	Altitude Feet	Geological substratum	Watershed
1	Summit Lake	Near Summit Lake, E. slope, Mt. Lassen, Shasta Co., Cal.	6,750	Probably basalt	Pit River
2	Sulphur Works	1 mile N. of Sulphur Works, S. side Mt. Lassen, Tehama Co., Cal.	6,800	Probably basalt	Mill Creek
3	Onion Valley	1 mile S. of Onion Valley, Plumas Co., Cal.	6,300	Tertiary andesite	Feather River
4	Pilot Peak	2 miles SW. of Pilot Peak, Plumas Co., Cal.	6,300	Tertiary andesite	Feather River
5	Plumas Co. line	Johnsville-Gibsonville Rd. just E. of Plumas-Sierra Co. line, Cal.	6,100	Tertiary andesite	Feather River
6	Gold Lake	½ mile NE. of Gold Lake, Sierra Co., Cal.	6,500	Glacial moraine, chiefly greenstone and rhyolite debris	Feather River
7	Yuba Pass	¼ mile W. of Yuba Pass, Sierra Co., Cal.	6,600	Granite	Yuba River
8	Martis Peak	W. slope of Martis Peak, Placer Co., Cal.	8,300	Tertiary andesite	Truckee River
9	Mt. Rose Road	About 4 miles W. summit Mt. Rose Rd., Washoe Co., Nev.	8,400	Granite	Truckee River
10	Echo Summit	¼ mile W. of Echo Summit, El-dorado Co., Cal.	7,300	Glacial moraine, chiefly granitic debris	American River
11	Carson Spur No. 1	About 4 miles W. of Carson Pass, Amador Co., Cal.	8,100	Tertiary andesite	American River
12	Carson Spur No. 2	About ½ mile E. of Carson Spur, Alpine Co., Cal.	7,850	Tertiary andesite	American River
13	Granite Lodge	2-3 miles W. of Lake Alpine, Al-pine-Calaveras Co. line, Cal.	8,100	Tertiary andesite	Stanislaus River
14	Sonora Pass	5 miles W. of Sonora Pass, Tuo-lumne Co., Cal.	8,100	Granite	Stanislaus River
15	Snow Flat	Near Snow Flat, Tioga Rd., Mari-posa Co., Cal.	8,600	Granite (perhaps some debris)	Merced River
16	Bridal Veil Creek	1.8 miles W. of Bridal Veil Cr., Mariposa Co., Cal.	7,500	Granite	Merced River

At all of the stations, a list of all the plant species present was made and notes taken on geologic, climatic, and biotic conditions affecting the life of the forest. At five of the stations (Nos. 4, 7, 10, 11, and 13) quadrat studies were made for quantitative analysis of the arborescent, frutescent, and herbaceous synusiae. All individual trees over two meters tall (or 1 inch d.b.h.) were counted, measured, and basal area calculated on quadrats 15 meters on a side (225 square meters). Fifteen such quadrats were laid out along compass lines in three rows of five, 20 meters apart, at each of the five selected stations. Shrubs and woody reproduction were counted on fifteen 25 square meter plots each in a corner of a larger quadrat. Herbs were counted in a one square meter plot in a corner of each of the shrub quadrats. Tree, shrub, and herb plots were always layed out in the same position relative to each other. All the data were analyzed for frequency, density, and presence, and the tree data for percentage of total basal area.

At these same stations, soil samples were taken in triplicate from the upper mineral soil and from the organic layers above the mineral soil. The pH of these samples was determined in the laboratory using a Beckmann pH meter with a glass electrode. Loss on ignition in a muffle furnace (3 hours) was used as a measure of organic matter content. For the mineral soil, only the fraction under 2 mm. in diameter was used for the ignition samples.

The percentages of material over 2 mm. in diameter and that under 2 mm. in diameter were determined on an air-dry basis after sieving. Mechanical analyses of the material under 2 mm. in diameter were made by the Bouyoucos hydrometer method.

GEOLOGY,[3] TOPOGRAPHY, AND SOILS

The present Sierra Nevada is the result of the uplift, beginning in late Tertiary times, of a peneplained ancestral mountain system. A series of faults exists along the eastern edge of the region. The uplift has occurred along these faults tilting the old peneplained surface toward the west and southwest and rejuvenating the streams. The result is a pronounced multiple escarpment on the eastern face and a long, gradual slope from the summit to the Sacramento and San Joaquin valleys on the west. The southern part of the Sierra is characterized by a high, narrow ridge of granitic rocks punctuated by sharp peaks. Toward the north, the summit ridge becomes lower, the peaks more rounded and in some cases wooded to the crest.

The range consists principally of granitic rocks (granodiorite and quartz monzonite) of late Jurassic age capped on the higher peaks and ridges in the north with volcanic rocks (andesitic tuffs, flows, and breccias) of the Tertiary system.

[3] The authors express their appreciation to Dr. H. E. Wheeler, Department of Geology, University of Nevada, for critical reading of the section on geology.

During the Pleistocene epoch, the range was capped with four successive glacial stages: McGee, Sherwin, Tahoe, and Tioga (Blackwelder 1931). The last mentioned is the Sierran counterpart of the Wisconsin of central North America, the two stages occurring simultaneously. The two earlier stages were very extensive but their moraines have been so eroded as to make it almost impossible to set any definite lower altitudinal limits. The Tahoe stage extended on the east slope down to 5,800 feet above sea level west of Truckee (Blackwelder 1931) and to 4,700 feet at Big Pine (Knopf 1918, cited by Blackwelder 1931). On the western slope, the Merced Glacier of the Tahoe stage terminated near Bridal Veil Falls at an altitude of 3,900 feet. The Tioga or most recent stage was not as extensive as its predecessors, its terminal moraines averaging about 7,500 feet above sea level on the eastern side or about 500 feet higher than those of the Tahoe stage. The lowest moraine of the Tioga stage is reported by Blackwelder (1931) at Big Pine at an elevation of 5,750 feet.

The result of these repeated glaciations has been the gouging of many U-shaped valleys and the deposition of prominent terminal and lateral moraines. In many places, the ice so scoured the granitic rock and disappeared so recently that vegetation has not yet invaded, except in patches where a shallow soil has been built up.

The red fir forest occurs principally at and around the summit of the range. In the south, where the divide is high and rocky, the type occurs as scattered stands in protected places or on plateaus where soil has accumulated and the winds are not too strong. In the northern Sierra, the stands are more extensive and in many places cover the summit ridge. They are not confined to soils developing from any particular parent material but are seemingly best developed on glacial moraines or on unglaciated sites with deep soil. Table 1 lists the parent materials for each of the 16 stations. Well-drained gentle slopes form the ideal topographical environment for the red fir association but in protected places it may be found on steep slopes and occasionally on level areas where melted snow water does not stand.

The soil is usually relatively shallow and somewhat rocky. The mineral soil proper is covered with a mat of organic debris averaging about 5 centimeters in thickness, which very gradually decays, adding organic material to the mineral soil. There is, however, a rather sharp line of demarcation between the inorganic soil and the organic horizons. During the dry summer after the snow has melted, the litter of twigs, cone scales, leaves, and the humus layer become so compacted that the whole can be lifted from the surface of the soil in large mats.

Samples of the surface mineral soil collected in triplicate from each of five stands ranged from 20 to 50 percent in material over 2 mm. in diameter, some pieces being as much as 4 or 5 centimeters across. The soil analyses (Table 2) show that the surface soils of these five representative stands are sandy loams or sands. While the texture of the subsoil is not known

with certainty, it can be presumed to be somewhat sandy also because of its shallowness, the nature of the parent material, and the climate. The drainage from such soils is good, as might be expected.

TABLE 2. Results of soil studies.

Station number	Horizon	Average percent over 2 mm.	Average total sands	Average percent total colloids	Average percent clay <0.005 mm.	Average percent silt 0.05-0.005 mm.	Average percent fine clay <0.002 mm.	Texture type	Average percent organic matter	pH range
4	A	45.6	81.17	6.08	2.64	16.19	2.52	sand	16.0	5.7 - 6.0
	Organic								71.4	5.2 - 5.7
7	A	27.6	81.64	3.79	2.16	16.20	2.02	sand	12.0	5.1 - 5.4
	Organic								69.2	4.6 - 5.3
10	A	22.3	86.07	4.60	2.16	11.77	2.02	sand	6.8	5.6 - 6.1
	Organic								81.8	5.2 - 6.0
11	A	40.4	77.16	6.73	3.79	19.05	3.35	sandy loam	13.3	5.6 - 6.0
	Organic								64.9	5.1 - 5.7
13	A	34.1	80.63	5.10	3.62	15.74	2.52	sand	19.4	5.0 - 5.4
	Organic								67.4	5.3 - 5.9

Organic matter in the surface mineral soil (as loss on ignition of the 2 mm. fraction expressed as percentage of oven-dry weight) ranged from an average of 6.8 percent at Station 10 to 19.4 percent at Station 13 (Table 2). The average of all samples from five stations was 13.5 percent. The average loss on ignition from samples of the total organic layer ranged from 64.9 percent at Station 11 to 81.8 percent at Station 10. The average of all samples was 70.9 percent. These figures indicate a rather sharp break between the organic and mineral layers, the principal characteristic of a mor humus type (Heiberg 1937). It is difficult with the present data to assign the humus to a particular kind of mor. The long period of winter snow cover and the relatively long summer dry period combine to produce an unusual climate for the formation of humus. Most decomposition probably takes place in spring and late autumn and the rate is evidently slow. Some humus is leached into the mineral soil by melting snow water each spring, the unincorporated humus and litter drying out to form the organic mat.

As might be expected under a coniferous forest on soil developed from igneous rocks, the H-ion concentration of the soil is rather high. Fifteen samples from the surface mineral soil ranged between pH 5.0 and pH 6.1 (Table 2) while a like number from the decaying organic material ranged between pH 4.6 and pH 6.0. Of five stands examined, four exhibited a slightly greater hydrogen-ion concentration in the organic horizons than in the surface mineral soil. The single exception was at Station 13, west of Ebbetts Pass, where the mineral soil ranged between pH 5.0 and 5.4 and the organic layers between pH 5.3 and 5.9.

CLIMATE

The climate of the Sierra Nevada is greatly influenced by its general north-south axis lying across the path of the prevailing westerly winds. During much of the year, especially in the summer months, these winds bring little moisture to the mountains. This is due to the presence of a semi-permanent anti-cyclone off the California coast. During the colder part of the year, this thermal high pressure region may be pushed south of its usual position or at times may disappear completely. Storms coming in from the north Pacific during the absence of the anti-cyclone bring most of the precipitation of the Sierra Nevada. This results in rather clearly defined winter-wet and summer-dry seasons. The extent of this contrast in precipitation may be seen by comparing the distribution of the average warm season precipitation in Figure 1, E with that of the average annual precipitation in Figure 1, F. (Also see Table 3.) Approximately 80 to 85 percent of the total precipitation in the Sierra Nevada falls in the winter months from October to March, inclusive. The western slopes receive far more moisture than the region east of the summit. This is a result of the expansion and cooling to the condensation level of the moist air as it rises up the windward side of the range. Once over the crest, the descending air is warmed and its pressure increased, resulting in markedly lower precipitation.

Since most of the precipitation occurs during the winter months it is not surprising to find that, at the higher elevations, most of the moisture falls in the form of snow. Heavy snowfalls are a characteristic feature of the high Sierra Nevada. A maximum depth of 60 inches in one day has been recorded at Giant Forest, Tulare County, California. At Tamarack, Alpine County, California, the average seasonal snowfall is 449 inches while the maximum recorded for the same station is 884 inches, or more than 73 feet, during the winter of 1906-1907.

The red fir community reaches its best development in approximately the same altitudinal zone that snowfall reaches its greatest depths. This, however, is not the zone of maximum total precipitation. (See Figure 1, F.) The greatest precipitation (actual and potential) occurs lower down the western slope where pressure, temperature, and moisture content of the air are all most favorable for maximum precipitation. The zone of greatest precipitation thus coincides with the luxuriant mixed coniferous montane forest between 5,000 and 7,000 feet elevation and is of primary importance in its expression.

Because of geographical isolation, there are few precipitation stations clearly within the red fir zone. The records of three of these appear in Table 3. The annual precipitation varies from 41.02 inches at Twin Lakes, Alpine County to 46.75 inches at Tamarack in the same county. By locating the 16 fir stations on the map in Figure 1, F, the annual precipitation within the type may be roughly interpolated. This precipitation seems to range principally between 40 and 50 inches per year, although some stations in southwestern Plumas and Sierra counties receive close to 60 inches while some of those east of the Sierran crest receive not more than 30 inches per year.

Since few total precipitation records are available and since snow makes up such a great proportion of the actual precipitation, reference to snow survey data may give a more adequate picture of the source of moisture for the red fir. Survey records for 1940 and 1941 obtained from the California and the Nevada Cooperative Snow Surveys are listed in Table 4 for the snow courses nearest the stations used in this study. The figures represent the depth of the snow pack as of April 1 and the amount of water stored in this snow. Disregarding the data from Mount Lassen where the nearest snow course is much higher than the fir stations and the snow accumulates in great drifts, the average depths on April 1, 1940 ranged from 66 inches at Eureka Lake to 132 inches at Snow Flat. Eureka Lake represents conditions at Stations 3, 4, and 5 while Snow Flat is only a few hundred yards west of Station 15. On April 1, 1941, the snow depths varied from 46.8 inches at Big Meadow to 129.9 inches at Snow Flat. The water contents varied from 28.2 to 55.4 inches in 1940 and from 22.6 to 63.2 inches in 1941. The normal water contents for the entire winter snowfall up to April 1 have been determined for only

TABLE 3. Climatic summary for three weather stations within red fir zone.*

Station	Length of record—years	Temperature in °F.				Length of record—years	Killing frost average dates			Length of record—years	Average precipitation (inches)												
		January average	July average	Maximum	Minimum		Last in spring	First in fall	Growing season (days)		January	February	March	April	May	June	July	August	September	October	November	December	Annual
Soda Springs, Nevada County	40	27.2	60.1	90	−28	33	June 20	Aug. 27	68	8	7.65	8.62	5.76	4.01	1.95	1.60	0.18	0.16	0.54	3.05	4.12	8.00	45.64
Tamarack, Alpine County	22	25.7	57.9	88	−29	20	June 28	Aug. 13	46	26	10.14	6.91	6.82	2.94	2.36	1.20	0.72	0.51	1.07	2.67	4.90	6.51	46.75
Twin Lakes, Alpine County	13	24.1	58.4	91	−26	12	July 2	Aug. 11	40	20	7.64	7.45	5.81	3.90	1.54	1.39	0.50	0.30	0.72	1.90	3.15	6.72	41.02

*Adapted from "Climate of California" in "Climate and Man," U. S. D. A. Yearbook 1941.

TABLE 4. Snow depths and water contents as of April 1 at snow courses nearest stations*.

Station Number	Snow course nearest station	Altitude of station	Altitude of course	Depth of snow in inches 4/1/40	Water content in inches 4/1/40	Depth of snow in inches 4/1/41	Water content in inches 4/1/41	Normal water content in inches for entire season to April 1 †
		Feet	Feet					
1	Mount Lassen..	6,750	8,400	233.2	110.9	287.1	134.5	100.5
2	Mount Lassen..	6,800	8,400	233.2	110.9	287.1	134.5	100.5
3	Eureka Lake....	6,300	6,300	66.0	30.4	109.4	44.4	——
4	Eureka Lake....	6,300	6,300	66.0	30.4	109.4	44.4	——
5	Eureka Lake....	6,100	6,300	66.0	30.4	109.4	44.4	——
6	Church Meadows	6,500	6,700	89.0	36.8	102.3	40.7	53.6 surveys started 1937
7	Yuba Pass....	6,600	6,700	81.5	36.6	72.7	34.4	
8	Big Meadow.	8,300	8,800	72.0	31.7	46.8	22.6	28.1
9	Marlette Lake....	8,400	8,000	81.3	33.7	59.6	26.7	27.8
10	Echo Summit..	7,300	7,500	87.6	39.5	87.2	37.2	40.0‡
11	Twin Lakes...	8,100	7,900	78.7	30.3	70.2	30.0	——
12	Twin Lakes...	7,850	7,900	78.7	30.3	70.2	30.0	——
13	Lake Alpine.	8,100	7,500	100.2	45.5	115.0	49.6	44.8
14	Sonora Pass....	8,100	8,800	74.6	28.2	73.1	29.7	29.0
15	Snow Flat....	8,600	8,700	132.0	55.4	129.9	63.2	48.4
16	Ostrander Lake....	7,500	8,200	91.6	35.8	113.7	48.6	——

*Compiled from: (1) California Cooperative Snow Surveys, Snow Surveys and Precipitation Data and Seasonal Forecast of Stream Flow, April 10, 1940 and April 10, 1941. Sacramento.
(2) Nevada Cooperative Snow Surveys, Seasonal Snow Survey and Forecast of Stream Flow. Part I – Central Sierra Quadrangle. April, 1941. Reno.
† "Snow season." ‡Tentative normal.

FIG. 2. Depth of snow pack (about 90 inches) on March 1, 1942 at Station 10, Echo Summit.

seven of the courses listed here. The range is from 27.8 inches at Marlette Lake to 53.6 inches at Church Meadows with an average of 38.8 inches. Assuming this water content of the snow to be 80 percent of the total annual water falling in the type, an average annual precipitation of 48.50 inches is obtained. This figure approximates within reasonable limits the figures obtained from precipitation stations and map interpolation. A yearly precipitation of 40 to 50 inches for the best development of the red fir seems to be fairly well substantiated.

The deep winter snow pack is a constant characteristic of the fir community and supplies on melting about 80 percent or more of its effective water. Figure 2 shows the pack at Station 10, Echo Summit, as it appeared on March 1, 1942, measuring about 90 inches. This is about normal for the station since the records for April 1, 1940 and April 1, 1941 show averages of 87.6 and 87.2 inches, respectively. The snow pack is more than just a water supply. It provides insulation and protection from the cold dry winter winds and thus protects the woody reproduction until it is firmly established. Much reproduction never exceeds the height of the pack, for unless the terminal shoot can grow vigorously in a single season from below the pack to a height of a foot above the pack, it may never reach that height. Snow blast and

low temperature within the first few inches above the glazed surface of the snow would, in many cases, kill back the terminal shoots of saplings already weakened by the intense competition for light and water. The great abundance of woody vegetation below 2 meters in height seems to be greatly influenced by and possibly caused by the presence of the snow pack.

The last of the snow pack has usually melted by mid-July, although in some places it may have disappeared by late June. Frosts may be expected until late June or even into July and usually appear again by the middle of August (see Table 3). The growing season as ordinarily considered is thus very short, varying from about 40 to 70 days throughout the type. This seems too short for luxuriant forest vegetation, and probably is, for it really indicates the growing season for cultivated crops and not for the native vegetation. Most of the herbaceous plants are perennials and begin growth very early, probably before all the snow has melted and certainly before the last frost. Many of these plants seem able to withstand very low temperatures both at the beginning and the end of the normal growing season without any apparent ill-effects. It is also probable that the frosts may not be as severe under the forest cover as in the open where observations were made.

The temperature is low in the winter but not excessively low. The January averages of 24.1°, 25.7°, and 27.2° for the three stations listed are high compared to those of most of the northern United States and similar locations in the Rocky Mountains. Even the extremes of −26°, −28°, and −29° are not exceptionally low. The thick mantle of snow must provide excellent insulation and it is improbable that soil temperatures ever go below 0°F. The summer temperatures are cool, the July averages being 57.9°, 58.4°, and 60.1° with extremes of 88°, 90°, and 91°. Even during the growing season, it is probable that night minima are always below 50°F.

The climate of the red fir zone can be summarized as cool, with clearly defined winter-wet and summer-dry seasons, the precipitation being mainly in the form of heavy snowfall. Late and early frosts in combination with the deep snow pack cause the growing season to be short and coincident with the driest portion of the year.

VEGETATION

The results of the vegetational analyses of Stations 4, 7, 10, 11, and 13 are presented in Tables 5 to 8, inclusive, and in Figure 3.

FIG. 3. General appearance of interior of red fir forest.

The characteristic appearance of this association (Abietum magnificae)[4] is shown in Figure 3. The massive fir trunks scattered over a relatively open forest floor are typical. Data concerning the arborescent strata (Table 5 and 7) show that only 5 species of tree size occurred on 75 plots at the 5 quadratted stations. Only two of these species, *Abies magnifica* and *Pinus monticola,* occurred in the quadrats at all five stations. *Abies magnifica,* of tree size, was present in every plot at all five stations thus having a frequency index of 100 for each station. *Pinus monticola* was its nearest competitor in homogeneity of distribution but had frequency indices of only 60 at the two stations (4, 13) where most abundant.

The density of individuals of *Abies magnifica* over 1 inch d.b.h. per 225 square meter plot ranged from

4 The concept of the association as used in this paper is that of the Swiss-French school encompassing both the phytocoenosis and its environment. The name Abietum magnificae is here given to this association following the long-established custom of adding the suffix "-etum" to the root of the generic name of the controlling species and changing the specific name to the genitive case.

TABLE 5. Analysis of tree population over 1 inch d.b.h. at five quadratted stations.

Species		Station 4	Station 7	Station 10	Station 11	Station 13
Abies magnifica	d	15.7	12.0	35.0	16.1	10.6
	f	100	100	100	100	100
	b	277.14	423.09	423.19	351.71	295.42
	%b	97.6	99.3	98.3	98.8	94.4
Pinus monticola	d	0.8	0.13	0.5	0.5	1.0
	f	60	20	30	40	60
	b	6.08	0.01	1.30	3.34	13.68
	%b	1.1	0.01	0.3	0.94	4.4
Abies concolor	d	0.9	1.5			
	f	33	33			
	b	0.93	1.89			
	%b	0.9	0.44			
Pinus contorta	d		0.13	0.85	0.07	0.4
	f		13	35	7	27
	b		0.06	6.15	0.91	3.82
	%b		0.06	1.4	0.25	1.2
Tsuga mertensiana	d					0.06
	f					7
	b					0.2
	%b					0.6

d = density per 225 sq. m. plot, f = frequency percentage of 15 plots, b = basal area in square feet, %b = percentage of total basal area.

10.6 at Station 13 to 35.0 at Station 10. The average seems nearer the former figure. The density of individuals of the other four species was much lower. At all stations, individuals of *Abies magnifica* constituted over 80 percent of the arborescent population and usually this figure was 90 percent or more.

The total basal area per 3,375 square meters ranged from 284.1 square feet at Station 4 to 430.6 square feet at Station 10 (that is 340.64 to 516.29 square feet per acre). Of the total basal area at each station, *Abies magnifica* constituted from 94.4 percent at Station 13 to 99.3 percent at Station 7.

From the above figures, it can be seen that this forest type is clearly and completely dominated by *Abies magnifica.* This domination by a single species is even more strikingly brought out by the phytographs in Figure 4 showing the relative importance of the tree species at each station.

The reproduction data in Table 6 show *Abies magnifica* reproducing heavily at all stations. As might be expected, the number of seedlings under ½ meter in height is much greater than the number of saplings from ½ meter to 2 meters high. Many coniferous seedlings still in the cotyledon stage are present in all stands. These could not be positively identified as either Pinus or Abies and so are listed merely as coniferous seedlings. From the abundant seed source and in view of the percentages in older reproduction classes, it is very likely that most of these unidentified seedlings are those of *Abies magnifica.*

Table 6. Analysis of woody reproduction and shrubs*.

Species		Station 4	Station 7	Station 10	Station 11	Station 13
½ - 2 m.	d	6.3	6.6	27.9	4.5	6.3
	f	33	27	65	13	80
Abies magnifica						
under ½ m.	d	220.5	30.6	49.5	7.2	220.5
	f	100	93	85	27	100
½ - 2 m.	d	0.6				4.5
	f	7				27
Pinus monticola						
under ½ m.	d	1.2		0.9	0.6	8.1
	f	7		10	7	40
½ - 2 m.	d			0.45		4.5
	f			5		20
Pinus contorta						
under ½ m.	d					4.5
	f					13
under ½ m.	d					0.9
Tsuga mertensiana	f					7
Coniferous seedlings	d	49.5	211.5	39.6	17.1	18.9
	f	93	100	90	80	73
Quercus vaccinifolia	d			0.45		
	f			5		
Lonicera conjugialis	d			11.7		
	f			15		
Ribes viscosissimum	d	40.5		12.6		
	f	20		25		

*Densities, determined on 25 sq. m. plots, have been raised to 225 sq. m. basis for better comparison with tree counts.

The younger reproduction of red fir seems to be fairly evenly distributed throughout the stands as evidenced by the rather high frequency indices. On the other hand, the saplings are densely aggregated under the openings in the crown cover. It is very likely that only a few seedlings in the shade of the overstory trees ever live more than a year or two. Enough survive to maintain an uneven-aged stand; most of these survivors gain dominance over their associates in the first few years of growth. Reproduction in the openings caused by the death of an old tree, however, survives until the opening is crowded with young saplings all of approximately even height. Most of these saplings never grow to a height exceeding the average depth of winter snow. Only those exceptional few which, in spite of competition, make sufficient height growth in a single season to raise their terminal buds well above the winter snow line can survive and eventually fill the gap in the overstory. Figure 5 illustrates this aggregation of saplings under an opening.

The reproduction of the other tree species is scattered and nowhere abundant. Seedlings and saplings of *Pinus contorta* are almost lacking within the association but are fairly abundant around the edges of stands especially where the ground is marshy and in places where snow is heavy and late in melting. The lodgepole pine thus seems to be the principal pioneer tree in the expansion of the Abietum.

That the arborescent pattern as presented above is likely to be long continued is indicated by Table 7.

Table 7. Number of arborescent individuals by size classes on 75 quadrats of 225 square meters each.

Species	Under 1.0″ d.b.h. under ½ m. high	Under 1.0″ d.b.h. ½-2 m. high	1.0-2.0″ d.b.h.	2.1-6.0″ d.b.h.	6.1-12.0″ d.b.h.	12.1-24.0″ d.b.h.	Over 24.0″ d.b.h.
Abies magnifica	7,956	1,647	165	766	239	123	195
Pinus monticola	153	72	15	12	9	4	3
Pinus contorta	63	63	4	8	4	4	1
Abies concolor	0	0	9	24	3	0	0
Tsuga mertensiana	18	0	0	0	1	0	0

The abundance of individuals of *Abies magnifica* in every size class from seedlings to maturity seems to insure the long-continued dominance of this species and to indicate the climax nature of the type under present environmental conditions. Of the other four species, *Pinus contorta* is successional at this altitude, preceding red fir especially on moist sites and often on dry sites; *Abies concolor* occurs only in the warmer places, *Tsuga mertensiana* is a straggler from higher altitudes, and *Pinus monticola* is a constant species but always of low abundance in the type.

Relatively few shrubs occurred on the 75 reproduction plots. Only three species: *Ribes viscosissimum*, *Lonicera conjugialis*, and *Quercus vaccinifolia* were represented. Their densities and frequencies (Table 6) show that the individuals tend to be aggregated in the few places where they occur. The shrub flora of the association is rather sparse, only 12 species being present at the 16 stations. *Ribes viscosissimum* and *Symphoricarpos rotundifolius* are the most abundant species and are typical of the low, spindly shrubs scattered under the undisturbed red fir.

Herbs are nowhere abundant in the association. Thirty-two species occurred on the 75 herb plots at the 5 stations intensively studied, and most of these species were found in only one or two plots out of the 75 (Table 8). The two most abundant species, *Chrysopsis breweri* and *Hieracium albiflorum*, were found on only 16 and 12 plots, respectively. Like the shrubs, the herbs tend to be grouped in small colonies of one or two species. These may range from one foot to several feet in diameter and are scattered at wide intervals through the stand in places where the organic litter is not too deep and the shade not too dense.

The species pattern characterizing the red fir association is well shown by the presence list of all 16 stations in Table 9. These data are of utmost importance in the synthesis of the Abietum magnificae. Of the six species of trees, two, *Abies magnifica* present at all 16 stations and *Pinus monticola* present at 14 stations, might be said to be constantly present. Following the scale of Braun-Blanquet (1932), these two species would be of presence degree 5. *Pinus contorta*

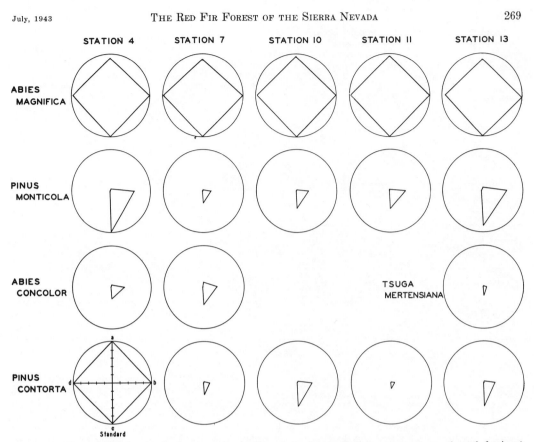

FIG. 4. Phytographs for the dominant species in the red fir forest. Radius o-a percentage of total dominant abundance; o-b frequency; o-c percentage of size classes; o-d percentage of total basal area.

is of degree 4 or mostly present, occurring at 12 stations, while *Tsuga mertensiana* and *Abies concolor* are often present or degree 3. The remaining arborescent species, *Acer glabrum,* is rare, being found at only three stations.

Of the shrubs, the only constant is *Ribes viscosissimum* which was present at 13 of the 16 stations. *Symphoricarpos rotundifolius* is often present, occurring at 7 stations, while the remaining frutescent species are seldom present or rare.

Five herbaceous species can be considered as constants of the association. These are *Chrysopsis breweri, Pedicularis semibarbata, Gayophytum ramosissimum, Pirola picta,* and *Monardella odoratissima.* The last species is usually found only around the edges of the stands where light conditions are better than within the stand. Eight herbaceous species are mostly present, occurring in 60 to 80 percent of the stands. These are *Phacelia hydrophylloides, Poa bolanderi, Arabis platysperma, Corallorrhiza maculata, Thalictrum fendleri, Kellogia galioides, Erigeron salsuginosus* var. *angustifolius,* and *Hieracium albiflorum,* all of which may be expected to be usually present in any normal stand of the association. Only five herbaceous species show a presence of degree 3, while

eleven species are of degree 2. Degree 1, with 46 species, exhibits the greatest number of species in any one presence class. These figures indicate relatively few characteristic herbs with a large number of accidentals derived from lower or higher elevations.

One of the true constants of the association is the lichen, *Evernia vulpina* (L.) Ach. This fruticose species, present in abundance at every station, is characteristic of the flora of the red fir forest, being rare above or below this zone. The lichen evidently starts growth on the young fir branches, working gradually back toward the trunk. Shortly after the lichen has become established on the trunk near the branch, the branch through natural causes dies and falls away. Since the fir branches are borne in whorls, this leaves a ring of the bright yellow-green *Evernia* well established on the bark of the trunk at every point at which a whorl existed. If it may be assumed that one whorl of branches is produced during each growing season, the age of a standing fir can be closely estimated from counting the existing whorls plus the rings of *Evernia* down almost to ground level. This is workable for trees up to two or three feet in diameter. Beyond that diameter, the older bark of the fir becomes so furrowed and corky that

FIG. 5. Rreproduction under opening in crown cover of red fir forest.

favorable spots in the red fir forest adding their weight to presence class 1.

While successional areas within the red fir zone are

TABLE 8. Analysis of herbaceous population.

Species		Station 4	Station 7	Station 10	Station 11	Station 13
Hieracium albiflorum	d	1.2	0.2	2.6		7.3
	f	20	7	30		13
Chrysopsis breweri	d	2.3		0.6	0.07	0.07
	f	66		20	7	7
Chimaphila umbellata	d	1.4	0.45			
	f	40	13			
Lupinus andersoni	d	0.3			0.2	
	f	37			13	
Pirola dentata var. integra	d	0.7				
	f	7				
Pedicularis semibarbata	d	0.07				0.3
	f	7				20
Arabis platysperma	d	0.13			2.9	
	f	7			20	
Pirola picta	d	0.4				
	f	20				
Gayophytum ramosissimum	d		0.4		0.7	
	f		7		7	
Poa bolanderi	d		0.27			1.3
	f		7			27
Polemonium sp.	d		0.6			
	f		7			
Mitella breweri	d			0.05		
	f			5		
Valeriana sylvatica	d			1.5		
	f			30		
Kelloggia galioides	d			0.65	0.07	
	f			30	7	
Anaphalis margaritacea	d			0.75		
	f			5		
Erigeron salsuginosus var. angustifolius	d			0.35		4.1
	f			5		40
Pteris aquilina var. lanuginosa	d			0.1		
	f			5		
Silene lemmonii	d			0.15		
	f			5		
Viola purpurea	d				0.4	
	f				20	
Phacelia hydrophylloides	d				0.4	0.07
	f				13	7
Brodiaea ixioides	d					0.3
	f					13
Silene montana	d					1.5
	f					40
Delphinium decorum var. patens	d					1.0
	f					13
Viola praemorsa	d					1.4
	f					33
Collinsia parviflora	d					4.1
	f					20
Plagiobothrys hispidus	d					2.5
	f					20
Carex brevipes	d					0.6
	f					13
Thalictrum fendleri	d					0.2
	f					13
Galium sp.	d					0.07
	f					7
Mimulus tilingi	d					3.1
	f					7
Ligusticum grayi	d					1.3
	f					40

most of the Evernia disappears or in moist situations becomes so abundant on the trunk as to obscure any individual rings.

Presence on a unit area basis is termed "constance" by Braun-Blanquet (1932). Constance is measured by the same scale as that used for presence. If the number of species in each class is plotted, a constancy diagram is derived. This diagram is very unlike a frequency diagram in that the greatest number of species is in the lowest class and there is a sharp drop to the higher classes with no secondary peak in that region of the curve. Presence diagrams which are not based on unit areas resemble constancy diagrams but usually show a higher percentage of casual or accidental species and thus are very high in class 1. If the 97 species encountered in this study are plotted according to their presence classes, 9 will occur in each of classes 3, 4, and 5; 16 will occur in class 2; and 54 in class 1. This is in conformity with presence diagrams in general. The use of a unit area at each station would probably result in a substantial decrease in class 1 due to the greater odds on the discovery of a single plant of an accidental species on the smaller area. Floristically, then, the Abietum is not rich, but through its relatively few species exhibits a high degree of uniformity. However, because of its intermediate altitudinal position, a number of species more typical of the yellow pine forest or the true semi-open subalpine zone are found here and there in

TABLE 9. Species of the Abietum listed in decreasing order of presence at sixteen stations.

Species	1	2	3	4	5	6	7	8	9	10	11	12	13	14	15	16
TREES																
Abies magnifica	x	x	x	x	x	x	x	x	x	x	x	x	x	x	x	x
Pinus monticola	x	x	x	x		x	x	x	x	x	x	x	x	x		
Pinus contorta			x			x	x	x	x	x	x	x	x	x	x	x
Tsuga mertensiana		x						x	x	x	x	x	x	'x	x	x
Abies concolor			x	x	x	x	x	x						x		x
Acer glabrum			x	x										x		
SHRUBS																
Ribes viscosissimum			x	x	x	x	x	x	x	x	x	x	x			x
Symphoricarpos rotundifolius			x		x			x	x		x			x		x
Ribes montigenum								x	x	x	x	x		x		x
Sambucus racemosa				x				x					x		x	
Ribes cereum	x						x	x	x							
Spiraea densiflora			x			x				x						x
Arctostaphylos nevadensis	x		x			x		x								
Symphoricarpos mollis			x							x						x
Lonicera conjugialis							x			x	x					
Quercus vaccinifolia				x					x							
Amelanchier alnifolia			x	x												
Rubus parviflorus				x												
HERBS																
Chrysopsis breweri	x	x	x	x	x	x	x		x	x	x	x	x	x	x	x
Monardella odoratissima	x	x	x	x	x		x	x	x	x	x		x	x	x	
Gayophytum ramosissimum		x	x	x	x	x	x	x	x	x	x	x	x	x	x	x
Pedicularis semibarbata			x	x	x	x	x	x	x	x	x		x	x	x	x
Pirola picta	x	x	x	x	x	x	x	x			x	x		x		x
Phacelia hydrophylloides			x	x	x		x	x	x		x	x	x	x	x	x
Poa bolanderi	x	x	x			x	x	x			x	x	x	x	x	
Arabis platysperma	x		x	x	x	x	x	x		x	x			x	x	
Corallorrhiza maculata	x	x	x	x		x	x	x		x	x		x	x	x	
Thalictrum fendleri		x	x	x	x	x	x	x		x	x	x				
Kelloggia galioides		x	x	x	x	x	x	x	x			x		x		x
Erigeron salsuginosus var. angustifolius	x	x		x		x	x			x	x		x		x	x
Hieracium albiflorum		x		x	x	x	x			x	x		x		x	x
Lupinus andersoni var. fulcratus	x	x	x	x	x				x	x			x	x		
Viola purpurea		x	x			x		x			x		x	x	x	
Chimaphila umbellata	x	x	x	x	x	x	x	x			x		x	x	x	
Pentstemon gracilentus		x	x	x	x	x	x	x								
Plagiobothrys hispidus		x			x	x				x	x					
Eriogonum nudum	x						x	x						x	x	x
Collinsia parviflora		x	x			x					x			x		
Smilacina amplexicaulis		x			x			x	x		x			x		
Viola praemorsa			x	x	x	x				x						
Erysimum asperum var. perenne		x					x	x		x	x		x			
Castilleia pinetorum								x	x	x	x		x			
Silene lemmonii					x				x		x			x	x	x
Arceuthobium campylopodum	x	x	x	x	x											
Pirola dentata var. integra		x	x	x	x		x									
Potentilla douglasii		x	x	x	x											
Calyptridium umbellatum			x				x	x		x						
Aquilegia formosa				x			x		x					x		
Sarcodes sanguinea		x				x		x								
Brodiaea izioides var. lugens			x						x					x		x
Ligusticum grayi				x			x		x							
Wyethia mollis					x	x				x					x	
Sitanion hystrix					x	x		x			x					
Pterospora andromedea		x	x				x									
Delphinium decorum var. patens								x	x		x					

TABLE 9 (Continued).

Species	1	2	3	4	5	6	7	8	9	10	11	12	13	14	15	16
Arabis repanda											x	x	x			
Carex sp.											x	x	x			
Mertensia ciliata var. stomatechoides													x		x	x
Aquilegia truncata	x						x									
Stellaria jamesiana		x				x										
Paeonia brownii		x				x										
Pteris aquilina var. lanuginosa			x				x									
Osmorrhiza obtusa					x		x									
Mitella breweri			x				x									
Polemonium sp.				x											x	
Chaenactis sp.						x										x
Phacelia heterophylla var. compacta						x										x
Lupinus lyallii									x						x	
Senecio triangularis									x	x						
Pirola secunda									x		x					
Mimulus tilingi									x		x					
Antennaria dioica var. rosea									x		x					
Lupinus sp.		x														
Pentstemon newberryi			x													
Erythronium citrinum			x													
Apocynum androsaemifolium var. pumilum					x											
Habenaria elegans			x													
Aster sp.									x							
Gilia aggregata									x							
Erysimum sp.							x									
Valeriana sylvatica									x							
Anaphalis margaritacea									x							
Lilium parvum									x							
Arnica nevadensis											x					
Silene montana												x				
Galium sp.												x				
Sidalcea glaucescens												x				
Lewisia triphylla												x				
Luzula subcongesta												x				
Eriophyllum lanatum var. integrifolium														x		
Castilleia affinis														x		
Potentilla glandulosa														x		
Streptanthus tortuosus var. orbiculatus																x
CRYPTOGAMS																
Evernia vulpina (L.) Ach.	x	x	x	x	x	x	x	x	x	x	x	x	x	x	x	x
Fomes sp.	x	x		x	x		x	x			x		x			x
Polytrichadelphus lyallii Mitt.		x			x								x	x		
Coral fungus											x			x		

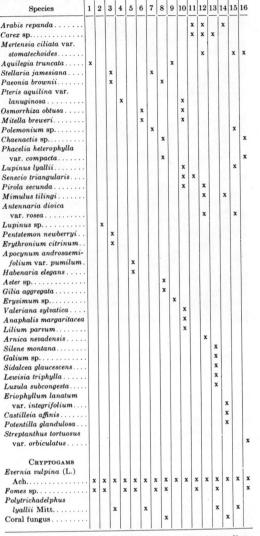

beyond the scope of this paper, it might be well to list a few of the outstanding species in such areas. Open sandy areas not yet colonized by Abies are abundant in many places throughout this altitudinal zone. These dry meadows are characterized chiefly by *Gilia aggregata, Wyethia mollis, Stipa occidentalis, Sitanion hystrix, Calyptridium umbellatum, Paeonia brownii, Phacelia heterophylla* var. *compacta,* and *Monardella odoratissima.* Moist or marshy places support *Senecio triangularis, Lilium parvum, Athyrium filix-foemina* var. *californicum, Alnus tenuifolia,* and usually the pioneer *Pinus contorta. Arctostaphylos nevadensis* is often found on rocky outcrops and in many places where burning has occurred. It seems to be a rather good indicator of past fires if present in abundance.

INTEGRATION

Compared to many forest associations, the Abietum magnificae is relatively simple in structure. The number of synusiae present, outside of the cryptogamic, is usually not more than three: herbaceous, low shrubby, and dominant arborescent. The unions involved in these three synusiae can be fairly well delineated except in the herbaceous stratum.

The overstory is always characterized by the presence of the *Abies magnifica* union. This union is overwhelmingly dominated by *Abies magnifica*. In addition to the type species, small percentages of *Pinus monticola*, *Pinus contorta*, *Abies concolor*, and *Tsuga mertensiana* may occur in almost every stand. This is the controlling union of the association; unless it is fully developed, a stand cannot be typical of the association.

Most of the shrubs of the association belong to the Ribes-Symphoricarpos union. This group is typified best by *Ribes viscosissimum*, *Ribes cereum*, *Ribes montigenum*, *Symphoricarpos rotundifolius*, and *Symphoricarpos mollis*. In the closed Abietum, this union is represented by low, spindly shrubs but during the formation of the association or after cutting, the frutescent union is dominated by very robust individuals, especially of *Ribes cereum*. The presence or absence of a closed crown cover has a pronounced effect on the development of this very plastic union. While an essential part of the climax Abietum, the Ribes-Symphoricarpos union is certainly not at its optimum stage of development in this climax.

Whether the herbaceous synusia consists of a single union or whether two are represented is not obvious. Like the shrubby layer, the development of the herb stratum is greatly enhanced by a little light. Under semi-open stands, plants such as *Monardella odoratissima*, *Chrysopsis breweri*, and *Gayophytum ramosissimum* are very abundant and robust but under closed stands the plants of these same species are scattered and rather delicate. Under closed stands, in dense shade, other species such as *Corallorrhiza maculata*, *Pirola picta*, and *Chimaphila umbellata* are conspicuous. At first inspection, no clear line of demarcation seems to exist between these extremes, since other herbaceous species, apparently indifferent to shade and organic matter, are found with Monardella and Chrysopsis and also with Corallorrhiza.

Seven species are especially characteristic of the dark closed red fir forest with its deep dry mat of twigs and needles. These are *Corallorrhiza maculata*, *Chimaphila umbellata*, *Pirola secunda*, *Pirola picta*, *Pirola dentata* var. *integra*, *Pterospora andromedea*, and *Sarcodes sanguinea*. All are ericads except Corallorrhiza; three of the seven are non-green saprophytes (Corallorrhiza, Pterospora, and Sarcodes), the remaining four are low evergreen, rhizomatous plants. *Chrysopsis breweri*, *Monardella odoratissima*, *Gayophytum ramosissimum*, *Pedicularis semibarbata*, *Phacelia hydrophylloides*, *Erigeron salsuginosus*, *Hieracium albiflorum*, *Thalictrum fendleri*, *Kelloggia galioides*, and *Poa bolanderi* are typical of those species

found under semi-open conditions or at times in deep shade. Most of these are somewhat robust when growing in rather open stands and are spindly but surviving in shade and deep organic litter. All are green plants; none are evergreen.

A study of the ranges (as of Jepson 1925) of the 17 species mentioned above yields some pertinent information. Of the seven species of deep shade, four are widely distributed throughout northern North America ranging to the Atlantic coast in New England and the Maritime Provinces. These species are *Corallorrhiza maculata*, *Pterospora andromedea*, *Pirola secunda*, and *Chimaphila umbellata*. The latter two are more or less circumboreal, being found also in Europe and Asia. *Pirola picta* ranges north to British Columbia and east to Wyoming while *Pirola dentata* var. *integra* occurs as far north as Washington in the Cascade Range. The remaining species, *Sarcodes sanguinea*, is confined to the Siskiyou Mountains, the southern Cascades, the Sierra Nevada, and the mountains of Southern and Lower California. The saprophyte-evergreen group is represented principally by wide-ranging species.

Four species (*Chrysopsis breweri*, *Gayophytum ramosissimum*, *Pedicularis semibarbata*, and *Phacelia hydrophylloides*) of the 10 in the second group are confined to the Sierra Nevada and the mountains of Southern California. The remaining six species are all restricted to the Pacific slope of North America, although some range east as far as New Mexico, Colorado and Utah and northward to Washington and British Columbia.

Thus the herbs fall into two distinct groups whether considered on the basis of ranges or life-forms and the presence of two herbaceous unions in the association is indicated. The first is designated the Pirola-Corallorrhiza union. It is made up of saprophytic or low evergreen herbs most of which are ericaceous. The principal species are widely distributed in the coniferous forests of the cooler portions of North America with at least two occurring in similar situations in Europe and Asia. The union is adapted to dense shade and deep accumulations of coniferous litter. It is thus the dominant herbaceous union in the closed or climax Abietum. The Sierran example is probably a variant of this continental or perhaps circumboreal union. The second herbaceous union can be called the Chrysopsis-Monardella union. It is typified by a number of autophytic non-evergreen plants whose ranges are confined to the mountains of the Pacific slope of western North America. This essentially Sierran union is best developed in open forests or around their margins where mineral soil is exposed and sunlight reaches the soil surface. Probably the majority of the 75 herbaceous species encountered in this study belong to this union.

Both herbaceous unions may occur together in the same stand, although when one is strongly developed the other is usually weakly developed or absent. In the semi-open or developmental Abietum magnificae, the Chrysopsis-Monardella union is present and optimally developed while the Pirola-Corallorrhiza union is usually absent. As the forest becomes closed and organic matter begins to accumulate because of changing environmental conditions, the plants of the Chrysopsis-Monardella union become more scattered and spindly. Plants of the Pirola-Corallorrhiza union invade and in the mature closed Abietum make up the characteristic herbaceous flora. The individuals of this predominantly ericaceous union are seldom abundant but their presence is indicative of the mature fir association. It is very likely that neither of these herbaceous unions is confined to this particular association. The Pirola-Corallorrhiza union is widely distributed throughout coniferous forests in North America and is probably developed to some extent in the mixed coniferous montane forest at lower elevations in the Sierra. The Chrysopsis-Monardella union, while primarily Sierran, may also occur in other Pacific slope ranges. Even in the Sierra, it tends to occur also in semi-open forests both above and below the red fir zone and is especially prominent in the successional stages of the Abietum.

Little is known of the cryptogamic unions. The abundant *Evernia vulpina* probably represents a Sierran variant of a northward-ranging, more complex, corticolous lichen union characteristic of more moist coniferous forests. Mosses are rather rare except in very damp places and occasionally on rotting wood and the bases of trees. Fungal unions were not studied but a species of Fomes on old fir trunks broken off by the wind was noted as almost universally present.

From the present studies, it can be concluded that the virgin Abietum magnificae represents a true climatic climax. The regional limits of this climax are not definitely known but it attains its best development in the central and northern Sierra Nevada of California. Its altitudinal limits in this range vary with topography, latitude, and longitude between 6,500 feet and 9,500 feet. Perhaps in any one location, this zone in which the closed red fir association may develop is not much more than 1,000 feet of elevation.

The overstory is almost entirely composed of very large old trees of *Abies magnifica*. Ring-counts on some average-sized stumps of this species near Station 9 in Washoe County, Nevada, showed between 400 and 500 rings. These indicate a time-period of approximately 500 years for the development of a mature dominant arborescent synusia. Table 7 shows that this arborescent stratum is constantly reproducing under its own cover in essentially the same pattern. *Abies magnifica* is present in every stand in all age classes from seedlings to large dominant trees. *Pinus monticola* seems also to be a constituent, though minor, of this climax. The other tree species appear to be successional or accidental in relation to the true climax red fir forest.

This slowly developing red fir association is climax in a montane region of recent glaciation. At the present time, it is best developed on lateral moraines or unglaciated spots because of a deeper soil. Heavily glaciated areas in the zone are still largely without the closed Abietum although developmental stages are plentiful. The absence of the true climax on such areas is not surprising considering the length of time involved in building up a soil sufficient to support a mature red fir community.

The red fir forest as a climax is particularly adapted to a region where snowfall is heavy and the summers are dry. This forest is very unlike its nearest relatives, the subalpine forests of the Cascades and Rocky Mountains and the transcontinental boreal forest. In these latter forests, higher moisture conditions during the summer season result in the development of a true mor humus and accompanying podsol. There is little evidence of this in the summer-dry Abietum magnificae of the Sierra. Over 80 percent of the moisture of the red fir forest falls in the form of winter snow. When this has melted, by late June, little more precipitation can be expected until October. Evaporation and transpiration are probably high during this period. The organic horizons dry out to form a rather impenetrable mat. These factors and others produce a distinctive variant of the spruce-fir forests of other regions both in dominant and secondary vegetation. There is no spruce in the overstory, a single species of Abies making up almost this whole stratum. The shrubs are few and poorly developed. Of the two herbaceous unions, the Pirola-Corallorrhiza union does show relationship with spruce-fir forests in general but the Chrysopsis-Monardella union is the product of this particular Sierran climate. The Abietum magnificae is thus shown to be a climax forest association bearing an ancestral relationship with the spruce-fir boreal forest but adapted to the winter wet-summer dry climate of the central Pacific coast of North America.

SUMMARY

1. Geobotanical studies were carried on during the summer of 1941 on the virgin red fir (*Abies magnifica*) forest of the Sierra Nevada of California.

2. Sixteen representative stations were selected for study which ranged from the south rim of Yosemite Valley to Lassen Peak. Presence lists and field notes were made at all stations. Five stations were intensively studied in regard to vegetation and soils by means of quadrats, soil samples, and subsequent laboratory analyses.

3. The distribution of the association is shown to be related to a geological history involving uplift and glacial activity that have produced relatively new sandy soils at medium-high altitudes, the subalpine zone of the Sierra Nevada. The best development of the type is on the deepest soils.

4. The limitation of the association to this zone is shown to be correlated with the prevailing climate, using climatological and snow survey data. The climate is characteristically winter wet with extremely

heavy snowfall and summer dry with a short cool growing season.

5. The association is designated the Abietum magnificae consisting of at least three and usually four unions of vascular plants and at least one cryptogamic union represented by a corticolous lichen.

6. The arborescent stratum is occupied by the *Abies magnifica* union, of which other species *(Pinus monticola, P. contorta, Tsuga mertensiana, Abies concolor)* make up a fairly constant but minor part. The frutescent stratum consists of the Ribes-Symphoricarpos union represented by few individuals of still fewer species.

7. Two unions make up the herbaceous layer. These are (1) the Chrysopsis-Monardella of autophytic, non-evergreen plants, a union which is essentially Sierran or western American and is best expressed in semi-open stands, and (2) the Pirola-Corallorrhiza of saprophytic or evergreen herbs, a predominantly ericaceous group of species which is a Sierran variant of a transcontinental union and is best developed in closed red fir stands.

8. The Abietum magnificae is the Sierran example of the spruce-fir boreal forest. As such, it constitutes the true climax at medium-high elevations in the Sierra Nevada with a vegetational constitution adapted to wet winters with heavy snowfall and to a short growing season which is dry and cool.

LITERATURE CITED

Blackwelder, Eliot. 1931. Pleistocene glaciation in the Sierra Nevada and Basin ranges. Bul. Geol. Soc. **42:** 865-922.

Braun-Blanquet, J. 1932. Plant sociology. Translated by G. D. Fuller and H. S. Conard. New York. 439 pp.

California Cooperative Snow Surveys. 1940 and 1941. Snow surveys and precipitation data and seasonal forecast of stream flow. April 10. Sacramento.

Clements, F. E. 1920. Plant indicators. The relation of plant communities to process and practice. Carnegie Inst. Wash. Publ. **290.** 388 pp.

Harlow, W. M., and E. S. Harrar. 1941. Textbook of dendrology. 2nd ed. New York. 542 pp.

Heiberg, S. O. 1937. Nomenclature of forest humus layers. Jour. Forestry **35:** 36-39.

Jepson, W. L. 1925. A manual of the flowering plants of California. 1st ed. Berkeley. 1238 pp.

Muir, John. 1911. My first summer in the Sierra. Boston. 272 pp.

Nevada Cooperative Snow Surveys. 1941. Seasonal snow survey and forecast of stream flow. Part 1—Central Sierra quadrangle. April. Reno.

Pratt, M. B. 1907. California red fir in the Tahoe forest reservation. For. Quart. **5:** 159-165.

Smiley, F. J. 1915. The alpine and subalpine vegetation of the Lake Tahoe region. Bot. Gaz. **59:** 265-286.

Sprague, Malcolm. 1941. Climate of California. In Climate and Man. U. S. D. A. Yearbook.

Sudworth, G. B. 1908. Forest trees of the Pacific slope. Forest Service, U. S. Dept. Agr. Washington. 441 pp.

9

FOREST VEGETATION OF NORTHERN IDAHO AND ADJACENT WASHINGTON, AND ITS BEARING ON CONCEPTS OF VEGETATION CLASSIFICATION

R. DAUBENMIRE
State College of Washington

TABLE OF CONTENTS

INTRODUCTION

In 1942 the writer published the results of a preliminary survey of the semi-desert and grassland vegetation of that part of the Columbia Plateau lying east of the Columbia River in Washington, and north of the Clearwater River in Idaho (Daubenmire 1942). The plan of the present study was to begin where the former survey ended, and consider the coniferous* forests to the north and east of the Plateau. Field work at first extended northward to the Canadian border, eastward to the Idaho-Montana boundary, southward to the divide between the Clearwater and Salmon Rivers in Idaho, and westward to the Columbia River in northeastern Washington (Fig. 1). Since practically all of the forest associations in this area extend beyond its limits, an attempt was later made to increase the utility of the study to workers in bordering areas, by including notes obtained during brief excursions beyond the limits of the main study area just described.

When the project was initiated the writer thought of the area selected as being characterized by a high degree of phytogeographic homogeneity. However, as field work progressed considerable north-south differentiation of vegetation was found and a significant east-west boundary line was discovered, which cuts the total area studied nearly in half. The line appears to fall a little south of the 47° parallel. Hereafter the two parts of the study area thus defined will be

* Climax forests dominated by the deciduous *Alnus sinuata* (Regel) Rydb. are also to be found in this predominately evergreen forest region, but these occupy small area in the Thuja-Tsuga and Picea-Abies zones and have not been included in the present account.

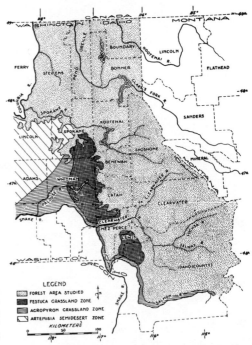

FIG. 1. Map of northern Idaho with adjacent parts of Oregon, Washington, Canada, and Montana, showing the forest area studied and its relation to contiguous grasslands and semidesert to the southwest.

referred to as the northern and southern sectors. It is to be assumed that each plant association occurs in both sectors unless otherwise stated.

This work is purely descriptive, but its stimulation was derived in large measure from the absolute necessity for a critical delimitation of vegetation units in order to organize and relate studies of the causes of vegetation distribution which the writer began concurrently.

Within the area studied most intensively, numerous isolated investigations of a phytosociologic nature have been made, but few comprehensive accounts based on original investigation have been published. Weaver (1917) studied a very small segment of the above-delimited forest area as it occurs in the vicinity of Pullman, Washington, and considered all vegetation in it as seral to a "*Thuja-Tsuga* association." Larsen (1930) described a large area briefly, but did not draw sharp distinctions between seral and climax forests. In the present study succes ional relations are held to be far more complicated than Weaver considered them, and much more significant than Larsen believed. Halliday (1937) has described the forests of the adjoining area north of the Canadian border, but because all of Canada is encompassed in the publication his descriptions of pertinent communities are too generalized to permit comparison with the present study, which is of a more intensive scope. Kujala (1945) has described and classified some forest types in Canada which may be essentially identical with certain types described in this report. Specific references to these and other pertinent researches will be made at appropriate places in the text. In addition, certain fundamental problems of vegetation classification and the practical application of the results will be discussed following the presentation of the vegetation descriptions.

TERMINOLOGY AND PRINCIPLES OF CLASSIFICATION OF THE VEGETATION

Several systems of phytosociologic terms and concepts have each gained a following among different groups of ecologists, and as the field work progressed, the writer sought to apply all of the series of concepts and terms known to him to the data at hand. The system advocated by no one "school of thought" seemed adequate in all respects, but with slight modification it has been possible to combine such favorable features of different ones as appears to be applicable, and thus arrive at a system that appears well suited to the vegetation of this particular area. The hybrid nature of the terminology demands a preliminary definition of terms and explanation of concepts before they are used in the subsequent descriptions.

DYNAMIC STATUS OF COMMUNITIES

A community is considered *seral* if interrupted age-gradients in species populations show that the sociologic status of at least some of the species is temporary, but *climax* if it appears to be self-regenerating and there is no concrete evidence that it is followed by a different subsequent community. Actually, no vegetation is absolutely permanent, but a reasonable and useful line of demarcation can be drawn between succession that is demonstrable and predictable, and that which is brought about by unforseeable events and therefore is unpredictable.

The term *association* is applied only to climax communities.

PHYTOSOCIOLOGIC ENTITIES

The *union* (or *synusia*) is considered the smallest structural unit in the organization of vegetation, each union consisting of a population of one species, or of several species that are closely similar in ecology (i.e., in microenvironmental requirements) as indicated by similarity of local environmental amplitude, phenology, and frequently by similarity of life-form as well. Of these criteria the last is believed to be the least important; thus in the *Pachistima* union there is included such diverse life-forms as a tall evergreen shrub (Taxus), a low deciduous shrub (Lonicera), a grass (Bromus), and a forb (Disporum).

The species comprising a union vary from place to place, but the variations lie within rather narrow limits and for the most part seem a result of historical factors or chance dissemination rather than of present variation in environmental factors. It is to be noted that the same species may participate in different unions, but ordinarily the abundance and/or life-form are different. *Amelanchier alnifolia*, for example, was found in 11 of the 12 forest associations analyzed, varying from a low shrub to a small tree, and from rare to ubiquitous. Also, one union may participate in the structure of different associations. The *Pachistima* union, for example, exhibits identical physiognomy and floristic composition under the Picea-Abies and the Thuja-Tsuga unions.

The association is considered the basic unit of vegetation classification. It embraces all unions that are superimposed on the same area, and each distinctive combination of vascular plant unions is ordinarily considered a separate association. Such an aggregation of union constitutes a *phytocoenosis*. A designation such as *Thuja-Tsuga/Oplopanax* association is intended to signify a dominant union typified by Thuja plus Tsuga, combined with a subordinate union typified by Oplopanax, this combination serving to characterize a particular association without indication which other unions are also present. The shortest name that seems adequate in the light of present information has been used, although further refinement in nomenclature may well be needed later. Such a binomial system of nomenclature has the merits of constituting a brief diagnostic description and conveying some notion of physiognomy, inasmuch as the unions are named after dominant species.

Association is a concept embodying those characters of all actual *stands* among which differences in species composition are attributable to historical events or chance dissemination rather than to inherent differences in environments. Widely separated dissimilar stands may eventually be shown to be connected by an intermediate series exhibiting continuous variation that is correlated with a gradual macroclimatic gradient. Such an indivisible unit might have to be viewed as a single association simply for convenience even though considerable floristic and ecologic diversity is involved. But this problem has scarcely presented itself within the limits of the study area. When surrounding regions are analyzed in detail, it may become important, although not incapable of solution. This opinion is based on reconnaissance westward to the Pacific Coast in 1949, eastward to the plains of central Montana in 1948, and northward through the Canadian Rockies in 1951, all of which excursions have been made subsequent to the accumulation of most of the data reported here, and earlier studies by the writer farther southward in the Rockies.

ECOLOGIC CLASSIFICATION OF CLIMAX ASSOCIATIONS

That type of climax vegetation which is characteristic of undulating topography and loamy soils that are moderately drained, is called the *climatic climax* (*sensu* Tansley 1935, not Clements). Within a small region those climaxes which differ from the climatic climax because of some stable but peculiar and ecologically important soil condition are called *edaphic climaxes*. On still other habitats where mature vegetation differs from the climatic climax because rough topography results in excessive exposure to, or protection from, direct solar radiation, or where wind manifestly modifies vegetation, such vegetation is considered *topographic climaxes*. Frequently soil peculiarities accompany strong topographic influence, and in such cases the term *topoedaphic climax* seems appropriate.

Most of the associations considered in the present study play different climax roles in different parts of their areas, e.g., climatic climax in one place, and topographic climax in another.

Vegetation which is peculiar because it has attained an equilibrium with a particular frequency and intensity of burning is called a *fire climax*, and that which has been differentiated by and adjusted to a particular type and degree of animal influence is a *zootic climax*. Clearly each climatic, edaphic, and topographic climax may have its own series of fire and zootic climaxes. The former group may be considered *primary climaxes*, and the latter two *secondary climaxes*, or *disclimaxes*. Climatic, edaphic and topographic climaxes are no more difficult to distinguish in the field than zonal, azonal and intrazonal soils, and the significance of the former for synecology is much the same as the latter group for pedology.

THE CLIMAX ASSOCIATION AS THE BASIS OF AREAL CLASSIFICATION

For several important reasons, it has become abundantly clear that the most significant biogeographic classification of the land surface is that based on climax conditions exclusively. One important fact that supports this conclusion is that seral communities are not well organized. It may frequently be observed in the northern Rockies that the composition of forests that appear on different parts of the same habitat type after burning is governed chiefly by the kind of plants which happened to remain on, or at the margin of, the burned area. Also the timing of the holocaust in relation to cyclic variations in seed production exerts a powerful influence upon the composition of seral forests. Seral species are thus replaceable to a large extent, and relatively few of the potential occupants of a disturbed habitat are usually present. Only after sufficient time has elapsed for extensive exchange of disseminules from place to place, and further time has allowed selection and adjustments, does a large unit of a given habitat type tend to attain uniformity, as well as conformity with ecologically similar habitats elsewhere. Finally, the same seral species is often found equally well represented on areas that support different climaxes. For these reasons climax vegetation is much more significant as a criterion of habitat potentialities than are seral communities.

Disturbance almost never eliminates all the species characteristic of a given habitat. Consequently it seems best to take the philosophic viewpoint that the habitat (soil, macroclimate, and topography) is the most durable component of an *ecosystem*,[*] and that disturbed vegetation presents varied appearance owing to differences in the degree to which the ecosystem has been thrown out of balance, or to accidents of minor ecologic importance that so strongly determine the details of recovery.

As suggested above the same climax type extends into distinctly different climatic regions, and in so doing occupies very diverse soils and topography by way of compensation. Nevertheless, each stand contains essentially the same galaxy of subordinate species, plants and probably animals too, which depend either upon intrinsic characters of the habitat or upon microenvironmental conditions created by the dominant trees. Because of this, diversity with respect to ecologic role must be subordinated in biogeographic classification to the fact that all stands of an association have a high degree of biotic similarity. Such biologic homogeneity among the stands is of much greater significance than are the macroclimatic and geologic variations among them, especially to sciences concerned with the management of wild lands.

In the interest of clarity, it is desirable to make a distinction between vegetation and the area it occupies. The collective area which one association occupies, or will come to occupy as succession advances, is called a *habitat type*. Considerable variation of intrinsic factors may be encompassed but the ecologic sums of the different sets of conditions are essentially equivalent with respect to the nature of the climax. Each time the forest is destroyed, as by fire or logging, plant succession leading toward the same climax association is initiated once more because the fundamental characters of the habitat type are not permanently affected by disturbance.

---

[*] Although this paper concerns plant associations, the writer considers plants and animals, plus climatic and edaphic factors, as inseparable constituents of one interrelated unit, the ecosystem (Tansley 1935). Vegetation is simply the most evident component of such an entity.

TABLE 1. Constancy and distribution by unions and associations, of shrubs and herbs that attain a constancy of 5/6 or 6/6, i.e., 83% or more, in at least one association.

Unions	Species	*Pinus/Purshia*	*Pinus/Agropyron*	*Pinus/Symphoricarpos*	*Pinus/Physocarpus*	*Pseudotsuga/Physocarpus*	*Abies grandis/Pachistima*	*Thuja-Tsuga/Oplopanax*	*Thuja/Pachistima*	*Thuja-Tsuga/Pachistima*	*Picea-Abies/Pachistima*	*Picea-Abies/Xerophyllum*	*Picea-Abies/Menziesia*
Agropyron spicatum	Balsamorhiza sagittata (Pursh) Nutt.	5	2	1									
	Agropyron spicatum (Pursh) Scrib. & Sm.*	6	6	5	2	1							
	Achillea millefolium L. var. lanulosa (Nutt.) Piper	6	6	6	4	4							
	Festuca idahoensis Elmer	4	6	2									
	Epilobium paniculatum Nutt.	5	6	4	2			1					
	Carex geyeri Boott		1	4	2	5	3	1	1				
Poa secunda	Stellaria nitens Nutt.	3	5	1	2	1							
	Bromus tectorum L.	4	5	4	2	3							
	Collinsia parviflora Lindl.	6	5	6	4	3	2						
Purshia tridentata	Purshia tridentata (Pursh) DC.	6											
Symphoricarpos rivularis	Perideridia gairdneri H. & A.		1	5	1								
	Prunus virginiana L. var. melanocarpa (A. Nels.) Sarg.	6	3	1	6	3							
	Spiraea betulifolia Pall.		1	4	6	6	6	1	1				
	Rosa spaldingii Crepin & R. ultramontana (Wats.) Heller	4	3	6	6	4	1						
	Symphoricarpos rivularis Suks.	5	3	6	6	6	4	1	2				
	Berberis repens Lindl.				2	5	2						
	Galium aparine L. & var. vaillantii (DC.) Koch	1	2	5	5	4	1						
	Galium boreale L.	1		4	5	2	1						
Physocarpus malvaceus	Hydrophyllum capitatum Dougl.	1		2	5								
	Physocarpus malvaceus (Greene) Kuntze		1	2	6	5	4						
	Erythronium grandiflorum Pursh		1	5	6	3	2						
	Trillium petiolatum Pursh			2	6	1							
	Fragaria cuneifolia Nutt. & F. bracteata Heller	3	4	4	4	4	6		2	2			
	Amelanchier alnifolia Nutt.	5	1	4	6	6	6	1	2	1	2	1	
Pachistima myrsinites	Arenaria macrophylla Hook.		2	3	3	6		3	1	4	1		
	Thalictrum occidentale Gray (incl. vars.)		1			6	2	5		5	2		
	Osmorhiza spp.				4	2	3	5					
	Bromus vulgaris (Hook.) Shear.				4	3	3	6	3	2	3		
	Arnica cordifolia Hook.			3	3	5	3	6	3	5	2		
	Galium triflorum Michx.					1	2	5	5	3	6	1	1
	Mitella stauropetala Piper					2	5	6	3	6	5	1	
	Rosa gymnocarpa Nutt.					1	5	6	2	1	4		
	Smilacina stellata (L.) Desf.					1	5	2	2	1	2	1	
	Actaea arguta Nutt.					1	3	6	2	1	5	1	
	Trillium ovatum Pursh						5	2	1				
	Lastrea dryopteris (L.) Slosson						4	6	4	2	4		
	Coptis occidentalis (Nutt.) T. & G.						6		2				
	Adenocaulon bicolor Hook.						3	2	6	2	4		
	Tiarella unifoliata Hook.						2	6	6	3	1		
	Asarum caudatum Lindl.						2	6	6	3	1		
	Anemone piperi Britt.						1	6	5	1	3		
	Linnaea borealis L. var. americana (Forbes) Rehder						5	3	6	2	5	1	
	Clintonia uniflora (Shult.) Kunth						4	3	3	6			1
	Pachistima myrsinites (Pursh) Raf.						2	5	6	6	5		
	Lonicera ebractulata Rydb.						2	6	6	5	1	1	
	Viola glabella Nutt.					3	1	5	3	6		1	1
	Goodyera oblongifolia Raf.			1	1	3	6	6	5	6	1	1	2
	Pyrola secunda L.					1	3	5	4	5	1	3	
Oplopanax horridum	Athyrium filix-femina (L.) Roth.						1	6	5	1	1		
	Oplopanax horridum (Sm.) Miq.						6	6					
Xerophyllum tenax	Vaccinium membranaceum Dougl.						1	3	4	6	6	6	6
	Xerophyllum tenax (Pursh) Nutt.									3	5	6	6
Menziesia ferruginea	Menziesia ferruginea Sm.							2	2	2		2	6
	Luzula glabrata (Hoppe) Desv.											5	5

*Includes also *A. inerme* (Scribn. & Sm.) Rydb.

The area occupied or potentially occupied by a closely related group of associations is called a *zone*.

METHODS

No statistical studies were attempted until several seasons of reconnaissance had permitted a tentative subjective classification of the climax communities. During this time it became apparent that these forest associations are best defined as combinations of two or more unions of vascular plants inhabiting the same area simultaneously. Eventually analyses were made to compose more accurate descriptions of the unions and to discover which species are of most constant occurrence in the unions comprising a given association, and therefore serve best as indicators. General impressions are useful in formulating descriptions of unions with respect to dominance and physiognomy, but something more objective is needed to draw up reasonably accurate floristic descriptions of the associations.

For each of the associations recognized on the basis of reconnaissance, six representative stands were selected for analysis, these being as widely distributed over the main study area as possible. In selecting representative locations for the sample plots, the aim was to choose climax or near-climax stands of average densities, for in any forest association all degrees of intergradation may be found from open stands in which the ground is populated with an abundance of heliophytic herbs or shrubs to heavy stands of timber where the shade is so dense that only sciophytic species are represented. In stands of intermediate denseness the majority of species of sun and shade plants that characterize the forest association are represented.

The sample plot in each stand was 5 x 25m in dimensions and was visited sufficient times during one growing season that a list of all the vascular plants growing on it could be compiled. Because no attention was paid to the numbers of individuals of each species in a plot, the influence of moderate grazing in altering the relative abundance of species in some of the stands at low altitude had little effect upon the statistics. The summary of floristic lists from a series of large sample plots widely scattered over the geographic area of an association yields *constancy* data. By this method of analysis information of statistical character was obtained which shows distinct differences and similarities in the kinds of plants and their constancy of occurrence in the various unions. However, it should be pointed out that these constancy data are wholly inadequate to suggest the sharp differences in physiognomy which are at least equally as characteristic of the unions. No single type of statistic should be expected to portray more than one characteristic of a community.

When the data were accumulated and tabulated, it was decided to disregard all those species which do

TABLE 2. Distribution and roles of tree species in relation to habitat types.

Zones	Habitat Types	Pinus ponderosa Laws.	Populus tremuloides Michx.	Pseudotsuga taxifolia var. glauca (Mayr.) Sudw.	Larix occidentalis Nutt.	Abies grandis (Dougl.) Lindl.	Pinus contorta var. latifolia Engelm.	Pinus monticola Dougl.	Betula papyrifera Marsh. (incl. vars.)	Populus trichocarpa Torr. & Gray	Thuja plicata Donn.	Tsuga heterophylla (Raf.) Sarg.	Tsuga mertensiana (Poir.) Britt.	Picea engelmanni Parry	Abies lasiocarpa (Hook) Nutt.	Pinus albicaulis Engelm.	Larix lyallii Parl.
Picea engelmanni-Abies lasiocarpa	Picea-Abies/Menziesia.....				S			S						C	C	S	c
	Picea-Abies/Xerophyllum...		s	s	S		S	s						C	c	S	
	Picea-Abies/Pachistima.....	s		S	S	S	s	S						C	C		
Thuja plicata-Tsuga heterophylla	Thuja-Tsuga/Pachistima...	s	s	s	S	s	S	S	s	s	C	C			s	s	
	Thuja-Tsuga/Oplopanax....					s		S			C	C			s	s	
	Thuja/Pachistima.........	s		S	S	S	S	S	s	s	C				s		
	Abies grandis/Pachistima...	S	s	S	S	C	S	s	s	s							
Pseudotsuga taxifolia var. glauca	Pseudotsuga/Physocarpus...	S	s	C	s		S										
	Pseudotsuga/Calamagrostis..	S	s	C	S		S										
Pinus ponderosa	Pinus ponderosa/ Physocarpus............	C	s														
	Pinus ponderosa/ Symphoricarpos.........	C	s														
	Pinus ponderosa/Agropyron.	C															
	Pinus ponderosa/Purshia...	C	s														

C — major climax species   S — major seral species
c — minor climax species   s — minor seral species

These evaluations are all based upon the author's experience. In reviewing the manuscript for this publication, Mr. Charles A. Wellner, a forester with intimate knowledge of the area, added the following notes from his observations:
Abies lasiocarpa minor seral species n *Thuja/Pachistima.*
Populus trichocarpa, Pinus contorta, Larix occidentalis, Pseudotsuga and *Betula* minor seral species in *Thuja-Tsuga/Oplopanax.*

not have very high constancy in at least one of the
associations studied. Specifically those species which
did not occur in at least 5 of the 6 plots in any one
of the associations were excluded. The list of herbs
and shrubs thus rigorously purged is given in Table 1.

The species of trees, their relative abundance in
the dominant union, and their successional roles on
the different habitat types are data obtained by examination of extensive area and are not of quantitative character. A summary of these observations are
presented in Table 2.

After the characters of the climax associations had
been worked out, by the above techniques, it became
a relatively simple matter to assess the direction of
succession in seral stands and thus extend the use of
the data to predict the nature of the ultimate climaxes
and classify habitat types with considerable confidence. The significance of the climax thus does not
reside in its ubiquity, for absolutely virgin climax
stands are difficult to find, but rather in the fact that
different kinds and degrees of disturbance result in
an almost infinite variety of plant communities all of
which immediately begin to evolve in the direction of
the climax. The concept of climax therefore is fundamental to the prediction of direction of succession,
and thus essential as a means of relating widely different types of existing stands into a common pattern of development (see: Ilvessalo 1929, Lippmaa
1939).

In an effort to fortify general impressions of soil
characters and so expand the descriptions of the ecosystems a little beyond the purely vegetal component,
soils were collected at three of the six stations chosen
as representative of each association. After remov-
ing the A_0 layers (litter, duff, and leafmold), the
first five decimeter-horizons of mineral soil were sampled. These were passed through a 2mm screen and

analyzed for moisture equivalent (original centrifuge
method), pH (glass electrode method), capacity for
adsorbed cations, and degree to which this capacity
is satisfied by hydrogen ions. The latter two values
were estimated after the techniques of Brown (1943).
In the absence of confirming tests, the degree to which
Brown's technique yields quantitatively accurate values for soils in the northern Rockies is unknown, but
the results should have considerable comparative
utility. Details of all analyses are contained in the
Appendix. Graphic summaries of certain of the data
are presented in Figure 2, where the horizontal bars
opposite each association name indicate the total
range of variation found in the 1st dm (upper bar)
and 5th dm (lower bar) of the three stands selected
for exploratory analysis. A separate study of the
details of profile development in each of the associations has already begun, but until that work is completed the cruder data obtained as described above
shall have to suffice.

[*Editor's Note:* Material has been
omitted at this point.]

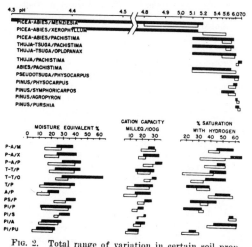

FIG. 2. Total range of variation in certain soil properties encountered in samples taken in three widely separated representative stands of each association. Upper
bar represents upper dm. of mineral soil, lower bar the
5th dm. Summary based on data contained in Appendix.

EVALUATION OF CONCEPTS IN FOREST CLASSIFICATION

If due consideration is given to the fact that each
stand has had a more or less distinctive developmental
history, but by the age-class distribution of its woody
plants can be related successionally to one of several
easily recognized climax associations, the forest vegetation of the northern Rockies does not appear as a
continuum with strong gradients extending in all directions. This is proven by the similarity in the
basic vegetation divisions recognized by different ecologists and foresters who have approached the subject
of vegetation classification from widely divergent
viewpoints (e.g. Larsen, Ilvessalo, J. E. Weaver, and
the writer). Neither is the vegetation composed of
sharply defined units that can easily be described in
terms of discontinuities in the distributions of separate species. However, there are broad pieces of
vegetation characterized by a relatively low degree of
variation or gradient (the stands), that are separated
by narrow areas of sharp gradients (the ecotones)
wherein the relative dominants of species-groups (the
unions) changes rapidly. It is possible to compose
generalized descriptions based upon similarities and

differences among stands which permit the grouping of ecologically equivalent, but by no means identical, stands into the same association. Certain of the associations and unions are well defined and easily characterized (e.g. the *Thuja-Tsuga/Oplopanax* association, or the *Pachistima* union), whereas other vegetation units are poorly defined and may be characterized in only a very generalized fashion (e.g. herbaceous vegetation accompanying the *Physocarpus* union). On account of the complexity resulting from floristic richness, wide ranges in soil and topographic conditions, unequal opportunities for ecologically equivalent species to occupy a given area, strong climatic differentiation, and frequence of disturbances of varied types and intensities, the existing forest vegetation presents an almost endless array of variations. All of these can, however, be related successionally to 13 basic ecosystems.

Although vegetation classification has long been a matter of basic importance in both theoretical and applied botany, the methods used in classification still exhibit wide variation among different workers in different geographic areas. The results of the present study provide a material contribution toward a critical evaluation of the different systems that are in use. If a universally applicable approach toward vegetation classification is possible, it must be so developed as to be applicable to the vegetation of the Rocky Mountain areas as well as other regions.

Forest classifications have been made on four principal bases. Some have used the arborescent union exclusively to characterize the vegetation units, e.g. the "cover type" or "forest type" classification of foresters in the U.S.A. Others have ignored the trees and used the structure and composition of the undergrowth shrubs, herbs, and cryptogams to distinguish forests, e.g., the Finnish "forest type" or "forest (site) type" classification. A third group concentrate their attention on soils and topography in the belief that vegetation is not equally as good an indicator of biologically equivalent areas. And finally there are those who consider it necessary to take into account the characteristics of all of the above components of ecosystems in defining the units of forest classification.

The ecosystem concept,[*] which to the writer seemed increasingly more natural and useful as this study advanced, is not new, finding clear expression as far back as Morosov's classification of Russian forests published in 1904. Although Morosov's work actually represents no more than the culmination of a trend of thought that had begun at least half a century earlier, the 1904 paper crystallized the concept, presenting it in such a fashion as to attract wide interest in Europe. It is significant that this system of vegetation classification was developed concurrently with the system of soil classification with which it is in close harmony, and although the soil concepts have been widely accepted the vegetation concepts are as yet unknown to the majority of plant ecologists outside of Russia. Morosov considered the dominant union, subordinate

* Now recognized as the "geobiocoenosis" by Russians (Sukatchev 1950).

vegetation, animals, climate, topography and soil all as parts of a natural entity, a concept now becoming known in the English language as the ecosystem. A second notable feature of Morosov's concept was that the reproductive abilities of different tree species on each habitat type were considered important, thus leading to a clear distinction between climax and seral communities. Much subsequent work by Russians has followed Morosov in a general way, but numerous modifications of the concept have been attempted. Thus Sukatchev (1928) arranged forest ecosystems into a series somewhat comparable to the present catena concept of North American soil scientists,— each member of one series finding its counterpart in a homologous series in a different region. Recently Sukatchev would modify Morosov's concept by abolishing the distinction between climax and seral stands; to him the succession concept is "unnecessary" (Sukatchev 1950). The latter of Sukatchev's modifications definitely cannot be accepted for the northern Rocky Mountains. Without the succession concept here, silviculture (Forest Practices Committee 1948) and phytosociologic classification would both become chaotic.

Ecosystems are recognized most easily in the field as combinations of unions. Russian forest ecologists long ago recognized this, and some have used the same system of nomenclature as the writer has used, i.e., diagnostic names which indicate the combinations of unions that characterize the associations.

Several bases have been proposed to define the smallest units of vegetation, which are here called unions. A. Kerner in 1863 and R. Hult in 1881 were responsible for the development of the concept of "layers" or "layer societies." In 1918 H. Gams recommended that subdivisions of the phytocoenosis be defined on the basis of Raunkiaerian life forms. It should be obvious from earlier discussions that the writer has found neither of these modes of defining the union acceptable, although he tried to use the layer concept in the 1942 study. In contrast, the more flexible concept expressed by Gleason in 1926 has proven to be very useful: "It has often appeared to the writer that much of the structural variation in an association would disappear if those taxonomic units which have the same vegetational form and behavior could be considered as a single ecological unit" (Gleason 1926). This was Gleason's entire contribution to the concept; he made no further comments, nor did he test this idea in the field. The writer's studies have indicated that more weight should be accorded ecologic "behavior," interpreting this as similarity of ecologic amplitude within a limited study area, giving less weight to "vegetational form." If Linnaea, Adenocaulon and Taxus have essentially identical autecologies throughout the study area, this fact should not be undervalued on account of the great diversity in life form and stature represented by these subordinate, sciophytic mesophytes which are intolerant of highly leached soils.

The concept of the association (*sensu* phytocoenosis) as a group of distinctive unions has been taken seriously by relatively few North American workers, among whom may be mentioned Cain & Penfound (1938), Cain & Sharp (1938), Billings & Drew (1938), Oosting & Billings (1943), *et al*, although essays on the merits of the concept have been published in North American literature by Cain (1936) and by Lippmaa (1939). In most work on North American vegetation, analyses are not made in relation to a careful preliminary division of the entire association into its component sociologic groupings, or if the unions are distinguished, one is given far too much attention at the expense of the others. There is a strong tendency to lay out plots with emphasis on the mechanics involved and simply annotate their contents, without giving much hint of the existence of important phytosociologic relationships that lace a vegetation matrix together.

Returning briefly to Gleason's remark quoted above, the implication is clear that there are two levels of variation in vegetation. One is a result of chance dissemination and survival, or unequal vegetation spread, that produces unpredictable variation from place to place in a community even though the environment is essentially uniform. The greater the floristic richness of the community the greater this type of heterogeneity. The second level of variation results from intrinsic differences in habitat types and determines the areas occupied by different associations, their seres, and their derived disclimaxes. When any aspect of this type of variation is known, the remainder is predictable. Variations of the first category must be taken into account in describing the structure and composition of the unions, whereas variation at the higher level is important in describing the associations that comprise the vegetal matrix.

The union concept is basically subjective, but the constancy studies reported above show that it is possible to make objective analyses that are of great value in reducing the personal error involved in making the initial subjective delimitations of the unions.

Statistical studies have little if any value in discovering the limits of vegetation entities, "but it may properly be pointed out that the intimate knowledge of vegetational structure obtained in this way may easily lead to a much fuller appreciation of synecological structure, entirely aside from any merits of the actual statistical results" (Gleason 1926).

The delimitation of natural sociologic entities in a complex and largely disturbed vegetation is by no means an easy task that can be resolved to simplicity in a short time; even a small area of vegetation may contain thousands of species which at first seem to form a chaotic pattern. However, if attention is confined to a reasonably uniform climax or near-climax stand and the species are first grouped according to similarity in ecology as indicated by phenology and life form or stature, a reasonably good segregation of unions results. The preliminary list for one stand of a union is generally long and unwieldy, and of definitely limited significance because it includes species uncommon in the community. Therefore a series of similar-stands should be compared next to ascertain

which of the species occur regularly, i.e., have high constancy, in stands of the same type. In this way a shortened list of infinitely greater value in characterizing the community is obtained, and at the same time the investigator has opportunity to note the amplitude of variation that is encountered before a significant hiatus is encountered. From such studies of climax or near-climax stands one can proceed to interpret disturbed vegetation with relative ease, at least under conditions now existing in the area of study.

To the taxonomically-minded the inclusion of other species in each stand along with the characteristic species is disconcerting. Most species occur in several communities, and some (e.g. *Amelanchier alnifolia*) may attain equal abundance in more than one, yet this phenomenon does not obliterate community limits if a series of characters including dominance and succession are used jointly as the basis of delimitation. In reporting the results of analyses of different stands of a union these incidental species should be included where found, although they contribute nothing to the diagnostic characterization. A plant species likewise possesses many characters which are of no taxonomic consequence. And as pointed out earlier the most characteristic species of a union may be locally absent, yet the situation does not defy classification. One can identify specimens of *Calypso bulbosa* before the flowers, upon which taxonomy is based, appear, simply by taking all other characters into account.

Another common phenomenon that emerges as the study of unions progresses is that contiguous habitat types (e.g. *Picea-Abies/Xerophyllum* and *Picea-Abies/Menziesia*) may present critically different conditions to one union (*Menziesia*) but not to another (*Xerophyllum*). In progressing up a north facing slope across a Picea-Abies/Menziesia stand which also contains the *Xerophyllum* union, the species of the *Menziesia* union thin out at about the same point on the crest of the ridge although the other two vascular unions continue far down the south slope.

Where these two associations come into contact practically all species may occur on both sides of the ecotone, and from a simple taxonomic viewpoint differentiation would seem to rest on a slender foundation,—the presence or absence of Menziesia being the only absolute floristic difference which separates a stand of Picea-Abies/Menziesia that occupies the north half of a mountain from a stand of Picea-Abies/Xerophyllum which occupies the south half. To the ecologist this essential identity of the two floras is a minor matter compared with the facts that the two stands differ also as to (1) microclimate, (2) degree of podsolization, (3) abundance of game, (4) stature of trees, (5) possibility of supporting *Pinus monticola* subsequent to burning, and (6) physiognomy. The situation is comparable to a subspecies distinguished by less than 1% of its thousands of genes, but these few so change the appearance and geographic distribution that recognizing the entity is very desirable.

While the foregoing discussion is offered to substantiate statements as to the importance of close study of subordinate unions in vegetation classification, one must not go so far as to assign superior indicator significance to forest undergrowth everywhere as is done by the Cajander school. Considerable significance must be attached to the fact that over the ecologic range of the *Pachistima* union there are subtle habitat differences that seem not to affect the structure or composition of this union, but which are sufficient to throw the environmental balance in a direction favoring a climax dominated by *Abies grandis* in one place, by Thuja in another, by Thuja plus Tsuga in another, and by Picea plus *Abies lasiocarpa* in still another. The contrast between homogeneity in undergrowth and divergence in soil characters between the *Thuja/Pachistima* and *Thuja-Tsuga/Pachistima* associations is especially noteworthy.

The work of Cajander (1909) and others who have followed his methods (Ilvessalo 1929, Heimburger 1934, Kujala 1945, etc.) represents a distinctive viewpoint in which the workers base their vegetation classification entirely upon the subordinate vegetation, and accord practically no importance to plant succession. It is not difficult to find a reason for the custom of long standing in certain parts of northwestern Europe of virtually ignoring the tree union in forest classification. Forestry there is based primarily on *Pinus sylvestris* and *Picea abies* which are the only coniferous tree species. Where these occur on the same habitat the Pinus is seral to the Picea, but on land managed intensively for timber production there may be no opportunity for the latter to assert its superior competitive ability. For the most part trees are cut singly as soon as they are large enough to be useful (in some instances when diameter at breast height reaches 2in.!), a process which perpetually keeps the forest so open that insolation at the ground remains above the high minimum requirements of the Pinus. In such a simply developed zootic climax the two species are ecologic equivalents and their theoretical successional relations can be safely ignored without jeopardizing the land use system. Consequently attention is largely directed from the tree union to subordinate unions.

These conditions stand in direct contrast to the forests of the northern Rockies where nearly all stands owe their origin to fires, therefore tend to be evenaged with a result that most of the trees mature at about the same time. Most of the valuable species are seral and when one of these is old enough to cut, an understory of slow-growing climax species has generally become well developed. Cutting therefore serves only to hasten the course of succession so that a forest of low commercial value but of unlimited tenure comes into full possession of the habitat. Until recently, when a market for low-value species has developed, fire has been the only method of perpetuating these seral trees, and its practicality is attested by the continued productivity of the land in the face of natural burning that has been taking place for

millennia. With a complex flora of seral and climax trees, an intricate mosaic of small fragments of different habitat types, and a process of stand renewal that involves complete destruction, successional relations cannot be ignored regardless of the aims of management. The sociology of the arborescent union is at least equal in importance to that of subordinate unions.

If Cajander's method were used in the writer's study area one could not distinguish the shrub thickets on north-facing slopes in the grassland (Daubenmire 1942) from the *Pinus/Symphoricarpos* association, nor distinguish between *Pinus/Physocarpus* and *Pseudotsuga/Physocrapus* associations, nor differentiate the several associations where the *Pachistima* union alone may form the undergrowth. Westveld (1951) also notes this phenomenon in the forests of northeastern U.S.A.

In 1931 Kujala (1945) spent about three months in the interior of British Columbia in an attempt to classify the forests there on the same basis used in his native Finland. Despite the fact that Kujala and the writer may have dealt with some identical associations, approaches and results have been somewhat different.

The frequence with which Kujala's stands were dominated by *Pinus contorta* var. *latifolia* or by *Populus tremuloides* shows that he did not attach much importance to secondary succession, in accord with the viewpoint prevalent in northeastern Europe. Although it is true that many plants of subordinate unions sprout from underground organs following a fire, the altered conditions also allow many other species to enter upon the area temporarily and thus give the vegetation a distinctly different aspect for some time so that climax or near-climax stands alone provide the critical data essential to the delimitation of fundamental vegetation units. Failure to take this into account prevents drawing clear-cut distinctions, and multiplies complexity by separating the ephemeral conditions of a union from the climax condition. The writer agrees fully with Lippmaa (1939) and others that disturbed areas can be interpreted correctly only after the undisturbed are understood.

Whether a result of ignoring the disturbance factor or not, Kujala has provided sample plot data for as many as 32 kinds of forest in an area with size comparable to eastern Washington and northern Idaho for which but 13 entities are recognized by the writer. The *Picea-Abies/Vaccinium scoparium* association of the writer is treated as two types and four phases by Kujala. Seven of Kujala's entities contain Pseudotsuga plus *Abies lasiocarpa*, a combination recognized but once in the present study. The *Pachistima* union of the writer is indistinguishable from Kujala's *Vaccinium membranaceum*, *Rubus pedatus-Vaccinium membranaceum*, *Tiarella-Rubus pedatus*, *Mitella-Tiarella-Rubus pedatus*, *Pachistima*, and *Tiarella-Pachistima* types. The writer has observed many variations in species dominance within the *Pachistima* union, but as these seem endless and unrelated to either the physical environment or the kinds and sociologic relations of associated trees, he sees little justification for giving them nomenclatorial recognition in the Washington-Idaho forests. It is significant that six of Kujala's 32 forest types are based on single stands which he encountered, a practice which seems hazardous in view of the wide variation in vegetation composition.*

Although a few other attempts have been made to apply the Finnish viewpoint to North American forests, there has been an overwhelming tendency both in forestry and theoretical ecology to delimit vegetation types according to species of the dominant union alone. This too is impractical in the northern Rockies. Climax stands of *Pinus ponderosa* differ markedly in physiognomy, soil, and climatic relations, response to fire, growth rate of trees, etc., and these differences can be accounted for only in a classification scheme that accords subordinate vegetation a position of equal importance with the dominant union.

Still another viewpoint is exemplified by Coile (1938) who considers vegetation, especially undergrowth, as unimportant in forest classification. He started with the premise that in general tree roots extend deeper than those of shrubs and herbs, an assumption that stands in opposition to the only special study that has been made of the problem (Kivenheimo 1947), and then reasoned that trees should be more sensitive to the fundamental habitat differences because they are influenced by the nature of soil horizons lying too deep to affect the structure or composition of the undergrowth. Also, he believes that trees are more directly affected by intrinsic climatic factors than is the undergrowth, for the latter occupies a microenvironment the nature of which is strongly determined by the tree union. These deductions led Coile to conclude that undergrowth is not to be trusted as an indicator of soil conditions that affect trees.

In partial substantiation of this viewpoint it might be pointed out that in some cases at least, there seems to be no evident variation in undergrowth accompanying environmental variations which affect tree vigor within one habitat type. In the region under study there may be observed distinct variations in the maximum possible heights of trees among different stands of the same association. Others, such as Viro (1947), have reported the same for different regions.

* There is a good possibility that European ecologists in general tend to attach too much importance to variations in the specific composition of a union. Their stands are for the most part so discontinuous and interrupted by artificial communities that normal migration of species from stand to stand is impeded, if not absolutely prevented. This is important because each of the relic stands has been subject to extreme disturbance approaching obliteration at least once in their past history, so that their present condition reflects only such an approach to primeval conditions as the drastically impoverished floras can achieve without receiving immigrants from outside floristic reservoirs. If disturbance eliminates the serious competitors of one species in one stand, leaving a different species without much interference in another stand, the consequent difference in dominance is far more conspicuous than significant. Even where it can be demonstrated that the microenvironments differ according to these differences in floristic composition, we may be dealing either with residual differences in the types of disturbances some decades ago, or with differences induced by the varied influences which these dominants have exerted upon their environment.

Furthermore, the continuity of the *Pachistima* union beneath four distinctive tree unions appears to be supporting evidence.

Even though these two aspects of forest communities tend to bear out Coile's conclusion, the present study has brought forth some important factual evidence which shows that the premise on which his conclusion is based is not valid in the northern Rockies. For example, *Pinus ponderosa* plays exactly the same ecologic role in four kinds of habitats each supporting distinctive subordinate unions. This shows that the seedlings of this tree species are not noticeably sensitive to certain environmental variables that are very critical for herbs and shrubs. The same phenomenon is reflected again in the Picea-Abies zone where undergrowth varies markedly under the same pair of climax tree species.

Another point which Coile offers to minimize the significance of the vegetation-indicator approach is his contention that differences in environment that are of sufficient magnitude to be of silvicultural importance are obvious without close inspection of undergrowth. He maintains that the direction and degree of slope, together with the texture and thickness of the major soil horizons are more reliable as guides to the vegetative potentialities of environment. The writer's experience in the northern Rockies does not support this opinion. Many climax herbs and shrubs survive holocausts that bring about a complete replacement of trees by a temporary community of deciduous shrubs, as well as a loss of the distinctive characters of the mature soil profile. The visible characters of the profiles vary more with the stages of vegetative regeneration subsequent to fire than does the composition of the subordinate unions that characterize the habitat, so that the latter is really the more permanent visible feature of the ecosystem. Furthermore, on the same slope one can often pass from one forest type to another within a few hundred meters without finding any changes in soil character other than those which are due to, and therefore no more permanent than, the existing vegetation types. It is true that differences in texture, slope, and direction of exposure usually produce pronounced changes in habitat type in the northern Rockies, but there exist in addition many ecotones which cannot be related to these factors.

Lutz & Chandler (1946) criticize the vegetation-indicator concept because "in the *Oxalis-Myrtillus* type the characteristic plant *Oxalis acetosella* may be lacking." However, in another place (loc. cit.) one reads "The characteristic features (of podsol profiles) as seen in the field are the layer of unincorporated organic mater, the gray A₂ horizon and the brown to dark-brown B horizon. In less clear-cut profiles however, *the identity of the soil may not be obvious*, and then the silica-aluminum ratio of the colloids becomes critical." It must be inferred from these quotations that Lutz & Chandler require much more exacting standards in phytosociology than in pedology, —a view to which few botanists who have done much research in phytosociology will subscribe. Whole ecosystems are never susceptible of characterization by so simple a criterion as the presence or absence of a single species, but by a combination of biologic and physical characters, not one of which is absolute. When a community is named after a characteristic plant it is not to be inferred that this plant is always present.

In conclusion, the present study confirms the viewpoints of workers exemplified by Morosov, Braun-Blanquet (1932), Tansley, et al who have concluded that we should look upon complex ecosystems as the only natural units, and that macroscopic vegetation in its entirety comprises the best criterion of ecosystems.

It will be noted that differences in soil fertility, as indicated by capacity for adsorbed cations and degree to which this capacity is preempted by useless hydrogen ions, exhibit few differences of possible sig-nificance among forest associations of the northern Rockies. Nor can the associations be distinguished on the basis of water-holding capacity of the soil. However, some striking correlations between vegetation and soil pH are indicated by the analyses. The ecosystems as delimited are therefore expressions primarily of the reproductive abilities of vascular plants under intense competition, and secondarily of the pH of the upper soil layers. Perhaps other soil characters, animal communities, etc. may later be found correlated.

[*Editor's Note:* Material has been omitted at this point.]

LITERATURE CITED

Arnold, M. C. 1946. Cover type mapping for forest and wildlife management in Connecticut. N. A. Wildl. Conf. **11**: 330-338.

Baker, F. S., et al. 1945. Forest cover types of western North America. Washington, D. C.: Soc. Amer. Foresters.

Branteseg, A. 1941. (Vegetation types on forest soils and their importance in forest management.) Tidsskr. Skogbruk **49** (1): 3-11, (2): 45-50.

Braun-Blanquet, J. 1932. Plant sociology. New York, N. Y.: McGraw-Hill Book Co. 439 pp.

Brown, I. C. 1943. A rapid method of determining exchangeable hydrogen and total exchangeable bases in soils. Soil Sci. **56**: 353-357.

Billings, W. D. & W. B. Drew. 1938. Bark factors affecting the distribution of corticolous bryophytic communities. Amer. Midland Nat. **17**: 20: 302-330.

Cain, S. A. 1936. Synusiae as a basis for plant sociological field work. Amer. Midland Nat. **17**: 665-672.

———. 1947. Characteristics of natural areas and factors in their development. Ecol. Monog. **17**: 185-200.

Cain, S. A. & W. T. Penfound. 1938. Aceretum rubri; the red maple swamp forest of central Long Island. Amer. Midland Nat. **19**: 390-416.

Cain, S. A. & A. J. Sharp. 1938. Bryophytic unions of certain forest types of the Great Smoky Mountains. Amer. Midland Nat. **20**: 249-301.

Chase, W. W. 1949. Recent advances in forest game management. Jour. Forestry **47**: 882-885.

Coile, T. S. 1938. Forest classification: classification of forest sites with special reference to ground vegetation. Jour. Forestry **35**: 1062-1066.

Cajander, A. K. 1926. The theory of forest types. Acta Forest. Fenn. **29**: 1-108.

Cowan, I. M., et al. 1950. The effect of forest succession upon the quantity and upon the nutritive values of woody plants used as food by moose. Canad. Jour. Res. (D) **28**: 249-271.

Daubenmire, R. 1942. An ecological study of the vegetation of southeastern Washington and adjacent Idaho, Ecol. Monog. **12**: 53-79.

———. 1943. Vegetational zonation in the Rocky Mountains. Bot. Rev. **9**: 325-393.

Eastham, J. W. 1949. Personal correspondence.

Fenton, E. W. 1947. The transitory character of vegetation maps. Scot. Forest. **1**: 23-27.

Forest Practices Committee, Inland Empire Section, Soc. Amer. For. 1948. Forest practices for the Inland Empire. Jour. Forestry **46**: 550-557.

Gleason, H. A. 1926. The individualistic concept of the plant association. Torrey Bot. Club Bull. **53**: 7-26.

Hall, J. A. 1947. Possibilities in new forest products industries in the northwest. Northwest Sci. **21**: 101-108.

Halliday, W. E. D. 1937. A forest classification for Canada. Canada Domin. Forest Serv. Bull. 89. 50 pp.

Heimburger, C. C. 1934. Forest type studies in the Adirondack region. N. Y. (Cornell) Agr. Expt. Sta. Mem. 165. 89 pp.

Hilitzer, A. 1934. (Finnish forest types.) Czech. Acad. Agr. Ann. **9**: 288-290. (see: U. S. Forest Serv. Div. Silvics Transl. #231.)

Huberman, M. A. 1935. The role of western white pine in forest succession in northern Idaho. Ecology **16**: 137-151.

Humphrey, H. B. & J. E. Weaver. 1915. Natural reforestation in the mountains of northern Idaho. Plant World **18**: 31-47.

Ilvessalo, Y. 1929. Notes on some forest (site) type in North America. Acta Forest. Fenn. **34**(39): 1-111.

Kalela, E. K. 1948. (Root competition and regeneration of spruce woods.) Metsätaloudellinen Aikakausk. **11**: 320-323. (see: For. Abstr. 11, #983.)

Kivenheimo, V. J. 1947. (The root systems of spermatophytes of the ground vegetation of Finnish forests.) Soc. Zool. Bot. Vanamo Ann. Bot. **22**(2). 180 pp.

Koch. E. & R. N. Cunningham. 1927. Timber growing and logging practice in the western white pine type and larch-fir forests of the northern Rocky Mountains. U. S. Dept. Agr. Bull. 1494. 37. pp.

Korstian, C. F. 1919. Native vegetation as a criterion of site. Plant World **22**: 253-261.

Kujala, V. 1945. Waldvegetationsuntersuchungen in Kanada. Acad. Sci. Fenn. Ann. Ser. A, I. V. Biologica #7. 434 pp.

Larsen, J. A. 1923. Associations of trees, shrubs and other vegetation in northern Idaho forests. Ecology **4**: 63-67.

————. 1925. National forests of the northern district. In: Naturalist's guide to the Americas. Baltimore, Md.: Williams & Wilkins Co. 761 pp.

————. 1930. Forest types of the northern Rocky Mountains and their climatic controls. Ecology 11: 631-672.

Leiberg, J. B. 1900. Bitterroot forest reserve. U. S. Geol. Survey 20th Rept. Part 5: 317-410.

Leopold, A. S. 1950. Deer in relation to plant successions. Jour. Forestry 48: 675-678.

Lippmaa, T. 1939. The unistratal concept of plant communities (the unions). Amer. Midland Nat. 21: 111-145.

Lutz, H. J. & R. F. Chandler. 1946. Forest soils. New York, N. Y.: John Wiley & Sons. 514 pp.

Marshall, R. 1928. The life history of some western white pine stands on the Kaniksu National Forest. Northw. Sci. 2: 48-53.

Morton, J. N. & J. B. Sedam. 1938. Cutting operations to improve wildlife environment on forest areas. Jour. Wildl. Mangt. 2: 206-214.

Oosting, H. J. & W. D. Billings. 1943. The red fir forest of the Sierra Nevada. Ecol. Monog. 13: 259-274.

Pearson, G. A. 1931. Forest types in the southwest as determined by climate and soil. U. S. Dept. Agr. Tech. Bull. 247. 143 pp.

Pickford, G. D. & E. H. Reid. 1948. Forage utilization on summer cattle ranges in eastern Oregon. U. S. Dept. Agr. Cir. 796. 27 pp.

Puri, G. S. 1950. Soil pH and forest communities in t'e sal (Shorea robusta) forests of the Dehra Dun Valley, U.P., India. Indian Forester 1950 (July): 1-18.

Smith, N. F. 1948. Controlled burning in Michigan's forest and game management programs. Soc. Amer. Foresters Proc. 1947: 200-205.

Spilsbury, R. H. & E. W. Tisdale. 1944. Soil-plant relationships and vertical zonation in the southern interior of British Columbia. Sci. Agr. 24: 395-436.

Sprague, F. L. & H. P. Hansen. 1946. Forest succession in the McDonald Forest, Willamette Valley, Oregon. Northw. Sci. 20: 89-98.

Sukatchev, V. N. 1928. Principles of classification of the spruce communities of European Russia. Jour. Ecol. 16: 1-18.

————. 1950. Address to 7th Internat. Bot. Congr., Stockholm, July 19.

Tansley, A. G. 1935. The use and abuse of vegetational concepts and terms. Ecology 16: 284-307.

Tisdale, E. W. 1950. Grazing of forest lands in interior British Columbia. Jour. Forestry 48: 856-860.

Titus, H. 1945. Timber and game—twin crops. N. A. Wildl. Conf. Trans. 10: 146, 163.

Viro, P. J. 1947. (The mechanical composition and fertility of forest soil taking into consideration especially the stoniness of the soil.) Commun. Inst. For. Fenn. 35(2). 115 pp. (see: For. Abstr. 11, #3414.)

Weaver, H. 1947a. Management problems in the ponderosa pine region. Northw. Sci. 21: 160-163.

————. 1947b. Fire—nature's thinning agent in ponderosa pine stands. Jour. Forestry 45: 437-444.

————. 1951. Observed effects of prescribed burning on perennial grasses in the ponderosa pine forests. Jour. Forestry 49: 267-271.

Weaver, J. E. 1914. Evaporation and plant succession in southeastern Washington and adjacent Idaho. Plant World 17: 273-294.

————. 1917. A study of the vegetation of southeastern Washington and adjacent Idaho. Nebr. Univ. Studies 17: 1-133.

Webb, W. L. 1942. A method for wildlife management mapping in forested areas. Jour. Wildl. Mangt. 6: 38-43.

Wellner, C. 1946. Recent trends in silvicultural practice on national forests in the western white pine type. Jour. Forestry 44: 942-944.

Westveld, M. 1951. Vegetation mapping as a guide to better silviculture. Ecology 32: 508-517.

Whitford, H. N. & R. D. Craig. 1918. Forests of British Columbia. Canada Comm. Conserv., Comm. on Forests. 409 pp.

Young, V., et al. 1942. The influence of sheep grazing on coniferous reproduction and forage on cut-over western white pine areas in northern Idaho. Idaho Univ. Bull. 37(6): 3-45.

10

Reprinted from *Ecology* **36**(1):89–100 (1955), with permission of the publisher, Duke University Press, Durham, North Carolina

THE CLASSIFICATION OF TROPICAL AMERICAN VEGETATION-TYPES

J. S. Beard

P.O. Box 39, Pietermaritzburg, Natal, South Africa

In 1944, in a paper entitled "Climax Vegetation in Tropical America," the writer proposed a system of classification for the plant formations of the American tropics. In the decade since this paper was written, further experience has been acquired and our knowledge of this vegetation widened by the work of numbers of different authors. The writer has himself used the system in studies of the vegetation of Tobago (1944b), Trinidad (1946b), and the Windward and Leeward Islands (1949a). It has been put into use by Beebe and Crane (1947) in Venezuela, by Curtis (1947) in Haiti, and Wadsworth (1951) in Puerto Rico and, recently, (1952) has been followed by Fanshawe in his Preliminary Review of the Vegetation of British Guiana which constitutes its first application by another ecologist than its author on a large scale. Sundry interesting problems have been raised in the course of all this; definitions given and views expressed may require modification, and new plant formations can now be recognized. At this juncture, therefore, it is appropriate to make a review of the subject.

The broad principles of the original classification were laid down in the original paper and have been restated on two occasions (1945, 1949b). Briefly they are as follows. The basic unit is the plant association which is a floristic grouping, being the largest possible group with consistent dominants either of the same or closely allied species. Associations may be divided into minor floristic groups, to which it was proposed to apply the Clementsian terminology. Also, they may be termed consociations if they are single-dominant communities. Climax associations, that is, communities apparently mature, stable and integrated, may be further grouped together according to their structure and physiognomy into formations. The formation is a physiognomic group and is independent of flora. Communities which are situated in widely scattered parts of tropical America and for that reason are quite differently constituted may thus be classified into a single formation if of similar structure and physiognamy. The properties of the community express the habitat: the consistent dominance which defines the association expresses a constant local habitat within the regional flora and the consistent physiognomy which defines the formation expresses a constant essential habitat within the tropical zone. Types of habitat which are in general similar permit us further to group formations together into formation-series. There is thus a grouping at three levels: a floristic grouping—the association, a physiognomic grouping—the formation, and a habitat grouping—the formation-series. The association must bear a floristic name such as *"Eschweilera-Licania* association," the formation a physiognomic name such as "deciduous seasonal forest" and the formation-series a habitat name such as "montane formations." This rule may however be broken for formations in order to draw into the

RAIN FOREST

Fig. 1. "Type Specimen" of Rain Forest: the *Eschweilera-Licania* association of British Guiana, from the original diagram by Davis and Richards (1933).

classification well-established and understood names.

Twenty-six formations were described and defined in the original paper of 1944. One more, named "montane thicket" in English, was added to the Spanish translation published at Medellin in 1946. Another, "evergreen bushland," was defined in 1949 as a member of the dry evergreen formation-series. In the meantime also, it had appeared that one of the original formations, "palm-brake" might have to be dropped on the ground that it is a disturbance climax and not a natural climax of the same order as the other montane formations, work in the Lesser Antilles (Beard 1945b, 1949a) having led to this conclusion. To the twenty-seven climax types thus defined, Fanshawe (1952) has now added ten more, four in the dry evergreen, two in the seasonal-swamp and four in the swamp formation-series. A change of name in the seasonal-swamp series was also proposed. The discussion which follows below aims at consolidation of the classification to date.

An essential concept of the system is that rain forest (Fig. 1) stands alone, having both formation and formation-series rank, in the center, and that from it radiate the habitat groups in descending order from optimum to pessimum. The formation-series which has been best know hitherto is the seasonal group. A full series of five formations (six if "desert" is also illustrated) was defined and shown diagrammatically in the 1944 paper, repeated in 1946 and taken up by Richards (1952). This series is shown in Figure 2 herewith. It is envisaged that there is in reality one long unbroken series, in which the formations

are artificially delimited stages. Each stage shows a regular stepping-down of structure. Rain forest has three tree stories, the uppermost one continuous. Evergreen seasonal forest also has three stories but the uppermost is discontinuous and canopy is formed by the second layer. In semi-evergreen seasonal forest only two stories remain, of which the upper is closed, and in deciduous seasonal forest this upper layer becomes discontinuous. Thorn woodland has only one tree story left, and in cactus scrub we have only the bushes and succulents. Desert retains nothing but a ground layer. Other physiognomic changes of course accompany the reductions of structure throughout.

We ought also to look for a similar regularly descending series of types in the other formation-series: not necessarily, perhaps, in the inundated swamp and seasonal-swamp series owing to their greater specialization, but certainly in the dry evergreen and montane series. In 1944, the writer did not have the data to substantiate this, and could illustrate the structure of only one dry evergreen formation and three montane ones. Now, however, more can be done.

Fanshawe (1952) has recognized six dry evergreen formations in British Guiana: Wallaba forest, xeromorphic rain forest, xeromorphic woodland, xeromorphic scrub, heath and littoral woodland. The choice of the name "Wallaba forest" was unfortunate, since wallaba is the vernacular name for the *Eperua* spp. which dominate the local association, and it is not desirable to give a floristic name to a formation. Supposing that forests of similar structure and physiognomy but not constituted by *Eperua* spp. are now described

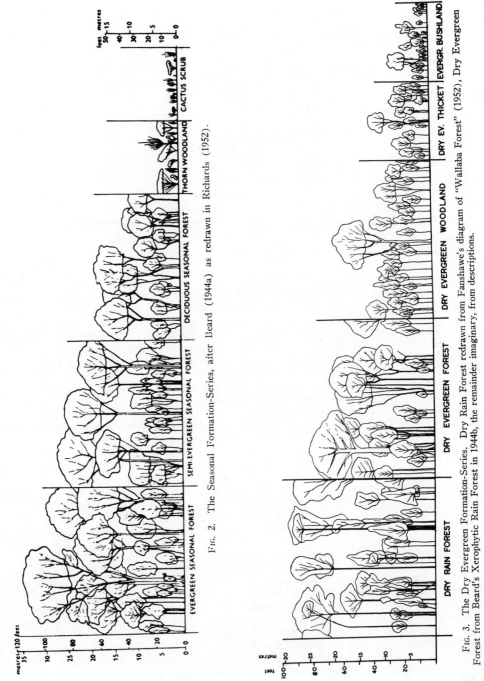
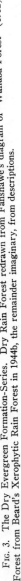

Fig. 2. The Seasonal Formation-Series, after Beard (1944a) as redrawn in Richards (1952).

Fig. 3. The Dry Evergreen Formation-Series. Dry Rain Forest redrawn from Fanshawe's diagram of "Wallaba Forest" (1952), Dry Evergreen Forest from Beard's Xerophytic Rain Forest in 1944b, the remainder imaginary, from descriptions.

elsewhere, we cannot justifiably call them wallaba forests and some other term will have to be found. "Dry rain forest" is proposed as the most convenient term, since this formation lies very close to the true rain forest but is differentiated therefrom by the underlying white sands which are unretentive of moisture and cause a paradoxical dryness of the habitat. Xeromorphic rain forest is stated to be the same as Beard's xerophytic rain forest, with a minor change of terminology that has much to commend it, but the structure does not agree with Beard's definition. Xeromorphic woodland is a new formation. Xeromorphic scrub is the same as the evergreen bushland which had already been brought into the terminology (Beard 1949a). Fanshawe's use of "heath" here for an "herbaceous vegetation of "xerophytic lily-like plants" is surely inadmissible as there is no resemblance to the true heath of temperate climates. The writer suggests "rock pavement vegetation" as a substitute.

In attempting to consolidate the dry evergreen formations, the writer suggests first the elimination of the awkward terms "xerophytic" and "xeromorphic" in favor of "dry evergreen." We can next proceed to delimit our formations structurally with the aid of Figure 3. Nearest to the optimum comes the *Dry Rain Forest*, Fanshawe's "Wallaba forest," characterized as follows (adapted from Fanshawe):

A three-storied forest with the canopy more or less closed between 25-35 meters, a discontinuous understory between 12-20 meters and a dense not very well-defined undergrowth from 6-12 meters. Illumination within the forest is relatively intense and the atmosphere dry. Stocking density is high and the trees are slender in relation to height. True under-growth species are few and the herb layers are poorly represented. Lianes are few and small. Epiphytes, especially sun epiphytes, are plentiful and descend low on the trees. Heavy buttressing is very rare. The proportion of semi-deciduous species is higher than in true rain forest. Leaves are mesophyllous, mostly compound in the canopy and simple in the understory.

Second in line of descent comes what we will now term *Dry Evergreen Forest,* equivalent to Beard's xerophytic rain forest but not to Fanshawe's xeromorphic rain forest. Reduction in structure now sees the canopy formed by a layer of crowded, slender trees reaching 12 to 20 meters in height above which an occasional emergent reaches 30 meters. Leaves are simple and ever-green, the majority being stiff and fleshy, a high proportion with latex or essential oil. Bark-shedding may be a conspicuous feature. The shrub and herb layers are poorly represented.

Thirdly, we have Fanshawe's xeormorphic rain forest, which we must now term *Dry Evergreen Woodland*. The understory of 6-12 meters is now the canopy-former, with emergents reaching 20 meters. Fanshawe describes this as:

"A 2 storied forest with the canopy formed of densely packed, attenuated trees, now larger than 18-20 in. diameter (45-50 cm.) and about 20-40 ft. high (6-12 meters). There is a discontinuous emergent layer between 60-80 ft. high, made up of larger trees. The forest is almost entirely evergreen."

Fanshawe's xeromorphic woodland, which it is preferred to term *Dry Evergreen Thicket,* is to be considered as a valid new formation, characterized as follows:

A 2 storied forest with a low, open, or dense canopy and a dense or sparse undergrowth. The canopy is between 20 and 40 ft. high (6-12 meters), and trees have slender stems not larger than 6-8 inches. (15-20 cms) diameter. There is an occasional, larger emergent.

Most reduced of all the woody formations is the *Evergreen Bushland,* Fanshawe's xeromorphic scrub, colloquially known as muri or moeri-moeri in the Guianas. This formation was detailed by Beard in 1949a.

Rock Pavement Vegetation takes the form of an irregular and open growth of herbaceous and woody plants less than two meters in height, growing in crevices or in mats of humus upon sheet rock, usually with much bare rock between. Such vegetation is frequently associated with outcrops of hard sandstone and granite bosses.

The 1944 formation *Littoral Woodland* must now be considered as identical with dry evergreen woodland, and the *Littoral Thicket* and *Littoral Hedge* (Beard 1949a) as dry evergreen thicket and bushland respectively. It may still be convenient to retain the names for particular studies, to differentiate dry evergreen vegetation under the influence of the sea, just as it is convenient to separate mangrove from freshwater swamp: but although the littoral habitat may be very different from that of inland sites carrying dry evergreen communities, we must recognize the essential physiognomic identity of these types. Differences such as the predominance of thickly cutinized salt-spray resistant leaves in littoral types must be regarded as of minor significance.

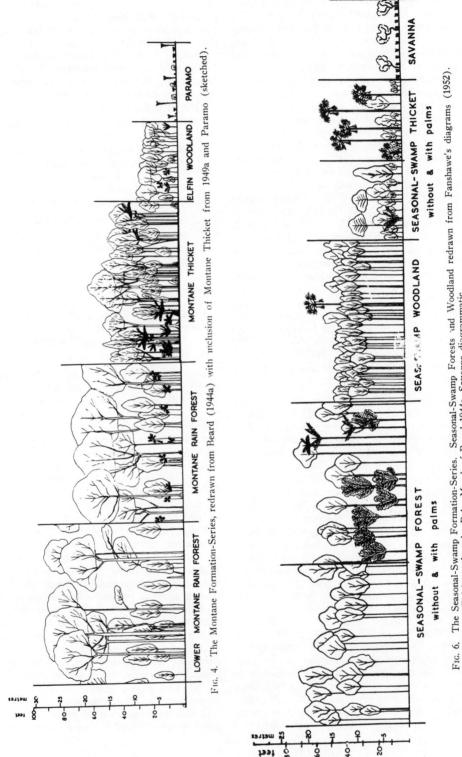

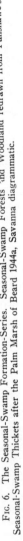

Fig. 4. The Montane Formation-Series, redrawn from Beard (1944a) with inclusion of Montane Thicket from 1949a and Paramo (sketched).

Fig. 6. The Seasonal-Swamp Formation-Series. Seasonal-Swamp Forests and Woodland redrawn from Fanshawe's diagrams (1952). Seasonal-Swamp Thickets after the Palm Marsh of Beard 1944a, Savanna diagrammatic.

It will be seen that the dry evergreen formations "step down" as their structure is reduced but not quite in the same way as the seasonal formations where successive strata become first discontinuous and then absent. In this way we seem always to have a canopy layer with emergent trees standing above it and little or no understory. As we descend the series, the height of these two layers is reduced.

We are now able to characterize the montane series of formations more fully than in 1944 (see Figure 4) and it will be seen that in this case the reduction of structure proceeds again somewhat differently. All forest-like montane formations present a continuous canopy layer without emergent trees and as we proceed through the series we find that the canopy layer is gradually lowered until finally it coalesces with the understory. On low mountains which do not rise above the upper limit of the mist belt (± 2,500 meters) the zonation is as shown in Figure 4. First and at lowest altitude, the *Lower Montane Rain Forest,* with two tree strata at 3-15 meters and 20-30 meters. Secondly, as the cloud belt is fully entered, comes the *Montaine Rain Forest* or *Cloud Forest* with the two strata reaching about 10 and 20 meters respectively (higher under favorable conditions). The above two formations featured in the 1944 account and there was an obvious discontinuity between the montaine rain forest and the elfin woodland, a gap which was filled in 1946a and 1949a by the inclusion of *Montane Thicket.* In this formation, the canopy layer is down to 10-15 meters and the understory is suppressed to the point of disappearance. Finally growth descends to the *Elfin Woodland* or *Mossy Forest,* a single-storied, impenetrable tangle, whose height varies from 10 meters down to one meter. Our terminology is not too good here, as we should correctly speak of montane woodland and elfin thicket; "elfin woodland" is however an old established name.

From studies undertaken (Beard 1945b, 1949a), it must appear that the palm brake (Fig. 5), although it may often appear stable, is a disturbance climax at an equivalent level to the montane

thicket. In the Lesser Antilles, there can be no doubt that it is a successional stage on areas bared by wind, landslides, or volcanic eruptions. Wadsworth (1951) supports this view for Puerto Rico. Cuatrecasas (1934) has described a "palmetum" of *Ceroxylon andicola* in the Andes of Colombia occurring as a society in the "Clethrion" which appears to be montane thicket. No opinion is expressed as to its status. In view of the fact that the physiognomy of the palm brake is at variance with that of the other members of the montane series, it seems very probable that it never does constitute a true climax.

On mountains rising to some extent above the mist belt, the elfin woodland gives place to the paramo. Closer acquaintance with this formation shows that it is in no sense comparable to an alpine meadow, as was suggested in 1944. It is much more similar to heath. The paramo finally gives way to the tundra at highest elevations. Where mountains rise so far above the mist belt as to carry a forest zone above it, we shall find a formation described as "frost woodland" in 1944 but which it is now preferred to call *High Mountain Forest. Mountain Pine Forest* is apparently a fire-climax derived from the latter and like the palm-brake must therefore be excluded from full rank in our classification. The writer has no structural data for high mountain forest. In zonation, it stands between cloud forest and elfin woodland. Further information on the *Bamboo Brake* proposed in 1944 indicates that *Chusquea* communities in the Andes are only of societal rank in the elfin woodland. There are *Guadua* societies also but these are riverain and at low altitude. It does not appear that we have any equivalent of the bamboo forest of East Africa and the Himalayas. The valid montane formations are thus reduced to seven.

The original four seasonal-swamp or "marsh" formations have been expanded to six by Fanshawe, who applies the name marsh to them. It is however highly debatable whether we should try to perpetuate the artificial distinction between "marsh" and "swamp" suggested by the present writer in 1944, in view of the fact that this distinction does not exist in common usage. While both represent inundated habitats, "swamp" is defined as having a perpetually waterlogged soil deprived of oxygen, whereas in "marsh" there is a fluctuating watertable permitting seasonal aeration of the topsoil. It seems generally preferable to substitute the term "seasonal-swamp" firmly for "marsh."

Four of his "marsh" formations are said by Fanshawe to occur in a regular zonation across the interfluves of the main rivers in the North-

Fig. 5. Palm Brake, from Beard, 1949a.

west District, in this order: palm marsh forest → marsh forest → palm marsh woodland → palm marsh. The first named of these is at the river bank, which in a delta region stands higher than the ground level further away between the main channels. Away from these, seasonal rise and fall of the watertable gradually decreases and the vegetation is reduced. Actually, the true order of progression of the marsh series is not exactly in the order of this zonation, since the marsh forest is clearly the least reduced type of all and should stand at the head of the series. Evidently conditions are at the most favorable not in the palm marsh forest nearest to the creek but in the marsh forest belt slightly further away.

Our knowledge of the seasonal-swamp series has been greatly extended by Fanshawe but certain comments must be made. It does not now appear to the writer that the presence and absence of palms is such an important diagnostic feature as was previously suggested. There is little structural difference, apart from the palm question, between Fanshawe's marsh forest and palm marsh forest, nor between marsh woodland and palm marsh. As Fanshawe states that there are also palm-dominated variants of rain forest, seasonal forest and swamp communities in British Guiana, and there is a palm type of savanna as well, it seems that we should regard the seasonal-swamp communities with palms as merely variants also.

Figure 6 illustrates some of the variants but the main classification is consolidated to seasonal-swamp forest, woodland and thicket followed by savanna. There is in this case a considerable discrepancy between the head of the series and the optimum, within which no intermediates are known. The forest-like seasonal swamp formations are characterized by highly irregular canopy. It may be possible to define strata, but they are erratic and the forest canopy is now high, now low as growth varies from short to tall, dense to open. Passage through the series witnesses a progressive reduction in the stature of this jumbled assemblage.

Seasonal-Swamp Forest (without palms) is somewhat similar to dry evergreen forest and is defined by Fanshawe as follows for his "marsh forest":

"A 2-3 storied forest in which palms only account for 5% of the stand. There is a low canopy of trees between 30-50 ft. (9-15 meters) and an emergent layer between 50-80 ft. (15-25 meters). Shrubs are more or less absent, but ground cover may be dense or sparse. Lianes are few, epiphytes fairly frequent. Buttresses are not marked.

Leaves are evergreen, mesophyllous, 75% simple."

Seasonal-Swamp Forest with Palms is the original marsh forest of Beard. Palms here enter the assemblage in great numbers and become dominant, providing 40-60% of all trees and 60-75% of the lower story forms the canopy at 5-10 meters, and emergents, both trees and palms reach heights up to 25 meters.

Seasonal-Swamp Woodland is, after Fanshawe: "A low woodland of small-stemmed, regularly spaced trees with scattered emergent palms or trees of peculiar habit. The canopy lies between 30-50 ft. (9-15 meters) with emergent palms to 60 ft. (18 m.). The shrub layer is virtually absent; the ground cover sparse or dense. The mean tree diameter is 4 in. (10 cm.); emergents are from 16-36 in. (40-90 cm.) in diameter. The stocking is high—between 300-400 stems per acre, 4 in. diameter and over—but the flora is restricted to about 20 tree species. Lianes are few, epiphytes rare and these mostly hemiepiphytes. Leaves are evergreen, mesophyllous, simple. Stilt roots are a common feature."

It is now clear that the *Amanoa* consociation described by Beard on the central plateau of Dominica (1949a) and tentatively classified as a "swamp phase of montane thicket" belongs to this formation.

The marsh woodland and palm marsh of both Beard and Fanshawe are now to be consolidated as seasonal-swamp thicket, without and with palms respectively. Savanna has been very fully dealt with by Beard in a recent paper (1953).

The original four swamp formations (swamp forest, palm swamp, herbaceous swamp, and mangrove woodland) have been expanded to nine by Fanshawe by the addition of four new formations and the recognition as full formations of the two former subdivisions of herbaceous swamp, now termed herbaceous swamp and semi-aquatic swamp. Not all of Fanshawe's treatment is, however, acceptable. At the pessimum he gives us aquatic swamp, "submerged trailing vegetation, attached to rocks in the shallow, fast running parts of the rivers." This would presumably include also the aquatic communities of lakes and ponds. All these are to be regarded as seral and unworthy of formation rank. We might very well, however, take up Fanshawe's term aquatic swamp for that division of herbaceous swamp which, at the true pessimum, consists of mats of grass rooted at the bottom but floating and rising up and down with the water level. This was rather inaptly termed swamp savanna in 1944.

Between herbaceous swamp and swamp forest, Fanshawe has given us two useful new formations, *Arborescent Swamp* and *Swamp Woodland*. *Arborescent Swamp*, which for the sake of uniform terminology it is preferred to call *Swamp Thicket*, stands in shallower water, marks the appearance of woody growth and "consists of a dense growth of shrubby vegetation, on sand banks or stretches of silt, submerged at high tide or at high water. Woody species are restricted, but gregarious; shrubs, perennial herbs and lianes are occasionally associated" (Fanshawe).

Swamp Woodland is defined by Fanshawe as follows:

"This term is proposed to connote the riparian fresh water fringe of low, spreading trees. Swamp woodland is remarkably homogeneous, comprises one or very few species and has a simple structure. There is one tree stratum 10-40 ft. high (3-12 meters) and the canopy is loose and open. A shrubby or herbaceous undergrowth may be present."

Swamp Forest and *Mangrove Forest* need no new definition are are the equivalents in fresh and brackish water respectively. They are to be regarded as fresh and brackish water phases of one and the same formation.

The above five formations constitute a harmonious series in reduction of structure. Neither of the remaining two types that we have to consider, Beard's palm swamp and Fanshawe's Mora forest, stands in a very satisfactory relation to the others and may have to be discarded. Palm swamp was not found by Fanshawe in British Guiana and in view of the probable modification by fire of the Trinidad swamps from which it was described, it is possible that it represents a fire climax, and is degradation of swamp forest or woodland, or of swampy normal forest. There may also be a confusion with seasonal-swamp types in some cases. Alternatively, as we have suggested that the presence of palms in seasonal-swamp types may not be diagnostic, "palm swamp" may be only swamp woodland or arborescent swamp in which palms happen to be present. Further data on this question are desirable.

Fanshawe's Mora forest is not a welcome member of this series. Apart from the nomenclature, which draws the same criticism as Wallaba forest, we cannot recognize it as a separate formation because the structure differs little if at all from rain forest and there is little or no other difference in physiognomy. The Mora may indeed be

associated with wet riverain sites, but it should be regarded properly as a type of rain forest. It is true that the formation swamp forest is far removed from the optimum in structure so that we might expect to find other types to fill the gap. Mora does not fill any gap, however, because in structure it *is* optimum. This same hiatus occurs also as we have seen with seasonal-swamp types and is probably due to their specialization. There is less room here for intermediate stages, because the site is either inundated or not and the distinction tends to be clear cut. In the Trinidad forests, there are societies of species tolerant of wet ground, tending to resemble swamp forest, on low-lying sites but there is always an abrupt change to true swamp where permanent waterlogging occurs (see Beard 1946b: 67).

The classification of American tropical vegetation types is thus consolidated as follows. For reasons to be discussed, the montane series is now placed before the dry evergreen and we follow Fanshawe in taking seasonal-swamp before swamp. On balance, the consolidation has only added two formations to the previous total of twenty-six.

A. Optimum Formation.
   Rain Forest.
B. Seasonal Formations.
   1. Evergreen Seasonal Forest.
   2. Semi-Evergreen Seasonal Forest.
   3. Deciduous Seasonal Forest.
   4. Thorn Woodland.
   5. Cactus Scrub.
   6. Desert.
C. Montane Formations.
   1. Lower Montane Rain Forest.
   2. Montaine Rain Forest or Cloud Forest.
   3. Montane Thicket.
   3a. High Mountain Forest.
   4. Elfin Woodland or Mossy Forest.
   5. Paramo.
   6. Tundra.
D. Dry Evergreen Formations.
   1. Dry Rain Forest.
   2. Dry Evergreen Forest.
   3. Dry Evergreen Woodland and Littoral Woodland.
   4. Dry Evergreen Thicket and Littoral Thicket.
   5. Evergreen Bushland and Littoral Hedge.
   6. Rock Pavement Vegetation.
E. Seasonal-Swamp Formations.
   1. Seasonal-Swamp Forest.
   2. Seasonal-Swamp Woodland.
   3. Seasonal-Swamp Thicket.
   4. Savanna.

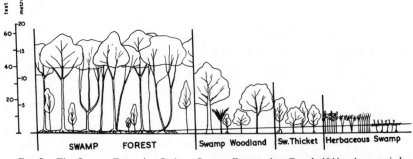

Fig. 7. The Swamp Formation-Series. Swamp Forest after Beard, 1944a, the remainder imaginary and diagrammatic.

F. Swamp Formations.
  1. Swamp Forest and Mangrove Forest.
  2. Swamp Woodland.
  3. Swamp Thicket.
  4. Herbaceous Swamp.

Space does not permit a recapitulation here of the physiognomic details of these formations and it is assumed that students of the question will be familiar with the earlier papers cited. The twenty-eight formations characterized should now be adequate to include all the possible types of vegetation in tropical America apart from "atypical formations" as envisaged in 1944a. Caution must be exercised in attempting to apply the system. It will not fit secondary or damaged forests, seral communities nor local societies, some of which, like the riverain *Guadua* bamboo groves of Colombia, may differ strongly from the climax association.

Certain observations on the habitat of the formations will now be made, since some of the original conceptions have to be modified. Fanshawe's work emphasizes in strong measure the overwhelming importance of soil drainage in controlling the vegetation of high rainfall areas. In dry and moderately dry areas, the most important character of the soil is its moisture-supplying ability, in which the drainage factor assumes a smaller and smaller role as the dryness of the climate increases. In wet areas, however, as the wetness increases only sites of increasingly favorable relief and physical structure of the soil are able to dispose of surplus water by run-off or infiltration and continue to support unspecialized vegetation. Even here, paradoxically, we can also find soils whose drainage is at the other extreme and too free. The climatic climax, in its narrow sense, can only exist on sites where drainage is optimum and neither too free nor impeded. In British Guiana, climate differentiates the rain forest from the seasonal forests of the eastern border

and from the montane forests of higher altitudes in the Pakaraimas. It is drainage which separates from these the enormous areas of dry evergreen, seasonal-swamp and swamp.

There is nothing to be added at this stage to our knowledge of rain forest and seasonal forests. These can only occur on well-drained sites and are invariably predominantly controlled by climate. Locally, sites of poor moisture supplying ability will carry vegetation of a drier type than the regional climax, and unusually favorable sites vegetation of a more luxuriant type. In such cases seasonal formations are still present and show only a move up or down in the formation-series.

Montane formations likewise reflect essentially the climatic zonation associated with altitude and have for this reason been placed next in the classification, rather than the dry evergreen formations which are less clearly climate-controlled. Generally speaking, rainfall and humidity increase progressively up to a level of some 2,500 meters, above which there is again a progressive decrease. Temperature falls steadily with altitude, but only becomes critical above about 3,000 meters. At this level and upward it must be responsible for the zonation of high mountain forest, elfin, woodland, paramo, and tundra below the snow line at about 4,500 meters. In different regions rainfalls vary widely within this zone without changing the essential character of the vegetation. There are both wet and dry paramos. Lower down in the cloud belt exposure becomes a factor of great importance and is responsible for alternations of lower montane and montane rain forest with montane thicket and elfin woodland. The more reduced formations are found to come in on ridge tops, very steep slopes or areas particularly exposed to the prevailing wind. Below 2,500 meters, the vegetation is still essentially tropical and it seems unlikely that temperature has any effect on zonation which is primarily due to the exposure factor. In previous papers the writer suggested that the exposure could result in excessive transpiration on clear days and that this was largely responsible for the reduction in stature and leaf size. It now seeems that this view is not tenable. "Exposure" is compounded of numerous effects. Most important of these is probably the mechanical effect of wind, which could break or blow down large trees on exposed ridges and slopes with shallow soil. The exposed situations in the mist belt are those where fog forms most consistently and where humidity and precipitation are highest. This may reduce transpiration to the point of stunting growth of the trees. A further limitation will be imposed by lack of root-room on steep slopes: however high the rainfall, shallow soil is bound to mean a lower forest. Again, in certain cases, the perpetual wetness of the climate sets up impeded drainage conditions even on mountain slopes, accentuated by pan-formation and the accumulation of a saturated surface peaty layer, with resultant reduction of the vegetation. This was found to be the cause differentiating montane thicket from lower montane rain forest in Puerto Rico by Wadsworth and Bonnet (1951). In such a case the montane thicket becomes close to marsh woodland.

Of the remaining three formation series, seasonal-swamp and swamp are clearly edaphic formations while the dry evergreen occupy a somewhat intermediate position and may be termed edapho-climatic. The swamp types are the most specialized, growing in permanently or more or less permanently inundated ground. The seasonal-swamp formations are in a sense intermediate between true swamp and dry evergreen, being periodically inundated on the one hand and associated with soils of obstructed internal drainage on the other, while the dry evergreen formations stand between seasonal-swamp and the truly climatic seasonal and montane types. These affinities between the habitat groups are shown by the certain degree of floristic interchange which takes place in a horizontal direction between them. Certain trees such as *Symphonia globulifera* are at home in both seasonal-swamp and swamp, while *Clusia fockeana* may be a dominant in both seasonal-swamp woodland and dry evergreen woodland in British Guiana.

Seasonal-swamp and swamp formations are in a true sense edaphic. Entirely under the influence of ground water, they will appear in substantially the same form under any climate where the edaphic conditions are similar. They stand moreover in a clear developmental relationship

to the climatic climax, occupying sites which are capable of evolving towards an optimum or have already devolved away from it. The dry evergreen formations are more difficult. Where littoral types are concerned they are controlled by climate, but it is a local and not the regional climate Dry evergreen forest which seems formerly to have existed in the Lesser Antilles at low altitude (Beard 1949a) would have been controlled by climate and climate induced soil conditions in "terras" and "shoal" soils which had developed drainage impedance. Dry evergreen formations in the interior of British Guiana are all differentiated from the climatic optimum by soil conditions, but these vary widely. In the white sands, we have a peculiar soil type of excessive drainage, which is not a climate-induced development, nor will it be capable of developing in any direction except by topographic senilty: there is no developmental relationship, therefore, between dry rain forest and the climatic climax. On the Kaieteurean sandstones, the dry evergreen vegetation occupies immature soils, where the rock has not yet decayed deeply: as it does so, there is presumably a development towards the climatic climax. Yet again, we find dry evergreen on ridges of lateritic ironstone, senile soils that have developed gradually the lateritic horizon: as they did so, there was presumably a development away from the climatic climax. Unlike seasonal-swamp and swamp formations, dry evergreen are always to some extent controlled by climate. Seasonal-swamp and swamp will be the same under any climate if groundwater conditions are constant whereas dry evergreen vegetation on a given soil type will vary according to the prevailing climate. The ecological status of dry evergreen formations may frequently appear to vary widely, but they are united by their essential habitat. An essential similarity of moisture relationships accounts for the fact that the same formation may appear on such widely different soil types, now deep white sands, now shallow sandstone, and now rocky laterite ridges.

There is little difficulty in assessing the habitat of swamp formations which are in general differentiated by the depth and duration of inundation, with salinity and aeration of the water and nature of the rooting medium as subsidiary factors. Seasonal-swamp vegetation is always associated with impeded drainage of the type which will lead to stagnation of the ground water and deprivation of oxygen, occurring on flat areas having a pan or impermeable subsoil or simply a high water table and deficient means of lateral drainage. In true swamps, water-logging is always present and the ground can never dry out. In seasonal-swamp there is at least a fall in the water table, allowing seasonal aeration of the topsoil. It is this factor which appears to control the segregation of these two major groups of edaphic communities. Under a strongly seasonal climate, waterlogging and stagnation of ground water take place on ill-drained sites during the rainy season while in the dry due to the nature of the soil there comes a period of desiccation. This acute alternation of moisture conditions is too severe for tree growth and is the commonest cause of the appearance of savanna. In higher rainfall areas, such severe seasonal desiccation may never occur, but there will be a relatively dry period of more favorable conditions in sites of bad drainage and stagnant ground water, giving us the woody seasonal-swamp formations, which are differentiated by the drainage factor into the series of formations, seasonal-swamp forest occurring where the drainage conditions are least severe and at the other extreme seasonal-swamp thicket standing for long periods in relatively deep stagnant water.

It is drainage variations which determine the zonation in and around savannas. With rainfalls of upwards of 2000 mm. a year, sedge savannas are most often found and occur typically as relatively small patches surrounded by a zonation of seasonal-swamp types from thicket to forest, these woody formations being enabled to appear due to the preponderance of the waterlogging factor. With rainfalls between 1000 and 2000 mm. a year, conditions on flats with impeded drainage are ideal for savanna, there being sufficient seasonal periodicity in the rainfall to maintain an intense alternation of the water conditions. This is the niche of tall bunch grass savanna which occupies wide areas, including sites which under higher rainfall would be constantly wet and carry woody seasonal-swamp vegetation. The transition to adjoining climatic forest is now usually abrupt, without ecotones or intermediate communities. An intermediate stage can however, sometimes be found where the rainfall is about 1500 mm. or less and the savanna is surrounded by semi-evergreen or deciduous seasonal forest. In such a case there may be a belt, called *cerradão* on the Brazilian plateau (Waibel 1948) and which we may perhaps term savanna woodland, constituted by a closing-up of the savanna trees to form a dense woodland 10-15 meters high. As the flora is still essentially that of the savanna, this must be considered as properly a type of savanna rather than as a separate formation. Under rainfalls below 1000 mm. a year, the waterlogging factor begins to fade out and is only now severe in depressions.

The low rainfall accentuates relief, so that the climatic climax—thorn woodland and cactus scrub—comes in on all the slight rises to form an intricate mosaic with short-bunch grass savanna which occurs as small patches in the slight depressions.

Dry evergreen communities not infrequently associate with savannas, appearing on reefs of white sand where rooting depth becomes very great, or on rock outcrops where rooting depth is greatly reduced: drainage is however good in both cases.

SUMMARY

The classification of plant formations in tropical America proposed by the writer in 1944 is reviewed in the light of subsequent work, particularly of Fanshawe's "Vegetation of British Guiana" (1952). The original twenty-six formations are expanded to twenty-eight, listed in the text on p. 96. The physiognomy and habitat of these formations are discussed.

References

Beard, J. S. 1944a. Climax vegetation in tropical America. Ecology 25: 127-158.

———. 1944b. The natural vegetation of Tobago. Ecological Monog. 14: 135-163.

———. 1945a. Some ecological work in the Caribbean. Empire For. Jour. 24: 40-46.

———. 1945b. The progress of plant succession on the Soufriere of St. Vincent. Jour. Ecology 33: 1-9.

———. 1946a. Los climax de vegetación en la América tropical. Rev. Fac. Nal. de Agronomía, Medellin, 6: 225-293. (Spanish translation of "Climax vegetation in tropical America.")

———. 1946b. The natural vegetation of Trinidad. Oxford Forestry Memoir No. 20.

———. 1949a. The natural vegetation of the Windward and Leeward Is. Oxford Forestry Memoir No. 21.

———. 1949b. Ecological studies upon a physiognomic basis. Actas del 2°. Congreso Sudamericano de Botánica, Lilloa 20: 45-53.

———. 1953. The savanna vegetation of northern tropical America. Ecological Monog. 23: 149-215.

Beebe, W., and J. Crane. 1947. Ecology of Rancho Grande, a subtropical cloud forest in northern Venezuela. Zoologica 32: 43-60.

Cuatrecasas, J. 1934. Observaciones geobotánicas en Colombia. Madrid.

Curtis, J. T. 1947. The palo verde forest type near Gonaives, Haiti, and its relation to the surrounding vegetation. Carib. Forester 8: 1-12.

Davies, T. A. W., and P. W. Richards. 1933. The vegetation of Moraballi Creek, British Guiana. Jour. Ecology, 21: 350-384.

Fanshawe, D. B. 1952. The vegetation of British Guiana. A preliminary review. Imp. For. Inst. Paper No. 29, Oxford.

Richards, P. W.   1952.   The tropical rain forest. Cambridge.

Schomburgk, R.   1847.   Travels in British Guiana. Leipzig (in German).   English translation by W. E. Roth, Georgetown, British Guiana, 1922. 2 vols.

Stehlé, H.   1945.   Forest types of the Caribbean Islands, Part. I.   Carib. Forester Suppl. to Vol. 6.

Wadsworth, F. H.   1951.   Forest management in the Luquillo Mountains, I.   Carib. Forester 12: 93-114.

Wadsworth, F. H., and J. A. Bonnet.   1951.   Soil as a factor in the occurrence of two types of montane forest in Puerto Rico.   Carib. Forester 12: 67-69.

Waibel, L.   1948.   Vegetation and land use in the Planalto Central of Brazil.   Geog. Rev. 38: 529-554.

11

Copyright © by Dr. W. Junk N. V.-Publishers

Reprinted from *Vegetatio* **20**:251–278 (1970)

PFLANZENSOZIOLOGISCHE BEOBACHTUNGEN AN ISLÄNDISCHEN DÜNENGESELLSCHAFTEN

von

Reinhold Tüxen

Arbeiten aus der Arbeitsstelle für Theoretische und Angewandte Pflanzensoziologie (42)

EINLEITUNG

Die Flora Islands ist dank der grundlegenden Arbeiten der dänischen und der isländischen Forscher, u.a. Jónsson, Ostenfeld, Stefánsson, Gröntved, Steindórsson 1948 (vgl. a. Gröntved 1942, p. 7-15) recht gut bekannt. Die pflanzensoziologische Erforschung folgte im allgemeinen den nordischen Methoden, wohl unbewußt beeinflußt durch die lange geltende Lehrmeinung, man könne bei der Einförmigkeit und Arten-Armut der nordischen Flora — Island hat nur etwa 430 Arten von Gefäßpflanzen — die Vegetation hier nur nach Dominanz und Frequenz einteilen. Wir haben diese Auffassung auf mehreren langen Reisen durch Fennoskandien seit langem widerlegen können. Selbst in Lapland wachsen nicht wenige Pflanzengesellschaften, die mit ihren Kryptogamen weit mehr Arten enthalten als viele mitteleuropäische! Andererseits wird die Zahl der Gesellschaften geringer, je weiter man nach Norden vordringt, und die Areale der einzelnen bedecken oft größere Flächen, was den monotonen Eindruck der Vegetation hervorruft oder verstärkt. Aber dennoch gibt es zahlreiche eigenartige Gesellschaften, besonders an Sonderstandorten (Quellen, Flußufern, Seen, Meeresküsten, Schneetälchen u.a.), und die vom Menschen und seinen Tieren geschaffenen Ersatzgesellschaften.

Dank der unvergleichlichen Gastfreundschaft von Herrn Direktor Gísli Sigurbjörnsson, Reykjavík, und der selbstlosen Hilfe von Herrn Dr. Sturla Fridriksson, Reykjavík, und Herrn Direktor Steindór Steindórsson, Akureyri, und nicht zuletzt durch die unermüdliche technische und wissenschaftliche Unterstützung meiner jungen Begleiter cand. Águst H. Bjarnason und cand. Sigurdur H. Richter konnte ich gemeinsam mit Herrn Dr. K. Höll, Hameln, dank der großzügigen Hilfe der U.S. Atomic Energy Commission, Biology Branch, und der Surtsey Research Society, die einen Wagen zur Verfügung stellten, vom 2.-13. August 1968 einige Pflanzengesellschaften SW-Islands eingehender untersuchen. Mein Mitarbeiter Herr H. Böttcher, der in Island ein Jahr weilte, um dort die Vegetation zu studieren, stellte mehrere seiner soziologischen Aufnahmen der hier beschriebenen Gesellschaften zur Verfügung und gab für die Auswertung der isländischen Literatur wertvolle Anregungen. Er fertigte auch die Reinzeichnungen der Abbildungen an.

Herr Dr. A. Scholz, Berlin-Dahlem revidierte die *Polygonum aviculare*-Kleinarten; Frau Dr. I. Markgraf-Dannenberg, Zürich, bestimmte die gesammelten *Festuca*-Belege und Herr Dr. Paul Aellen, Basel, übernahm die Durchsicht und

Benennung mitgebrachter Stücke der Gattung *Atriplex* von der Küste. Die Bestimmung der Moose übernahm Herr stud. K. DIERSSEN. Allen freundlichen Helfern darf ich herzlich danken für ihre Unterstützung, die sie bereitwilligst und erneut gewährt haben!

Wenn ich mir erlaube, einige Ergebnisse meiner Beobachtungen hier vorzu-legen, bin ich mir einer gewissen Zufälligkeit der Aufnahmen durch die Auswahl der besuchten Orte und der Unvollständigkeit der Darstellung voll bewußt, glaube aber angesichts des Strebens zur Vereinheitlichung der pflanzensoziolo-gischen Methoden nicht zögern zu dürfen, diese kurze Mitteilung zu geben, die ich zugleich als einen bescheidenen Ausdruck meiner tiefen Dankbarkeit für meine isländischen Freunde aufzufassen bitten darf.

Diese kleine Arbeit möchte nicht isländischen Untersuchungen vorgreifen oder solche stören. Sie könnte aber einige Anregungen geben durch Vergleiche mit anderen Gebieten und durch die Anwendung der Methode BRAUN-BLAN-QUETS, die bisher mit Ausnahme kurzer Notizen von A. STÄHLIN (1960) und N. KNAUER (1966) auf Island u.W. kaum erfolgt ist. Wir beschränken uns hier im wesentlichen auf die westliche S-Küste Islands und ihr Hinterland zwischen Eyrarbakki und Thorlákshöfn, wohin der Verf. mit den Herren STEINDÓRSSON, BJARNASON und RICHTER drei unvergeßliche Halbtags-Exkursionen machen konnte, und fügen nur einige Beobachtungen aus anderen Teilen Islands zum Vergleich hinzu.

1. ATRIPLICI - CAKILETUM ISLANDICAE

Der Strand SW-Islands wird von dunkelgrauem vulkanischen Sand gebildet, der aus zerriebener Lava und vulkanischer Asche entstanden ist und nur wenig Molluskenschalen enthält, also kalk-arm ist. Die groben Körner sind leicht, weil sie Lufträume enthal-

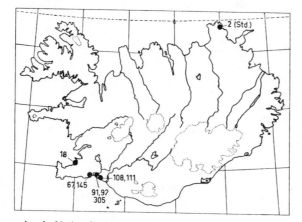

Karte 1. Atriplici - Cakiletum islandicae. Lage der Aufnahmen.

ten. Sie können darum von der Brandung und vom Wind stärker bewegt werden als der zwar feinere, aber schwerere helle Sand der meisten Küsten Europas. Der Auswurf an Algen ist in SW-Island ziemlich beträchtlich, so daß braunschwarze Bänder wechselnder Breite von toten *Fucus*-Arten, einzelnen *Laminarien* und Rotalgen den oberen Strand säumen, die entweder trocken und zerbrechlich auf dem Sande liegen oder — bei mächtigerer Schicht — im Sommer eine weiche stinkende Faulschlamm-Masse bilden.

Die vorjährigen dieser "Tangbeete" sind mit Therophyten bedeckt, unter denen Reinbestände der breitblättrigen *Cakile edentula* ssp. *islandica* weite Flächen einnehmen können, wo aber auch *Atriplex* vorherrschen kann.

Die *Cakile*-Bestände auf dem oberen Strand zeigen nicht selten zwei phänologisch verschiedene Zonen (Abb. 1). In der oberen ist Anfang August *Cakile* schon verblüht und beginnt zu fruchten und zu vergilben, während die strand-nähere grüne noch in voller

TABELLE I

Atriplici - Cakiletum islandicae (Jónsson 1900) Tx. 1968
   A. Typische Subass.
   B. Subass. von Polygonum heterophyllum (Ostenfeld 1905)

	A				B			
Nr. d. Aufnahme:	67	108	145	111	305	92	18	91
Veg.-Bedeckung (%):	70	70	.	100	30	15	30	.
Artenzahl:	2	6	6	5	4	6	7	6
Kennarten:								
Cakile edentula ssp. islandica	5.5	4.5	2.2	3.4	+	2.1	+	.
Atriplex glabriuscula								
(A.patula(?) et spec.)	.	.	3.3	4.3	3.3	2.2	1.1	1.1
Trennart:								
Polygonum heterophyllum	.	.	.	.	1.2	+	+	2.1
Begleiter:								
Stellaria media	.	r	r	2.3	2.2	1.2	+	3.3
Honckenya peploides	rK	rK	.	.	.	r	+	.
Elymus arenarius	.	+	.	.	.	.	+	1St
Tripleurospermum maritimum	.	.	+.2	+	.	.	.	+
Mertensia maritima	.	2.1	.	(+)	.	.	.	.
Capsella bursa-pastoris	.	+	.	.	.	r	.	.

Außerdem in Aufn. 145: Poa annua +, Rumex domesticus (Klg.) r; in 18: Potentilla anserina +; in 91: Festuca rubra ssp. 1St.

Fundorte:
67   7.8.68 Strand vor 4—5 m hohen Elymus-Dünen ö Thorlákshöfn
108 9.8.68 Nahe Stokkseyri (30 × 300 m).
145 12.8.68 N des Hafens Thorlákshöfn.
111 9.8.68 Beim Hafen von Stokkseyri.
305 26.8.67 Eyrarbakki (Aufn. BÖTTCHER).
92 9.8.68 Zwischen Eyrarbakki u. Stokkseyri
18 24.6.67 Álftanes sw Reykjavík (Aufn. BÖTTCHER)
91 9.8.68 Zwischen Eyrarbakki und Stokkseyri.

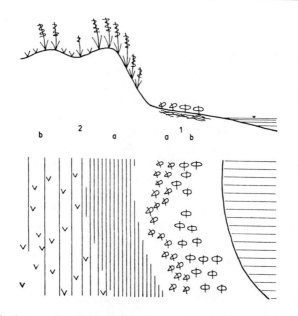

Abb. 1. Zonierung der Strandvegetation ö Thorlákshöfn (SW-Island). 1 :
Cakile-Fazies des Atriplici - Cakiletum islandicae auf Tangbeeten, in zwei
verschiedenen phänologischen Phasen (a Anfang August schon vergilbend, b
noch voll blühend). 2. Honckenyo diffusae - Elymetum arenariae —
Düne: a = Optimal-, b = Degenerationsphase der Ass.

Blüte steht. Die Erscheinung, die wohl nur durch die frühere
Erwärmung des trockeneren Hochstrandes gegenüber dem feuchte-
ren tieferen zu deuten ist, erinnert stark an die entsprechende Zonierung der Chenopodion fluviatile-Gesellschaften unserer
mitteleuropäischen Flußtäler.

Die in den neueren Floren und taxonomischen Arbeiten für die isländische
Cakile gebrauchten Namen erleichtern nicht gerade die Benennung unserer Assoziation:

Cakile maritima (Bigel.) Hook. var. latifolia Desf.
Cakile maritima f. islandica Gandoger 1900
Cakile edentula f. islandica (Gand.) O. E. Schulz 1923
Cakile maritima ssp. islandica Hylander 1955
Cakile edentula (Bigel.) Hook. ssp. islandica (Gandoger) L. et L. comb. nov.
  1961 (so auch in Flora Europaea)
Cakile lapponica Pobedimova 1963

Noch schwieriger ist nach Herrn Dr. AELLEN (schriftl.) die
taxonomische Beurteilung der Gattung *Atriplex* an den isländischen
Küsten, deren Studium noch nicht abgeschlossen ist. Außer *A.
glabriuscula* könnten noch weitere Taxa, z. B. *A. longipes* und *A.*

calotheca am Aufbau unserer Assoziation beteiligt sein[1]). Von den isländischen Floren wird *A. patula* genannt.

Wir ziehen daher vor, für die Benennung unserer Assoziation neben *Cakile *islandica* nur die Gattung *Atriplex* zu verwenden.

Unsere Tabelle I zeigt Unterschiede sowohl in der Artenzahl der Bestände als auch in der Dominanz der stetesten Arten. Vorläufig lassen sich zwei Ausbildungen erkennen, die vielleicht den Rang von Subassoziationen haben.

In der Typischen Subass. dominiert *Cakile*, auch *Mertensia maritima* ist hie und da als Rosette zu finden, die nicht zur Blüte kommt. Diese Gesellschaft ist in der Regel hohen *Elymus arenarius*-Dünen vorgelagert. Die andere Subass. wird durch *Polygonum heterophyllum* unterschieden. Zugleich ist die Menge von *Stellaria media* hier erheblich höher. Diese Ausbildung wächst auf mächtigen faulenden Spülsäumen (Aufn. 91, 92) hinter einem Vorland von Puccinellietum maritimae von folgender Zusammensetzung (zwischen Eyrarbakki und Stokkseyri):

5.5 Puccinellia maritima	r Polygonum heterophyllum
+.2 Plantago maritima	r⁰ Atriplex cf. glabriuscula

Wir glauben aus diesen Beobachtungen auf einen geringeren Salzgehalt im Substrat der Subass. von Polygonum schließen zu dürfen.

Schon JÓNSSON (1900, p. 45, vgl. a. JÓNSSON 1898, p. 363) gab als ausgesprochene Strandpflanzen *Cakile, Atriplex patula, Mertensia, Honckenya* und *Potentilla anserina* an und damit eine vollständige (wenn auch etwas komplexe) Liste unserer Assoziation und beschrieb die Zonierung in verschiedener Entfernung vom Meer: "Cakiletum" "Honckenyetum", "Potentilletum". OSTENFELD (1905, p. 120) teilte unter dem Namen "Cakile-Atriplex-Ass." eine Liste unserer Subass. von Polygonum mit. Auch THORODDSEN (1914, p. 319) hat in seiner inhaltsreichen Schilderung der isländischen Vegetationstypen von S-Island (wohl JÓNSSON folgend) diese auffallende Assoziation beschrieben (vgl. a. GRÖNTVED 1942, p. 43). Von STEINDÓRSSON (1936, p. 477) ist sie aus NO-Island als "Atriplex-Ass." erwähnt worden. Seine Aufn. 2 von Raufarhöfn ist wohl zu unserer Subass. von Polygonum zu stellen, wenn darin auch *Cakile* fehlt und *Puccinellia retroflexa* und *Cochlearia officinalis* auftreten, die in unseren Aufnahmen nicht vorkommen.

In seiner Übersicht über die isländischen Soziationen gibt STEINDÓRSSON (1951) 6 Soziationen (254—259) an, die alle zu unserer Assoziation gehören dürften. Nach THORODDSEN (1914, p. 338)

---

[1]) Wir danken Herrn R. VAN DER MEIJDEN, Leiden, für die Durchsicht unserer gesammelten *Atriplex*-Stücke.

und Gröntved (1942, p. 78) scheint sie, vielleicht in einer beson-
deren Ausbildung, auch an isländischen Bauernhöfen in Meeres-
nähe vorzukommen. Hadač (1949, p. 12), der im gleichen Gebiet
wie wir gearbeitet hat, nennt unsere Gesellschaft "Atriplicetum
glabriusculae" (n.n.).

Über die synsystematische Stellung der isländischen Cakileta-
lia-Gesellschaft wagen wir keine näheren Angaben zu machen,
weil immer noch zu wenig Vergleichsmöglichkeiten bestehen.
Eine nicht zu übersehende Verwandtschaft besteht zu der Atriplex
glabriuscula-Matricaria ambigua-Ass. (Ostenfeld 1908)
Tx. 1950 prov., die auf den Färöers vorkommt (Tüxen 1950, p. 103,
106).

Das Atriplici-Cakiletum islandicae ist wie alle Cakiletea-
Gesellschaften eine migratorische therophytische Initial- und
Dauer-Gesellschaft zugleich (Tüxen 1950). Sie kann zwar im
Flugsand-Gebiet von *Honckenya peploides-* oder *Elymus arenarius-*
Dünen und auf frischeren, nicht vom Winde bewegten Böden vom
Agropyro-Rumicion-Verbande abgelöst werden. Beide Folge-
gesellschaften können sich aber auch unmittelbar auf offenen
Böden einstellen, und viele, wenn nicht die meisten Bestände der
Cakile-Gesellschaft gehen nach Beendigung ihrer Lebensdauer
im Herbst spurlos zugrunde, um im nächsten Jahre auf neuen
Spülsäumen aus Samen wieder zu erscheinen.

Endlich sei der Hinweis erlaubt, daß die wenigen Stücke von
Cakile, die sich auf Surtsey gezeigt haben (Fridriksson 1964—1967,
Thorarinsson 1966, p. 38 u.a.) wohl als erste Vorposten unserer
Assoziation auf dieser neuen Insel aufzufassen sein dürften. Ihre ge-
naue taxonomische Untersuchung dürfte Aufschluß über ihre Her-
kunft (Island, Grönland oder Amerika) geben.

2. AGROPYRO - RUMICION

Meist sind die Grenzen zwischen der Therophyten-Gesellschaft,
und dem folgenden Hemikryptophyten-Rasen am Strande sehr
scharf, wie es die Lage der verschieden alten Tangwälle bedingt.
Nicht selten liegen junge Tangstreifen mit Therophyten höher
als die alten mit dem Agropyro-Rumicion bewachsenen (Abb. 2).

Wie Nordhagen (1940) eingehend beschrieben hat, nehmen die
Agropyro-Rumicion-Gesellschaften die Stelle alter Spülsäume
ein. *Potentilla anserina* ist hier einer der am schnellsten mit seinen
meterlangen oberflächlichen Ausläufern sich einstellenden Pioniere,
wie die folgende Aufnahme von Böttcher vom Strand bei Stokkseyri vom 3.9.1967 zeigt (Veg.-Bedeckung 70%, 20 m²):

4.5 Potentilla anserina
1.1 Senecio vulgaris
+.2 Puccinellia maritima
+ Cakile edentula ssp. islandica

+ Capsella bursa-pastoris
+ Tripleurospermum maritimum
+ Polygonum aviculare s.l.
1St. Taraxacum spec.

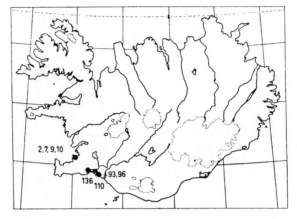

Karte 2. Agropyro-Rumicion (Tab. 2). Lage der Aufnahmen.

Eine ähnliche Liste hat auch STEINDÓRSSON (1936, p. 477, Aufn. 4) von Oddsstadir mitgeteilt. Das "Potentilletum" älterer isländischer Autoren gehört gewiß auch hierher. *Agropyron repens* folgt erst langsamer mit seinen Rhizomen, kann sich aber wohl auch unmittelbar einstellen und erweist sich als sehr konkurrenzkräftig und widerstandsfähig gegen Wellenschlag. Neben diesen

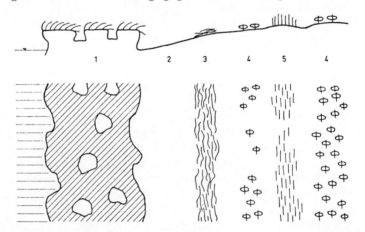

Abb. 2. Zonierung der Spülsaum-Vegetation hinter einem Puccinellietum maritimae am Strande bei Eyrarbakki. 1. Puccinellietum. 2. Offener Sand-Strand. 3. Junger Spülsaum ohne Vegetation. 4. Atriplici-Cakiletum, Subass. v. Polygonum heterophyllum. 5. Poa irrigata-Potentilla anserina-Ass.

beiden Arten ist *Poa pratensis* ssp. *irrigata* regelmäßig zu finden. Die meisten übrigen Hemikryptophyten und Therophyten, diese als Reste der Sukzession oder mit jungen Spülsäumen neu eingeschleppt, erreichen weder größere Menge noch Stetigkeit. Die Lebensdauer dieser Gesellschaft wird durch solche gelegentlichen Überflutungen, die neue organische Reste hinterlassen, eher verlängert. Sie ist eine echte Dauergesellschaft, die sich am Strande nicht weiter entwickeln kann.

TABELLE II.

A. Poa irrigata - Potentilla anserina-Ass. (Steindórsson 1936) Tx. 1968
B. Poa pratensis - Rumex domesticus-Ass. (Jónsson 1900) Tx. 1968

	A			B				
Nr. d. Aufn.:	93	96	116	7	9	10	136	2
Veg.-Bedeckung (%):	90	100	100	100	90	.	.	95
Größe d. Probefläche (m²):	8	8	.	10	5	10	30	50
Artenzahl:	9	6	11	4	5	5	10	15
Kennarten (Ass. A.):								
Agropyron repens	2.3	2.3	5.5	.	.	.	2.4	2.1
Potentilla anserina	5.4	5.4	2.2	.	.	.	.	.
Poa pratensis ssp. irrigata	1.2	1.2	1.2	.	.	.	.	.
Trennarten der Subass. (?) von Elymus								
Elymus arenarius	.	.	2.1	.	.	.	.	.
Lathyrus maritimus	.	.	1.3	.	.	.	.	.
Angelica archangelica	.	.	+	.	.	.	.	.
Kenn- u. Trennarten (Ass. B.):								
Rumex domesticus	.	2.2	+	2.2	.	3.2	2.2	2.2
Tripleurospermum maritimum	.	.	+	4.4	2.2	2.2	4.4	+2
Poa pratensis s. str.	.	.	.	2.2	5.4	2.3	1.2	1.1
Kennarten (Verb., Ordg. u. Klasse):								
Alopecurus geniculatus	1.2	.	.	.	.	.	+.2	+.2
Polygonum heterophyllum	1.1	.	.	.	.	.	2.2	.
Poa annua	.	+.2	.	.	.	.	1.2	.
Agrostis stolonifera	.	.	+.2	.	.	.	.	2.3
Ranunculus repens	.	.	.	.	.	.	.	4.3
Begleiter:								
Taraxacum spec.	.	+	.	.	+	+.2	.	1.1
Stellaria media	1.2	.	1.2	.	.	.	+	.
Leontodon autumnalis	.	.	.	.	.	+	+	1.2
Festuca rubra	2.3	.	.	.	+.2	.	.	.
Ranunculus acer	.	.	+	.	.	.	+	.
Poa trivialis	.	.	.	3.4	.	.	.	.
Pohlia nutans	.	.	.	.	.	.	.	2.4
Equisetum arvense	.	.	.	.	.	.	.	2.4
Deschampsia caespitosa	.	.	.	.	.	.	.	2.2

Außerdem in Aufn. 93: 1.2 Achillea millefolium, + Cerastium caespitosum; in 9: +.2 Rumex acetosella; in 2: +.2 Festuca pratensis, +.2 Ranunculus acer, +.2 Cephaloziella rubella.

Fundorte: 93: 9.8.68 Älterer Spülsaum auf Lava-Schicht mit Sand zwischen Eyrarbakki und Stokkseyri.
96: 9.8.68 Stokkseyri. Hinter dem Steinwall-Deich.
110: 9.8.68 W Eyrarbakki. Ausgedehnter alter Spülsaum hinter schmaler Elymus-Zone (Aufn. 109, Tab. III).
7: 2.8.68 Bauschutt vor einer Hauswand am Hafen Reykjavík.
9: 2.8.68 Straßenrand neben Mauer in Reykjavík.
10: 2.8.68 Dsgl.
136: 12.8.68 Thorlákshöfn. Am Hafen auf Schutt und Geröll.
2: 2.8.68 Aufgelassene Fläche nahe Flugplatz Reykjavík.

Diese Poa irrigata-Potentilla anserina-Ass. (Tab. IIA), die von STEINDÓRSSON (1936, p. 477, Tab. XIII A, Aufn. 4) und auch von GRÖNTVED (1942, p. 43) ohne Benennung erwähnt worden ist, kann nicht mit dem Agropyretum repentis maritimum (Störmer 1938) Nordhagen 1940 (vgl. TÜXEN 1950, p. 146) vereinigt werden. Sie scheint eine eigene Arten-Verbindung zu besitzen, die sich durch Poa pratensis ssp. irrigata und durch das Fehlen mancher Arten vom Agropyretum maritimum unterscheidet.

Am höheren Strande, der nur beweidet, aber nicht mehr überflutet wird, kann die Poa irrigata-Potentilla anserina-Ass. die hier herrschenden Weide-Rasen wie ein lückiger Teppich überlagern ("Teppich"-Gesellschaft), wodurch besondere Vegetationstypen vorgetäuscht werden können.

In einiger Entfernung vom Strande, wo Tier und Mensch meist in Hausnähe und an ähnlichen Standorten den Boden verdichtende und eutrophierende Wirkungen ausüben, die am Strande von Sturmfluten geschehen, wächst eine andere Agropyro-Rumicion-Gesellschaft, die Poa-pratensis-Rumex domesticus-Ass. (Tab. IIB). Sie ist in größeren Siedlungen sehr häufig und geradezu bezeichnend für diese Orte.

THORODDSEN (1914, p. 338) zeigt eine sehr klare Photographie dieser Assoziation und fügt eine Artenliste bei. Auch GRÖNTVED (1942, p. 77,80) hat sie unter Hinweis auf ältere Schriften von JÓNSSON (1900, p. 58) und von LINDROTH (1931, p. 457) erwähnt. Neben den herrschenden namengebenden Arten ist regelmäßig Poa pratensis zu finden. KNAUER (1966, p. 169) teilt drei Aufnahmen aus Island mit, die in den Agropyro-Rumicion Verband gehören.

Die isländische Assoziation zeigt wesentliche Abweichungen von der verwandten Rumex domesticus-Carduus crispus-Ass. Tx. et Becking 1950 aus N-Schweden (vgl. TÜXEN 1950, p. 175). Wir können hier leider nicht näher auf sie eingehen.

3. HONCKENYO DIFFUSAE - ELYMETUM ARENARIAE

Weil *Agropyron junceum* und *Ammophila arenaria* als südliche Arten Island fehlen und auch Koelerion albescentis - Gesellschaften nicht mehr vorkommen, die in Norwegen bis Stavanger (oder weiter nach dem Norden) und in Schottland fragmentarisch bis an die N-Küste reichen, zeigen die Zonierung und die Sukzession im Bereich der Sandküsten Islands einen ganz anderen Verlauf als an den Küsten W-Europas und der Britischen Inseln.

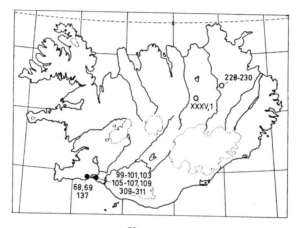

Karte 3.
Honckenyo diffusae - Elymetum arenariae. Lage der Aufnahmen.

Wir können für Island allerdings nur örtliche Verhältnisse schildern und rechnen damit, daß sie an anderen Dünenküsten dieser Insel nicht unerhebliche Abwandlungen oder Ergänzungen erfahren könnten, gab doch jede unserer kurzen Fahrten ungeahnte Überraschungen selbst auf kleinem Raume. Die Grundzüge unserer beobachteten Zonierung dürften aber eine gewisse allgemeinere Gültigkeit für Island haben.

Die oben (p. 254) dargestellte Zonierung findet sich nur an flachen Küsten, die nicht gegen das Meer vordringen. Dort, wo viel Sand angespült und durch den Wind weiter getragen wird, kann das Atriplici-Cakiletum entweder ganz oder nahezu ganz fehlen. Hier schmiegen sich dafür einzelne blaugrüne Stern-Rosetten der tiefblau blühenden *Mertensia* auf den violett-grauen Sand, auf dem sich hie und da dichte gelbgrüne Polster von *Honckenya* flach erheben. Bald folgen auch die ersten Pioniere des eigentlichen Dünenbauers *Elymus arenarius*. Seine Rhizome und graugrünen Halme und Blätter halten den groben, leichten Sand in locker bestockten Dünenkuppen fest, die sich allmählich zu einem mehrere Meter hohen welligen Dünenfeld zusammenschließen (Abb. 3).

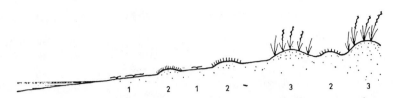

Abb. 3. Initial-Phase des Honckenyo diffusae - Elymetum arenariae auf einem anwachsenden Strand w Eyrarbakki. 1. *Mertensia maritima*. 2. *Honckenya peploides* var. *diffusa*. 3. *Elymus arenarius*.

Wo aber kürzlich das Meer vordrang und die jungen Vordünen zerstörte, erhebt sich das Elymetum als geschlossener 4—5 m hoher Dünenwall, der das dichte Cakiletum zu seinen Füßen, z.B. bei Thorlákshöfn, mit scharfer Grenze ablöst (Abb. 4, 1).

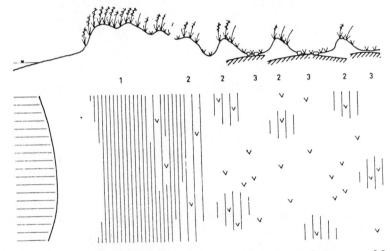

Abb. 4. Honckenyo diffusae-Elymetum arenariae auf einer vom Meer angerissenen Düne ö Thorlákshöfn. 1. Optimal-Phase. 2. Degenerations-Phase = Subass. von Silene maritima. 3. Silene maritima - Festuca *cryophila-Ass. auf Lava-Platten und -Geröll.

Diese Dünen-Gesellschaft ist in gewaltiger Ausdehnung zu beiden Seiten der Ölfusá-Mündung zwischen Thorlákshöfn und Eyrarbakki entwickelt. Sie erstreckt sich hier mehrere km tief in das ebene Land hinein, wo kahle Lava-Decken und flache Lava-Buckel teilweise mit grauem Sand überweht sind, der zu kleinen knie- bis mannshohen Dünenkuppen von *Elymus* aufgehäuft worden ist. Der Wind zerstört manche von ihnen (Parabolisierung: VAN DIEREN). Dann bedecken die hellbraunen, zähen Wurzeln von *Elymus* in vielen Metern Länge entweder als feines, offenes Netz die ausgeblasenen steilen Hänge, oder sie lagern locker auf ihnen wie ein dichter vom Winde aufgerollter Schleier (Abb. 5). Aber der frei gewordene

TABELLE III

Honckenyo diffusae - Elymetum arenariae (Regel) Tx. 1968
A. Typische Subass., B. Subass. v. Silene maritima

	A					B								
Nr. d. Aufn.:	109	68	311	310	99	105	309	100	107	137	103	69	106	101
Veg.-Bedeckung (%)	90	85	90	40	30	80	90	50	50	70	90	60	.	15
Artenzahl:	6	4	5	5	5	5	4	/4	5	5	6	8	6	6
Kennarten:														
Elymus arenarius	5.5	4.5	5.5	1.2	2.3	4.4	5.5	3.2	3.4	4.5	3.2	4.5	1.2	2.3
Honckenya peploides var. diffusa	3.4	+.2	2.3	2.3	2.3	2.2	2.3	3.3	2.3	+	5.4	1.3	2.3	2.2
Mertensia maritima	+	+	1.2	3.3	2.1	2.2	.	.	.	.	1.2	.	2.3	1.2
Inocybe spec.	.	+	.	.	.	.	.	.	.	.	.	.	.	.
Trennarten der Subass. u. Var.:														
Silene maritima	.	.	r°	r	r	1.2	+	+	+.2	+.2	.	.	.	.
Festuca rubra ssp. cryophila	.	.	.	.	.	.	+	+.2	+.2	1.2	+.2	2.2	2.2	2.2
Rumex acetosella	.	.	.	.	.	.	.	.	2.2	2.2	1.3	+.2	2.3	+
Begleiter:														
Cakile edentula ssp. islandica	.	2.1	1.1	2.1	r	.	.	.	r	.	.	.	.	.
Tripleurospermum maritimum	+.2	.	.	.	+	+	+	.	.	2.1	.	.	.	.
Atriplex glabriuscula et spec.	1.3	.	.	.	+	+	.	.	.	.	1.2	.	+.2	+

Außerdem in Aufn. 109: 2.3 Stellaria media; in 311: 1.2 Potentilla anserina; in 137: + Armeria maritima; in 69: + Potentilla anserina.

Fundorte: 109 9.8.68 2—4 m breiter junger Dünen-Saum vor altem *Agropyron repens*-Bestand (Tab. IIA, Aufn. 110) w Eyrarbakki nahe dem Hafen.

68 7.8.68 Erste *Elymus*-Düne 4—5 m hoch an *Cakile*-Zone (Aufn. 67, Tab. I) angrenzend ö Thorlákshöfn. *Elymus* bis 1,4 m hoch, reich fruchtend.

311 26.8.67 W Eyrarbakki, flache Sanddüne, 20 m von MHW-Linie (BÖTTCHER).

310 26.8.67 W Eyrarbakki, flacher Sandstrand, 15 m von MHW-Linie (BÖTTCHER).

99 9.8.68 W Eyrarbakki, flacher Sandstrand 40 m von Flutgrenze, 100 m²; neben der blauen Form von *Mertensia* auch eine grüne (+).

105 9.8.68 Etwas strandferner, 1,5 m hohe Düne, 20 m², neben Sileno - Festucetum *cryophilae im Dünental.

309 26.8.67 Nw Eyrarbakki, 2 m hohe Düne, 25 m von MHW-Linie (BÖTTCHER).

100 9.8.68 In der Nähe von 99, etwas weiter landeinwärts 1 m hohe flache Düne, 100 m².

107 9.8.68 W Eyrarbakki, 40 m vom Strand, flache Düne.

137 12.8.68 Stark zerblasenes Dünenfragment w Thorlákshöfn nahe der Lava-Kliff-Küste, 100 m².

103 9.8.68 W Eyrarbakki, 1 m hohe Düne ca. 100 m vom Strand, 100 m².

69 7.8.68 30 m vom Strand entfernt ö Thorlákshöfn. Elymus 1 m hoch, wenig fruchtend.

106 9.8.68 W Eyrarbakki unmittelbar hinter einem niedrigen Steinwall, 100 m².

101 9.8.68 W Eyrarbakki hinter Steinwall und Stacheldraht einzelne bis 3 m hohe *Elymus*-Düne. Rand einer *Sterna*-Kolonie im Sileno - Festucetum *cryophilae.

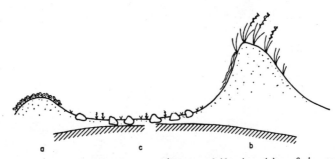

Abb. 5. Silene maritima-Festuca *cryophila-Ass. (c) auf einem Geröll-Feld zwischen einzelnen *Honckenya*- (a) und *Elymus*-Dünen (b), die z.T. ausgeblasen werden (offene *Elymus*-Wurzeln: "Dünen-Eingeweide").

Sand wird schnell wieder, sei es von *Elymus* oder *Honckenya*, aufgefangen, durchwurzelt und damit festgehalten. Hier zeigen sich Bilder, die in ihren Formen und in ihrer düsteren Farben-Harmonie überraschend an Dünen-Küsten Hokkaidos (N-Japan) erinnern.

Das Elymetum vertritt hier ganz die Rolle der Ammophile-

ten an südlicher gelegenen Küsten. Es kann bei starker Sandzufuhr nach einem Einbruch des Meeres und damit Düngung mit
einer Optimal-Phase beginnen, wo *Elymus* gleich einem Kornfeld so
reichlich fruchtet, daß seine Ähren in früheren Zeiten — in entle-
genen Gegenden noch heute — gedroschen und die Früchte zu
Mehl verarbeitet wurden, während das Stroh zum Hausdecken, die
Rhizome zu Packsätteln und die vielen langen zähen Wurzeln als
Bindfäden verwendet worden sind (Direktor STEINDÓRSSON mdl.)

In der Nähe des Strandes ist der Dünen-Gürtel geschlossener und
dichter (bis 80—90 %) bewachsen, in etwas weiterer Entfernung
vom Meer läßt nicht nur die Wuchskraft und besonders die Fruchtbarkeit von *Elymus* merklich nach, sondern hier treten zwischen
seinen Halmen und Blättern auch einige neue Arten auf, welche
die Degenerationsphase der *Elymus*-Gesellschaft anzeigen. Diese be-
siedelt meistens auch die isolierten sekundären *Elymus*-Dünen des
weiteren Hinterlandes, die durch Umlagerung der älteren Primär-
Dünen entstanden sind. Hier ist die mittlere Artenzahl etwas höher
als in der Initialphase, soweit sie nicht am flachen Strand mit
Strand-Therophyten durchsetzt ist, deren Samen der Wind mitge-
bracht hat (Tab. III, Aufn. 109).

Aufnahme 109 weicht durch ihre nitrophilen Begleiter von den
übrigen Beständen ab. Das erklärt sich aus ihrer Lage vor einer
älteren Erosionskante eines Bestandes (Tab. IIA, Aufn. 110) der
Poa irrigata - Potentilla anserina-Ass. mit dominierendem

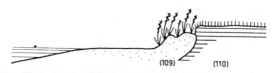

Abb. 6. Junges Honckenyo - Elymetum (109) in statu nascendi vor einer
Erosionskante einer *Agropyron repens*-Fazies der Poa irrigata-Potentilla
anserina-Ass. (110). W Eyrarbakki.

Agropyron repens, vor dem im Laufe sehr rezenter neuer Sand-Zufuhr
ein durch Arten des Atripici - Cakiletum "verunreinigter"
schmaler *Elymus*-Dünensaum sich aufzubauen im Begriff ist (Abb.
6). Die strandfernen *Elymus*-Dünen der Degenerationsphase möchten wir als Subassoziation von Silene maritima auffassen,
die zu der folgenden Gesellschaft, dem Sileno - Festucetum
*cryophilae (s.S. 268) überleitet. Die Differentialarten dieser Sub-
assoziation sind die Kennarten der Silene maritima-Festuca
rubra *cryophila-Ass., die sich bereits hier in der Degenerationsphase des Elymetum einzustellen vermögen, weil die Lebens-
und Konkurrenzkraft von *Elymus* (und *Honckenya*) mit der geringeren Sandzufuhr erheblich nachläßt. Die strandfernsten Bestände

der Subassoziation von Silene maritima können *Rumex acetosella* enthalten (Var. von Rumex acetosella), die vielleicht durch Seeschwalben-Kolonien begünstigt wird.

Junge sekundäre Initial-Bestände pflegen sehr artenarm zu sein, wie die folgende Aufnahme (78) in etwa 3,5 km Entfernung vom Strand nö Thorlakshöfn zeigt (Veg.-Bedeckung 30 %, 200 m²):

3.3 Elymus arenarius   1.3 Honckenya peploides

Nicht selten ist hier nur eine dieser beiden Arten auf diesen Sekundär-Dünen zu finden. Sie erinnern an die Inlands-Dünen nw Mödrudalur in etwa 450 m Meereshöhe, die sich als Ergebnis der starken Winderosion in pflanzenarmen Gebieten im Innern Islands bis zu 10 m hoch auftürmen. Herrn H. BÖTTCHER verdanken wir drei Aufnahmen aus diesem Gebiet (230, 228, 227 v. 10.8.67), denen wir eine weitere von STEINDÓRSSON (1967, Tab. XXXV, 1) von Sudurárbotnar (460 m NN) beifügen (Tab. IV).

TABELLE IV

Fragmente des Elymetum auf Inlands-Dünen

Nr.d.Aufn.:	230	228	227	XXXV,1
Veg.-Bedeckung (%):	40	40	35	.
Artenzahl:	2	2	3	3
Elymus arenarius	3.5	2.3	2.1	100
Festuca rubra	2.2	3.2	3.2	100
Silene maritima	.	.	+	.
Equisetum arvense	.	.	.	10

Diese Bestände darf man wohl als Binnenlands-Variante zur Degenerationsphase unserer Assoziation rechnen, der die Küsten-Arten *Honckenya peploides* und *Mertensia maritima* fehlen.

Bestände der Degenerationsphase der Elymus-Assoziation (Subass. von Silene maritima) können je nach Ausbildung der Dünen weiter entfernt vom Strande oder ihm näher liegen. Weitere Entfernung deutet auf neuerlichen Sandanwuchs (z.B. w Eyrarbakki), geringer Abstand oder gar unmittelbarer Kontakt bezeugt kürzlichen oder rezenten Dünenabbruch.

Die Verbreitung der Elymus-Gesellschaft in Island ist noch ganz ungenügend bekannt. Außer unseren Beobachtungen haben wir (abgesehen von den Floren) nur wenige Angaben finden können, so bei STEINDÓRSSON (1954, p. 205) von der N-Küste Islands (Gásar am Eyjafjördur). STEINDÓRSSON (1937, p. 127) erwähnt "tufts of *Elymus arenarius* and *Silene maritima*" aus Fjallabaksvegur, S-Island. (Vergl. auch eine komplexe Liste bei GRÖNTVED (1942,

p. 43, die nach Jónsson 1895 zitiert wird). Hadač (1949, p. 12) gibt das "Elymetum arenariae" von den Dünen der Reykjanes-Peninsula an. Es ist anzunehmen, daß sich die Verbreitung der Elymus-Gesellschaft mit derjenigen der Dünen, wenigstens an der Küste, deckt.

Die treffende Benennung der isländischen Elymus-Gesellschaft ist nicht leicht zu finden, weil auch außerhalb Islands erst wenig über die subarktischen Elymus-Gesellschaften bekannt ist. Immerhin lassen sich aus den bis jetzt vorliegenden Florenlisten und soziologischen Angaben gewisse Vorstellungen über die Soziologie der Elymus-Honckenya-Gesellschaften der gesamten Subarktis gewinnen. Vereinigt man alle diese Angaben zu einer Tabelle, die wir später an anderer Stelle vorlegen werden, und berücksichtigt man zugleich die Ergebnisse der Sippensystematik (vgl. Meusel, Jäger & Weinert), so erhält man das folgende Schema, aus dem neben den Arten-Verbindungen (ohne die wenig steten Begleiter) die Areale der verschiedenen Gesellschaften abgelesen werden können. Zugleich ergeben sich daraus auch die zweckmäßigsten Namen der verschiedenen Territorial-Assoziationen.

Tabelle V

Systematik der Honckenyo peploides - Elymetea

Soncho brachyotis-Elymetum mollis	..... — Elymetum mollis	Honckenyo robustae-Elymetum mollis	Honckenyo diffusae-Elymetum mollis	Honckenyo diffusae-Elymetum arenariae	Honckenyo latifoliae-Elymetum arenariae
N-Japan Kurilen	N-Pazifik	O-Kanada	Grönland	Island, Eismeer-Küste	Nordsee Ostsee

Elymus mollis

Elymus arenarius

Honckenya peploides

ssp. major

ssp. robusta

ssp. peploides
var. diffusa

var. latifolia

div. spec.
(Tx. 1966)

Lathyrus maritimus

Mertensia maritima

Unsere Übersicht erlaubt die kürzlich (Tüxen 1966) vorgelegten Honckenyo peploides-Elymetea-Klasse schärfer zu fassen und weiter zu gliedern.

Das Potentillo-Elymetum arenariae (Raunkiaer 1965) Tx. 1966 muß aus diesem Verband, wie schon vermutet, herausgenommen werden und beim Agropyro-Rumicion bleiben.

In der Gruppe der Elymus mollis-Assoziationen hebt sich das Soncho brachyotis-Elymetum mollis (Tatewaki 1931) Tx. 1966 durch seine japanischen Trennarten *Sonchus brachyotis*, *Artemisia montana* und *Oenothera* spec. sehr klar heraus[1]).

Die Elymus mollis-Gesellschaft der nördlichen Pazifik-Küsten kann wegen des Fehlens von Aufnahmen noch nicht benannt (oder zugeordnet) werden (vgl. jedoch R. Knapp 1965, p. 187/8).

Das kanadische Honckenyo robustae-Elymetum mollis (Dansereau 1949) Fernando Galiano 1959 zeigt durch die Verbindung namengebender Arten und sein Areal genügende Selbständigkeit.

Auf Grönland wächst das Honckenyo diffusae-Elymetum mollis, welches das Bindeglied zu den europäischen Elymus arenarius-Gesellschaften darstellt.

Unsere isländische Elymus-Assoziation, die wohl mit derjenigen der fennoskandischen Eismeer-Küste gleichzusetzen ist, kann wohl nur als Honckenyo diffusae-Elymetum arenariae bezeichnet werden. Die nahe liegende Verwendung von *Mertensia maritima* verbietet sich, weil diese Art in allen anderen bisher genannten Elymeten ebenfalls vorkommt. Sie ist keineswegs eine "typische Geröllpflanze", wie Nordhagen (1940, p. 78) sie wertet. Auch der von ihm erwähnte Name "Mertensietum", der allerdings nie durch eine Gesellschaftsdiagnose erläutert worden ist, sollte besser aufgegeben werden.

Auch der Name Elymo-Festucetum arenariae subarcticum (Regel) Nordhagen 1955 scheidet aus, denn die nur in der Degenerationsphase (Subass. v. Silene maritima), nicht aber in der Optimalphase auftretende *Festuca* ist sicher nicht *Festuca *arenaria* sondern eine andere Art, die im südwestlichen Island schon von Hadač nachgewiesene *Festuca *cryophila*. Im Bereich der Nordsee- und Ostsee-Küsten endlich bilden *Elymus arenarius*-Bestände mit *Honckenya peploides* ssp. *peploides* var. *latifolia* die letzten nach S vordringenden Vorposten unserer subarktischen Honckenyo-Elymetea-Klasse. Wir können diese verarmte Gesellschaft nur als Honckenyo latifoliae-Elymetum arenariae bezeichnen.

---

[1]) In der Übersichtstabelle (Tüxen 1966, p. 366) sind die Stetigkeits-Angaben für diese Ass. z.T. leider falsch eingesetzt worden. Sie können leicht nach der Original-Tabelle (p.363) berichtigt werden.

4. SILENE MARITIMA - FESTUCA *CRYOPHILA-ASS.
(Sileno-Festucetum cryophilae)

Die Degenerationsphase des Honckenyo diffusae-Elymetum arenariae. d.h. seine Subass. von Silene maritima, kann nur so lange bestehen bleiben, wie genügend frischer Sand durch den Wind herangeführt wird, der *Honckenya peploides* und vor allem *Elymus arenarius* ernähren kann. Wo die Sandzufuhr nachläßt oder ganz fehlt, gehen *Honckenya* und *Elymus* immer mehr zurück. Die

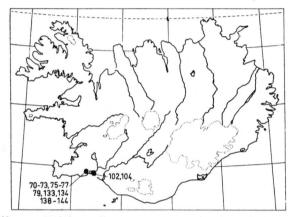

Karte 4. Silene maritima-Festuca *cryophila-Ass. Lage der Aufnahmen.

Trennarten der Subass. von Silene maritima, vermehrt um einige andere Arten (s.Tab. VI), bleiben allein in kleinen Vertiefungen der dunklen Lavadecke oder zwischen Geröll in geringen Mengen von Sand und Grus am Leben. Vor allem *Festuca *cryophila* vermag auch wohl die nicht mehr so zahlreichen vom Winde hergewehten Sandkörner zu winzigen Miniatur-Dünen zu sammeln. Die Vegetationsbedeckung ist hier zunächst nur sehr gering. Sie erreicht nicht oft mehr als ein Viertel der Fläche und kann nur ausnahmsweise bis zur Hälfte ansteigen. Nur die in der Nähe des

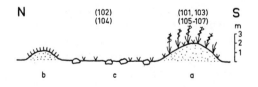

Abb. 7. Einzelne *Elymus*-(a) und *Honckenya*-(b) Dünen in der Silene maritima-Festuca *cryophila-Ass. auf einem Lava- und Geröll-Feld w Eyrarbakki. a. *Elymus*-Düne. b. *Honckenya*-Düne. c. Silene maritima-Festuca *cryophila-Ass. Die Zahlen geben die Lage der Aufnahmen an (Tab. III u. Tab. VI).

geschlossenen Elymetum-Gürtels noch einzeln auftretenden niedrigen *Elymus*-Dünenkegel, die scharf begrenzt auf dem äußerst locker bewachsenen dunkelgrauen Lava- oder Grob-Geröllfeld sich aufwölben, tragen ein dichteres Pflanzenkleid (Abb. 7).

Um diese Hügel findet *Rumex acetosella* stellenweise beste Wuchsmöglichkeiten. Sie bildet große kreisförmige Herden (bis 5 und mehr m²), die zu verschiedener Zeit blühen und fruchten, was sich schon aus der Ferne in der Farbe dieser Flächen zeigt: Während manche leuchtend rot sind, haben andere bereits ihre Früchte verloren und erscheinen nur noch im fahlen braungelb auf dem dunkelgrauen Sande. Diese Bestände, die ziemlich dicht geschlossen sein können, sind w der Ölfusá-Mündung das Brutgebiet von Küsten-Seeschwalben (*Sterna macrura*).

Die sorgfältige Ordnung der Tabelle VI, (die durch mehrere Teil-Tabellen der Trennarten vorbereitet wurde), zeigt nun eine Reihe offenbar gesetzmäßiger Zusammenhänge, die Einblick in die Struktur und die Syndynamik, die Synoekologie (und in Verbindung mit einer Karte, auch großen Maßstabs, in die Verbreitung (Synchorologie)) der Gesellschaft und ihrer Untereinheiten geben.

Es muß allerdings gesagt werden, daß unsere Aufnahmen von zu wenigen Orten stammen, um Endgültiges über diese Assoziation aussagen zu können. Vielmehr ist anzunehmen, daß in unserer Tabelle gewisse örtliche Eigenarten überbetont sind, welche die regionalen Züge überlagern. Wir wollen daher die Tabelle VI durchaus als *vorläufig* bewerten, die mehr das Methodische als das endgültige Gliederungsergebnis darstellen soll.

Im Großen sind die Aufnahmen nach ihrer Entfernung von der Küste oder mit anderen Worten nach der abnehmenden Stärke und Menge des Sandfluges geordnet. Richtiger gesagt läßt sich dieser Einfluß aus der getroffenen Anordnung in der Tabelle ableiten, denn die Ordnung der Aufnahmen erfolgte *lediglich* nach dem Auftreten oder Fehlen und nach dem Zusammengehen oder Sichausschließen von Arten. Danach lassen sich drei Untereinheiten (Subassoziationen) unterscheiden, die durch die Kennarten (Ch) und Trennarten (D) der Assoziation: *Festuca* *cryophila (Ch), *Silene maritima* (Ch) und *Armeria maritima* (D) zusammengehalten werden. Unter diesen Arten dürften auch die Arten der höheren Einheiten (Verband, Ordnung, Klasse) zu suchen sein, über die wir noch nichts angeben können. Wir wollen die Gesellschaft nach den beiden stetesten dieser Arten als Sileno maritimae-Festucetum cryophilae bezeichnen.

1. TYPISCHE SUBASSOZIATION

Die Typische Subassoziation hat keine eigenen Trennarten (Differential-Arten). Sie unterscheidet sich durch die geringe Arten-

TABELLE VI

Silene maritima-Festuca *cryophila Ass. (Thoroddsen 1914) Tx. 1968
A. Typische Subass.; B. Subass. v. Poa *irrigata (prov.); C. Subass. v. Rhacomitrium canescens

	A													B	C				
Nr. d. Aufn.:	70	76	72	71	75	73	77	79	138	139	102	104	141	134	133	140	142	143	144
Probefläche (m²):	·	50	30	50	50	50	15	30	40	40	·	100	50	70	100	4	·	70	·
Veg.-Bedeckung (%):	35	20	30	20	15	20	15	30	70	40	50	45	65	70	60	80	70	70	90
Artenzahl:	6	5	6	7	6	5	7	7	6	6	7	7	12	9	23	20	16	17	22
Kenn- u. Trennarten d. Ass.																			
Festuca cryophila	3.3	2.2	2.3	2.3	2.2	2.2	2.2	3.3	4.3	2.3	1.2	1.2	3.3	3.3	4.3	4.3	2.2	2.2	2.2
Silene maritima	2.3	1.2	2.3	2.2	2.2	2.2	1.2	1.2	2.2	2.2	2.2	3.2	2.2	2.2	1.2	+.2	2.2	1.2	+.2
Armeria maritima	+.2	2.2	·	2.2	2.2	1.2	+.2	1.2	1.2	1.2	·	·	1.2	+	+.2	+.2	2.2	+	+
Trennarten d. Subass. u. Var.:																			
Cardaminopsis petraea	·	·	·	·	·	·	·	+	·	·	·	·	+	·	+	(+)	·	·	·
Syntrichia ruralis	·	+.2	·	+.2	+.2	+.2	+	·	·	·	·	·	·	·	1.1	·	·	·	·
Agrostis stolonifera var. mar.	·	·	·	·	·	·	·	·	1.2	2.2	r.2	+.2	2.2	·	·	·	1.2	1.2	·
Plantago maritima et **fo.**	·	·	·	·	·	·	·	+	+	**2.2**	+	+.2	1.2	·	+.2	**1.2**	**1.2**	+	·
Poa pratensis ssp. irrigata	·	·	·	·	·	·	·	·	·	·	·	·	·	2.1	·	·	·	·	·
Rumex tenuifolius	·	·	·	·	·	·	·	·	·	·	·	·	·	1.1	·	·	·	·	·
Viola canina	·	·	·	·	·	·	·	·	·	·	·	·	·	2.1	·	·	·	·	·
Silene acaulis	·	·	·	·	·	·	·	·	1.2	·	·	·	2.2	·	·	2.2	2.2	+.2	1.2
Luzula spicata	·	·	·	·	·	·	·	·	·	·	·	·	1.2	·	+	+.2	+.2	1.2	1.2
Rhacomitrium canescens	·	·	·	·	·	·	·	·	·	·	·	·	·	·	+.2	+.3	+.2	5.5	5.5
Stereocaulon paschale	·	·	·	·	·	·	·	·	·	·	·	·	·	·	1.2	+.2	+.3	+.2	+.2
Empetrum spec.	·	·	·	·	·	·	·	·	·	·	·	·	·	·	+	+.2	·	+.2	+.2
Poa glauca	·	·	·	·	·	·	·	·	·	·	·	·	·	·	+	·	+.2	+.2	+.2

Begleiter:

Rumex acetosella	2.2	.	+	+	.	.	3.2	3.4	1.2	.	1.2	1.2	4.3	.	+
Elymus arenarius	+°	+	+	+°	.	+	.	2.1°	2.1°	.	+.2	+.2	2.3	.	.
Thymus drucei	.	1°	.	.	+°	2.1°	.	.	.	2.2	.	2.3	2.3	1.2	+.2
Honckenya peploides	+.2	+	.	+.2	+.2	.	.	.	.	+	.	+	+	1.2	+.2
Equisetum arvense var. alp.	2.1	.	.	.	2.1	.	1.3	1.2	.	+	.	+	.	.	+.2
Cornicularia aculeata	.	.	.	.	.	.	.	.	1.2	2.2	1.2	.	.	2.2	1.2
Trisetum spicatum	.	.	.	.	.	.	.	+.2	.	.	+.2	+.2	.	.	.
Pogonatum urnigerum	.	.	.	.	.	.	.	.	.	+.2	+.2	+.2	.	.	+.2
Polytrichum piliferum	.	.	.	.	.	.	.	.	.	1.2	+.2	.	.	.	+.2
Cerastium caespitosum	.	.	.	.	.	.	1.2	.	.	.	1St	+	.	+	.
Tripleurospermum maritimum	.	.	.	.	+	.	.	.	.	+.2	.	.	.	.	+.2
Rhacomitrium lanuginosum	.	.	.	.	.	.	.	.	.	+	.	+	+	.	+
Botrychium lunaria	.	.	.	.	.	.	.	.	.	+	.	.	+	.	+
Galium verum	.	.	.	.	.	.	.	.	.	2.1	.	.	.	2.2	+.2
Equisetum pratense	.	.	.	.	.	.	.	.	.	.	.	.	.	.	.
Festuca vivipara	.	.	.	.	.	.	.	.	.	.	+.2	+.2	.	2.2	.
Schistidium maritimum	.	.	.	.	.	.	.	.	.	.	.	.	.	2.4	2.3

Außerdem in Aufn. 133: +.2 Barbula fallax, + Cladonia furcata, 1.3 Peltigera canina, 1.3 Agrostis canina ssp. pusilla, +.2 Agrostis tenuis; in 140: 1.3 Musci, in 142: Encalypta ciliata; in 143: + Sagina nodosa; in 144: + Euphrasia frigida.

Fundorte:

70	7.8.68	Flache Delle in den Dünen ö Thorlákshöfn etwa 100 m vom Strand, Oberfläche schwach wellig.
76	7.8.68	Dünental mit Steinen in der Nähe zwischen *Honckenya-Elymus*-Kegeln. *Festuca*-Keimlinge.
72	7.8.68	Sandrücken weiter binnenlands.
71	7.8.68	Neben voriger ausgeblasenes Dünen-Tal mit etwas Sand auf Lava und Geröll.
75	7.8.68	Abhang einer Lava-Kuppe mit Lava-Brocken und Sand. Viele *Festuca*-Keimlinge.
73	7.8.68	Dünental in der Nähe von 72 mit Lava-Brocken und etwas Sand. Viele *Festuca*-Keimlinge.
77	7.8.68	4 km von der Küste nö Thorlákshöfn. Zwischen Basaltlava-Rundhöckern steinige Delle mit 15 % Sand. Alte Erosionsfläche.
79	7.8.68	4,5 km von der Küste nö Thorlákshöfn zwischen spärlichen Sandhügeln mit *Elymus* auf Basalt-Steinpflaster mit 85 % Steinen und 15 % Sand
138/139	12.8.68	Schwache Dünen- oder Flugsand-Decken auf Lava-Plateau w Thorlákshöfn in zunehmender Entfernung vom Kliff.
102	9.8.68	W Eyrarbakki in Ausblasungsebene zwischen Dünen mit Honckenyo diffusae-Elymetum (Abb. 7).
104	9.8.68	desgl.
141	12.8.68	Nahe 138. Übergang zur Subass. von Rhacomitrium canescens.
134	12.8.68	Flache Dünen (1 m hoch) am N-Eingang von Thorlákshöfn. *Festuca* ohne Blüten. Keine Moose.
133	12.8.68	Am nw Rand von Thorlákshöfn (gegenüber der Auto-Werkstatt). Niedrige flache Sand-Polster über Geröll-Ebene.
140	12.8.68	Flache Dünen auf Lava-Platte sw Thorlákshöfn.
142	12.8.68	In der Nähe zwischen kopfgroßen Lava-Blöcken auf Sand.
143	12.8.68	In der Nähe auf Lava-Platte mit wenig Sand.
144	12.8.68	In der Nähe des Dorfes Thorlákshöfn.

zahl ihrer Bestände, die nur ausnahmsweise höher als 5—7 liegt, von den übrigen Subassoziationen. Der Deckungsgrad der Vegetation ist hier gering, jedoch in den verschiedenen Varianten von wechselnder Dichte. In syndynamischer Hinsicht stellt die Typische Subass. die Initial-Phase der Assoziation dar, was sich schon im Gelände aus der strandnäheren Lage der Bestände und in der Tabelle aus dem Vorkommen der aus dem in der Entwicklung vorhergehenden Honckenyo-Elymetum noch übrig bleibenden Relikte, *Elymus arenarius* und *Honckenya peploides* erkennen läßt. Diese zeigen, daß noch immer eine gewisse Menge von Flugsand in die Wuchsorte der Typischen Subass. eingeblasen oder hier umgelagert wird, die sie zu ihrer Ernährung braucht. Allerdings ist die Vitalität und die Konkurrenzkraft dieser Arten hier nur noch gering. Sie werden hier durch die weniger sandbedürftigen Arten der Silene-Festuca-Assoziation verdrängt.

Innerhalb der Typischen Subass. lassen sich drei Varianten mit verschiedener Artenverbindung unterscheiden. Wir messen dabei dem Auftreten oder Fehlen einer Art im allgemeinen mehr Gewicht bei als ihrer Frequenz oder Menge, die als quantitatives Merkmal im Ganzen weniger aussagt als das qualitative der Arten-Verbindung.

Die Typische Var. (Aufn. 70, 76) hat keine eigenen Trennarten. Sie schließt unmittelbar an das Honckenyo-Elymetum an, das sich oft inselförmig fragmentarisch auf kleinen Dünen-Kegeln noch im Bereich dieser Variante erhebt (Abb. 7).

Wahrscheinlich ist der Untergrund (Geröll oder rissige Lava-Decken) die Ursache für das Zustandekommen der Variante von Cardaminopsis petraea. Sie kommt mit weniger Flugsand aus. Ihr Deckungsgrad ist dementsprechend im Ganzen eher niedriger (15—30 %). Mit zunehmender Entfernung von der Küste stellt sich *Equisetum arvense* in der prostraten var. *alpestre* ein (Subvariante?). Wir fanden die Var. von Cardaminopsis im Osten von Thorlákshöfn in sehr großer Ausdehnung, immer wieder von einzelnen letzten *Elymetum*-Dünen durchsetzt.

In der Agrostis stolonifera-Variante, die durch niedrige Horste dieses Grases und *Plantago maritima* unterschieden wird, erreicht die Vegetationsbedeckung die höchsten Werte in der Typischen Subass., die 50 % nicht selten übersteigen, obwohl die Sandzufuhr hier nicht größer ist, als in der vorigen Variante. Aber die dichtere Pflanzendecke und die Halophilie der Trennarten legen die Vermutung nahe, daß hier vielleicht durch Salzwasser-Spray eine gewisse Zufuhr von Nährstoffen stattfindet, die in der Fazies (Subvar.?) von Rumex acetosella (Aufn. 102, 104) noch durch Seeschwalben-Kolonien (*Sterna macrura*) gesteigert wird.

Mit wachsender Entfernung vom Strande tritt eine gewisse Alterung der Gesellschaften ein, die zur Bildung von saurem Humus führen kann, so daß azidophile Arten wie *Luzula spicata* und *Silene acaulis* einwandern können, welche die Degenerationsphase der Agrostis stolonifera-Variante einleiten (Subvar. v. Luzula spicata). Diese ist nichts anderes als der erste Schritt zur Bildung der Subass. von Rhacomitrium canescens.

2. SUBASS. VON RHACOMITRIUM CANESCENS

Diese Ausbildung des Sileno maritimae-Festucetum cryophilae ist auf dem Lava-Feld nw Thorlákshöfn weit verbreitet. Fragmentarisch sahen wir sie auch am n Rande des Flugsandfeldes n dieses Dorfes. Hier erreichen sowohl die Deckung als auch die Artenzahlen die höchsten Werte im Bereich der ganzen Assoziation. Dennoch ist das Minimum-Areal hier meist viel kleiner als in der

Typischen Subass., wo man gut tut, nicht unter 50 m² zu untersuchen, um die Artenverbindung vollständig zu erfassen. In der Rhacomitrium-Subass. können verschiedene Arten (*Festuca *cryophila*, *Rumex acetosella* oder *Rhacomitrium canescens*) vorherrschen und auch andere erhebliche Mengen erreichen. Hier kann zum ersten Male von eigentlichen sozialen Beziehungen der Arten gesprochen werden, die in der Typischen Subass. bei der geringen Deckung der Vegetation nur schwach erkennbar sind.

Gewiß lassen sich auch in dieser Subass. verschiedene Varianten unterscheiden, wovon wir aber aus Mangel an Aufnahmen von verschiedenen Orten absehen. Nur auf die Rhacomitrium-Fazies sei besonders verwiesen, in der die Gräser *Trisetum spicatum* und *Agrostis tenuis* wachsen, während *Cardaminopsis petraea* und *Polytrichum piliferum* dort fehlen.

Diese Subass. lebt nicht mehr unter so extrem harten Bedingungen wie die Typische, wie aus Deckungsgrad und Artenzahl hervorgeht. Sie nähert sich den Zwergstrauch-Heiden der küstenferneren Gebiete, in denen mehrere der Trennarten und verschiedene Begleiter regelmäßig vorkommen. Über die Weiterentwicklung dieser Subass. konnten wir keine Beobachtungen anstellen.

3. SUBASS. VON POA *IRRIGATA (PROV.)

In unmittelbarer Nähe von Thorlákshöfn bildet das Sileno maritimae-Festucetum cryophilae flache bis 1 m hohe Dünen-Polster von beträchtlicher Größe, die ziemlich dicht von Vegetation durchwachsen und bedeckt sind. Leider haben wir nur eine Aufnahe (134) dieser Ausbildung, die sich von allen anderen der Tabelle durch mehrere Differential-Arten (*Poa pratensis* ssp. *irrigata*, *Rumex tenuifolius* und *Viola canina*) unterscheidet. Diese floristische Abweichung, verstärkt durch die ökologische Sonderstellung starker Sandzufuhr (*Elymus*!) lassen die Möglichkeit zu, hier eine besondere Subass. aufzustellen, die allerdings durch weitere Aufnahmen bestätigt werden müßte. Diese Gesellschaft zeigt zu keiner der anderen Subassoziationen nähere Beziehungen. Auch von ihr kennen wir noch nicht die weitere Entwicklung.

Die Silene maritima-Festuca *cryophila-Ass. ist schon früher bemerkt, wenn auch nicht näher beschrieben worden. JÓNSSON (1900) nannte bereits unter Hinweis auf ältere Arbeiten die wichtigsten Arten, THORODDSEN (1914, p. 331) gab eine umfassendere Liste, ohne die Gesellschaft zu benennen, und STEINDÓRSSON (1951) führte die Namen mehrerer Soziationen an (249—252), die wohl zu unserer Ass. zu rechnen sein dürften. *Festuca *cryophila* wurde allerdings in allen Beschreibungen als *Festuca rubra* oder *Festuca arenaria* bezeichnet. Dieses südliche Taxon dürfte auf Island fehlen. HADAČ (1949) nannte *Festuca cryophila*.

Insgesamt vertritt die Silenemaritima-Festuca cryophila-Ass. in Island die dort fehlenden Sedo-Festucetalia-Gesellschaften, insbesondere den Koelerion albescentis-Verband. Es ist erwähnenswert, daß die wenigen Arten dieser Ordnung und dieses Verbandes, die Island erreicht haben, auch in dieser Assoziation leben, wenn auch nicht ausschließlich: *Syntrichia ruralis*, *Rumex tenuifolius*, *Rhacomitrium canescens*, *Polytrichum piliferum*, *Galium verum*, *Agrostis canina* ssp. *pusilla* (=*A.c.* var. *arida*) u.a.

Sie reichen aber, da sie nur ganz spärlich auftreten, bei weitem nicht aus, unsere Assoziation mit der südlicheren Ordnung der Festuco-Sedetalia synsystematisch zu vereinigen.

Sowohl aus wissenschaftlichen, aber auch aus wirtschaftlichen Gründen wäre nach einer Nachprüfung der von uns provisorisch aufgestellten Untereinheiten der Silene maritima-Festuca cryophila-Ass. durch weitere Aufnahmen und Tabellen aus anderen Gebieten die Erarbeitung einer Vegetationskarte etwa des Gebietes w und ö der Ölfusá-Mündung sehr wünschenswert. Sie würde klar die gesetzmäßige Zonierung der Untereinheiten erkennen und damit ihre syndynamische sowie synökologische Stellung besser verstehen lassen. Sie würde aber zugleich die beste Grundlage oder den Schlüssel für die gerade in diesem Gebiet notwendige und, soweit wir wissen, schon versuchte Festlegung des Flugsandes liefern!

Diese wird ebenso leicht wie sicher und schnell gelingen, wenn den soziologischen und synökologischen Tatsachen Rechnung getragen wird. Im anderen Falle bleibt sie Glückssache. Dabei ist zu beachten, daß die Pflanzung von *Elymus* nur dort Sinn hat und Erfolg verbürgt, wo genügend Sandzufuhr durch den Wind zu seiner Ernährung möglich ist. Eine NPK-Volldüngung nach dem Verfahren von Lux (1964) ist wertvoll, ja verbürgt erst vollen Erfolg an Stellen, wo weniger Sand hinkommt. Bodenanalysen müßten natürlich der Düngung vorausgehen um ihre Menge festzusetzen. Einheimisches Pflanzgut ist Vorbedingung und könnte leicht gewonnen werden.

Im Bereich des Sileno-Festucetum sollte *Festuca *cryophila* gesät werden. Unbedingt ist auch dafür einheimisches Saatgut zu sammeln und zu verwenden. Die Saatmenge braucht 1—3 gr/m² wohl nicht zu übersteigen. Ein vorübergehender Windschutz während der Keimzeit wäre anzustreben. Gedüngt wird wie oben angegeben. Die Beweidung der Ansaat-Flächen wäre unbedingt zu verhindern.

Zu empfehlen wäre im 1. oder 2. Jahr die Bearbeitung kleiner Probeflächen, um die Methoden der Samengewinnung und der Saatmenge, der Saat-Technik usw. (vergl. LEVSEN 1961, LOHMEYER 1961 a,b, 1964, LUX 1964, 1965) in die Hand zu bekommen.

Unser Beispiel zeigt die Fruchtbarkeit der Anwendung unserer pflanzensoziologischen Methoden auf Island, die in allen anderen

Pflanzengesellschaften des Grünlandes, der Heiden, der Wälder in entsprechender Weise, aber noch weit vielseitiger nutzbar zu machen sind. Möchten sie recht bald von jungen, isländischen Forschern zum Wohl ihres Heimat-Landes gebraucht werden und reiche Früchte tragen.

Aber auch für rein wissenschaftliche Vergleiche wird die Kenntnis der Strand-Vegetation SW-Islands von Wert sein. Auf der jungen Insel Surtsey, die 1964 durch einen Vulkan-Ausbruch im Meer sw der Vestmannaeyjar entstand und seither einer sehr sorgfältigen vielseitigen naturwissenschaftlichen Dauer-Beobachtung laufend unterliegt (FRIDRIKSSON 1964—1967), vollzieht sich der Einzug der Pflanzendecke. Bis jetzt sind die Pioniere des Atriplici-Cakiletum islandicae und des Honckenyo-Elymetum eingetroffen. Es wird von hohem wissenschaftlichen Interesse sein, die Weiterentwicklung dieser Erstbesiedler zu Gesellschaften zu verfolgen. Wir wissen dieses Problem bei unseren isländischen Freunden in besten Händen!

ZUSAMMENFASSUNG

Nachdem die Therophyten- und Hemikryptophyten-Spülsaum-Gesellschaften des Strandes geschildert worden sind, wird die Honckenya diffusa-Elymus arenarius-Ass. der isländischen Küste und der Flugsandflächen im Binnenlande beschrieben. Die weitere Entwicklung der Vegetation auf den Flugsand-Feldern führt zur Silene maritima-Festuca cryophila-Ass., die in verschiedenen Subassoziationen vorkommt.

Die Wünschbarkeit einer Nachprüfung der örtlich gewonnenen Ergebnisse wird dargelegt und die pflanzensoziologische Kartierung eines ausgedehnten Flugsandgebietes beiderseits der Ölfusá-Mündung in SW-Island empfohlen.

Die Festlegung des Flugsandes, die zur Sicherung der benachbarten Siedlungen notwendig ist, wird am sichersten, schnellsten und billigsten durch die Kenntnis und Beachtung der syndynamischen Gesetze der hier wachsenden Pflanzengesellschaften erreicht. Dabei sind Vegetation und Standort, d.h. die Flugsand-Dünen als eine lebendige Einheit im Sinne von VAN DIEREN (1934) zu betrachten und die praktischen Erfahrungen an der deutschen und dänischen Nordsee-Küste von Wert, auf die verwiesen wird.

Auf die Bedeutung der Kenntnisse von Soziologie, Syndynamik und Synökologie der s-isländischen Küsten-Vegetation für das Verständnis der Besiedlung der nahen jungen Vulkan-Insel Surtsey wird hingewiesen.

LITERATUR

DIEREN, J. W. VAN 1934 — Organogene Dünenbildung. Den Haag.

FRIDRIKSSON, ST. 1964 — Um adflutning lífvera til Surtseyjar. — The colonization of the Dryland Biota on the island of Surtsey off the coast of Iceland. *Náttúrufr.* 34(2). Reykjavík.

FRIDRIKSSON, St. 1965 — Fjörukál í Surtsey og fræflutningur á sjó. — The first species of higher plants in Surtsey the new volcanic island. *Náttúrufr.* 35(3): 97—102. Reykjavík.

FRIDRIKSSON, ST. 1966 — Melgresi í Surtsey. *Náttúrufr.* 36 (3): 157—158. Reykjavík.

FRIDRIKSSON, ST. & JOHNSEN, B. 1967 — The vascular flora of the outer Westman Islands. *Greinar* 4 (3). Reykjavík.

GELERT, O. & OSTENFELD, C. H. 1898 — Nogle Bidrag til Islands Flora. *Bot. Tidsskr.* 21 (3). København.

GRÖNTVED, J. 1942 — The Pteridophyta and Spermatophyta of Iceland. *Bot. Iceland* 4 (1). Copenhagen/London.

HADAČ, E. 1949 — The flora of Reykjanes Peninsula, SW-Iceland. *Bot. Iceland* 5 (1). København.

HANSEN, A. 1961 — Bibliographia phytosociologica: Island. *Exc. bot.* B. 3: 28—32. Stuttgart.

HYLANDER, N. 1955 — Förteckning över Nordens Växter. 1. Kärlväxter. Lund.

JÓNSSON, H. 1898 — Vaar – og Høst-Exkursioner i Island 1897. *Bot. Tidsskr.* 21 (3). København.

JÓNSSON, H. 1899 — Floraen paa Snæfellsnes og Omegen. *Bot. Tidsskr.* 22 (2). København.

JÓNSSON, H. 1900 — Vegetationen paa Snæfellsnes. Vidensk. Medd. fra den naturhist. Foren. i København 1900. København.

KNAPP, R. 1965 — Die Vegetation von Nord- und Mittelamerika und der Hawaii-Inseln. Stuttgart.

KNAUER, N. 1966 — Bericht über einige Pflanzengesellschaften Islands und deren Bedeutung für die Landwirtschaft. *Vegetatio* 13 (3): 148—171. Den Haag.

LEVSEN, P. 1961 — Die Begrünung der Dünen. Ein Erfahrungsbericht. *Angew. Pflanzensoz.* 17. Stolzenau/Weser.

LINDROTH, C. H. 1931 — Die Insektenfauna Islands und ihre Probleme. *Zool. bidrag från Uppsala* 13. 1930—1931.

LÖVE, Á. 1963 — Taxonomic botany in Iceland since 1945. *Webbia* 18. Firenze.

LOHMEYER, W. 1961 — Die pflanzensoziologischen Grundlagen für die Dünenbefestigung. *Angew. Pflanzensoz.* 17. Stolzenau/Weser (a).

LOHMEYER, W. 1961 — Kurzer zusammenfassender Bericht über die Exkursionen während der Tagung auf Amrum am 19. und 20. Juni 1958. *Ibid.* 17 (b).

LOHMEYER, W. 1964 — Über die künstliche Begrünung offener Quarzsandhalden im Bergbaugelände bei Mechernich. *Ibid.* 20. Stolzenau/Weser.

LUX, H. 1964 — Die biologischen Grundlagen der Strandhaferpflanzung und Silbergrasansaat im Dünenbau. *Ibid.* 20. Stolzenau/Weser.

LUX, H. 1965 — Flugzeugeinsatz zur Düngung der Amrumer Dünen. *Wasser ü. Boden* 12. Hamburg.

LUX, H. 1969 — Zur Biologie des Strandhafers (Ammophila arenaria) und seiner technischen Anwendung im Dünenbau. In: TÜXEN, R. (Edit.) Experimentelle Pflanzensoziologie. Ber. Int. Symp. Rinteln 1965. Den Haag.

McVEAN, D. N. 1955 — Notes on the Vegetation of Iceland. *Trans. Proc. Bot. Soc. Edinburgh* 36 (4). Edinburgh.

MEUSEL, H., JÄGER, E. & WEINERT, E. 1965 — Vergleichende Chorologie der Zentraleuropäischen Flora. Jena.

NORDHAGEN, R. 1940 — Studien über die maritime Vegetation Norwegens, I.

Die Pflanzengesellschaften der Tangwälle. *Bergens Museum Arsbok. Naturv. rekke* 2. Bergen.

NORDHAGEN, R. 1955 — Studies on some plant communities on sandy river banks and seashores in Eastern Finmark. *Arch. Soc. "Vanamo"*. 9. Suppl. Helsinki.

OSTENFELD, C. H. 1905 — Skildringer af Vegetationen i Island. *Bot. Tidsskr.* 27 (1). København.

POBEDIMOVA, E. G. 1963 — A review of the genus Cakile Mill. *Bot. J.* 48. Moskau.

STÄHLIN, A. 1960 — Beobachtungen an Grünland und Graslandansaaten in Island. *Das Grünland* 9 (12). Hannover.

STEFÁNSSON, ST. 1948 — Flóra Íslands. 3. Útg. Ed. St. Steindórsson. Akureyri.

STEINDÓRSSON, ST. 1936 — Om Vegetationen paa Melrakkasljetta i det nordøstlige Island. *Bot. Tidsskr.* 43 (6). København.

STEINDÓRSSON, ST. 1937 — Contributions to the plant geography and floristic conditions of Iceland. II. The flora at Fjallabaksvegur, South-Iceland. *Greinar* 1 (2). Reykjavík.

STEINDÓRSSON, ST. 1951 — Skrá um Íslenzk gródurhverfi. A list of icelandic plant associations. Akureyri.

STEINDÓRSSON, ST. 1954 — The coastline vegetation at Gásar in Eyjafjördur in the north of Iceland. *Nytt Magasin Bot.* 3. Oslo.

STEINDÓRSSON, ST. 1964 — Gródur á Íslandi. (Reykjavík).

STEINDÓRSSON, ST. 1966 — Bibliographia phytosociologica: Islandica. *Exc. bot.* B 7 (1): 1—4. Stuttgart.

STEINDÓRSSON, ST. 1967 — Um hálendisgródur Islands. 4. hluti. *Flóra* 5. Akureyri.

THORARINSSON, S. 1966 — Surtsey. Reykjavík.

THORODDSEN, TH. 1914 — An account of the physical geography of Iceland with special reference to the plant life. *Bot. Iceland* 1 (2), London.

TÜXEN, R. 1950 — Grundriß einer Systematik der nitrophilen Unkrautgesellschaften in der Eurosibirischen Region Europas. *Mitt. Flor.-soz. Arbeitsgem.* N.F.2. Stolzenau/Weser.

TÜXEN, R. 1966 — Über nitrophile Elymus-Gesellschaften an nordeuropäischen, nordjapanischen und nordamerikanischen Küsten. *Ann. Bot. Fenn.* 3 (3), Helsinki.

Manuskript eingegangen am 24. Oktober 1968.

Nach Abschluß dieser Arbeit sandte mr Prof. Dr. HADAČ, Průhonice, am 12. 6. 1969 ein Manuskript, in welchem er ebenfalls die Silene maritima-Festuca cryophila-Assoziation in zwei Subassoziationen, sowie das Honckenyo-Elymetum (in etwas anderer Ausbildung) beschreibt.

R. Tx.

11

PLANT SOCIOLOGICAL OBSERVATIONS OF ICELANDIC DUNE COMMUNITIES

Reinhold Tüxen

*This summary was prepared expressly for this Benchmark volume by Robert P. McIntosh from "Pflanzensozio-logische Beobachtungen an islandischen Dunengesell-schaften," Vegetatio **20**:251–278 (1970)*

The plant sociological research of Iceland generally followed the nordic methods probably influenced by the long-standing opinion that the vegetation could only be distinguished by dominance and frequency due to the lack of variety in the flora. Through several long journeys we have been able to repudiate this. Even in Lappland there are not a few communities which have, including cryptogams, far more species than many middle European ones. The number of communities decreases further north and the area of individual communities increases. Nevertheless, there are numerous original individual communities and the substitute societies created by man and his animals.

The therophyte and hemicryptophyte communities of the beaches (Atriplici-Cakiletum-Agropyro-Rumicion) are described, followed by descriptions of the Honckenya diffusae-Elymetum arenariae of the Icelandic coast and dune areas inland. For each the characteristic species (Kennarten) the differential species (Trennarten) and the accidental species (Begleiter) are given. Further evaluation of the vegetation in the dune fields leads to the *Silene maritima - Festuca cryophila* association which occurs in different subassociations.

The stabilization of the dune areas, which is necessary for the safety of the neighboring settlements, can be achieved, in the surest, fastest, and cheapest way by knowledge and recognition of the syndynamic laws of local plant communities. Vegetation and the dune location must be looked upon as a living unit and practical and valuable experience on other coasts have to be considered.

Part III

ORDINATION AND NUMERICAL CLASSIFICATION

Editor's Comments
on Papers 12 Through 17

The substance of the community type or traditional association concepts is that plant species occur together in more or less regularly repeating arrays that are homogeneous internally. The term valence was used in some early phytosociology studies to suggest the likelihood of combining the plant species to produce a limited number of the possible combinations of species. As in any biological classification, the problems faced by phytosociologists were the criteria by which the class or type was to be recognized and the degree of similarity required among members of the class or, alternatively, how much variation within a class was acceptable.

The familiar assumption of homogeneity of stands is that there is no trend of change within the boundaries of the stand or that there is a high degree of similarity from place to place given an adequate sample. The grouping of stands into an abstract association is also based on the similarity of the stands among themselves. Since any collection of samples or stands may differ in a large number of ways, such as by physiognomical, compositional, or habitat criteria, they may be studied and classified in an equally large variety of ways. Whittaker (1962) noted four criteria to be used in considerations of the association:

1. Relative similarity of stands.
2. Similarity of distributions of the component species.
3. Discontinuity between stands due to the similar distributions of component species.
4. Dynamic relations among species to account for the interrelations which bring about co-occurrence of the component species of the association.

Even so, the degree of similarity required among the stands of an association has largely been a matter of judgment.

One of the earliest approaches to quantitative study of similarity was the development of coefficients of community by Jaccard, Forbes and Gleason (McIntosh 1976). Such methods were developed independently by a number of ecologists (Curtis 1959) and have since formed the basis of extended quantitative studies. Among the most famous coefficients of similarity is that associated with Thorvald Sørenson and also independently developed by Polish phytosociologists (Kulczynski 1928). Sørenson's paper (Paper 12) reviews some of the earlier efforts in this field and develops the coefficient commonly associated with his name. Similarity coefficients include a variety of mathematical expressions, from the simple number of species in common between two stands or samples to more involved indices incorporating quantities of the species.

The measures of similarity should be correlated with the ecological similarity of the stands or samples (relevés) being compared. Goodall (1973a) reviews measures of similarity and correlation. Any similarity coefficient can be used to compare all possible pairs of samples or stands to produce a table of coefficients that may then be examined either graphically (McIntosh 1973) or in a variety of mathematical ways to arrange them into groups or orders.

Measures of similarity used in the ordination and classification of communities rest on the rationale that similar stands should be placed in the same classes, or close together in a system of ordination, since they represent similarity of distribution of species and of composition of stands. The converse of similarity is difference, and the overall problem of any classification or ordination system is to place similar things close together and dissimilar things apart to represent their ecological *distance.*

Vegetation samples or communities may be objectively related to each other in two ways: (1) Classification by grouping them into classes (community-types, formations, associations) which are discontinuous; (2) Ordination or arrangement of samples or communities in relation to one another on the basis of one or more gradients of environment, species, or physiognomic characteristics. Traditionally phytosociologists grouped communities on the basis of shared characteristics into community-types although many recognized the principle of species individuality and some arranged community-types into series related to the environment. The 1930s and 1940s saw increased interest in quantitative sampling and methods of data analysis (Goodall 1952). Phytosociologists continued their concern with sampling methods, quadrat number, size, shape, and placement; the interpretation of the resulting data, e.g., frequency, density, and cover measurements; and the synthetic considerations species-area curves, minimal area, interspecific association (co-occurrence), and correlation; and the difficult problems of spatial distribution of individuals and homogeneity of the community (Curtis and McIntosh 1950, Greig-Smith 1964, Goodall 1970).

The concept of vegetation as a continuum, the development of more rapid and adequate sampling methods, and the introduction of numerical methods for establishing similarity produced, in the late 1940s and the 1950s, a resurgence of interest in the largely submerged individualistic concepts of Gleason and Ramensky (Ponyatovskaya 1961, McIntosh 1967, Whittaker 1967, Sobolev and Utekhin 1973). In the English-language literature this was evidenced in the broad-scale criticism by Robert Whittaker (Paper 13) of the plant association and climatic-climax concepts which dominated American phytosociology. Whittaker based his criticism on his own thesis studies in the Great Smoky Mountains. In these studies he stressed the technique of *gradient analysis,* which distributes samples along environmental gradients (e.g., altitude, moisture) without regard to supposed associations. From these investigations he developed the idea that species population distributions form a series of separate but overlapping binomial curves along the gradient. Whittaker disagreed with the concept of the climax association as a regional climax and argued that the climax was better viewed as a climax pattern, changing from site to site, and forming a "complex continuum of plant populations" (cf. Whittaker 1953, 1956, 1967, 1972, 1973).

In the same year, 1951, and with strikingly similar phrasing to Whittaker's, Curtis and McIntosh (Paper 14) described what they called a "vegetational continuum." Curtis and McIntosh posed three alternative hypotheses about the upland forests of southern Wisconsin. They found that tree species occurred in a continuously shifting series of combinations and stated, "such a gradient of communities is here called a vegetational continuum" to differentiate it from the conventional concept of the plant association. These studies were based on an arrangement of stands according to their species composition which was called continuum analysis (McIntosh 1958) and subsequently was termed "indirect gradient analysis" by Whittaker (1967). The difference between the two approaches is that one, now called "direct gradient analysis," orders the phytosociological data on the basis of some external criterion, which is usually an environmental measurement. The other, now called "indirect gradient analysis," uses the phytosociological data themselves to establish an order showing the relationship among stands. Studies of either type are now familiarly described as *ordinations,* as proposed by Goodall (Paper 15).

Increasing interest in and use of the methods of statistical analysis have produced extended studies of numerical methods based on the presence and quantity of species in large numbers of samples or stands. The principal aim of these studies was to produce more objective methods of looking at phytosociological data. Also the early efforts at ordination produced linear one-dimensional axes on which the relations of species or species and environment were displayed. It was apparent that such unidimensional ordinations were inadequate (McIntosh 1957) and several phytosociologists explored methods of multidimensional analysis of vegetational data (Peet and Loucks 1977).

Goodall (Paper 15) introduced the use of factor analysis (specifically "principle axes") to produce an "arrangement of the vegetational data in a multidimensional series." It should be noted that he had previously (Goodall 1953) used the same data in studies of methods of numerical classification. Ordinations were widely regarded as distinct conceptually and methodologically from the traditional classifications of phytosociology and extended discussion ensued about their relative merits in analyzing vegetation. The development and use of numerical classification or clustering methods of analysis by Goodall and Williams and Lambert (1959), and others, paralleled the use of ordination methods and continues in present-day phytosociology (Lambert and Dale 1964, Van der Maarel 1969, Groenewoud 1965, Goodall 1973b). They are not mutually exclusive and may be complementary, although opinions differ on their relative merits.

Bray and Curtis (Paper 16) considered the application of factor analysis to their forest data but decided to use a more empirical geometric method of constructing ordination axes. They used the Sørensen coefficient as a basis of interstand compasrison and developed a technique for selection of pairs of "reference stands" for determination of the poles of an axis. Given the interstand similarities (or their complement, *distance*) the positions of other stands could be determined by a straightforward geometric triangulation procedure. The Bray-Curtis method was first widely used, then criticized for its mathematical naiveté as compared to more sophisticated methods of factor analysis. More recently, an extended series of papers have compared methods of ordination and often suggested that the relatively crude Bray-Curtis method has advantages over more involved methods (La France 1972, Beals 1973, Noy-Meir 1974, Kessell and Whittaker 1976). Current work is exploring nonlinear methods of ordination (Noy-Meir 1974, Austin 1976).

The parallel use of ordination and classification methods in vegetation study is illustrated in Austin, Ashton, and Greig-Smith (Paper 17). Other authors have also used multiple techniques (e.g., Groenewoud 1965, Moore et al. 1970). The relative merits of numerical methods of ordination and classification and the numerous variants of each are not yet agreed upon. Crovello (1970) provides an overview of numerical methods and their use in ecology and systematics, Goodall (1970) has reviewed statistical methods in plant ecology, Whittaker (1973b) has surveyed ordination and classification, and Orloci (1975) has written a handbook of methods of numerical analysis. Whatever their merits, numerical methods have been widely adopted, and an extensive literature has evolved about the methods, their validity, and the interpretation of vegetational data based on these quantitative techniques. The apparent confrontation between traditional methods and numerical methods has been partially circumvented as numerous studies suggest that the relevé methods of Braun-Blanquet can be adapted to numerical clustering techniques (Janssen 1975, Lieth and Moore 1971, Whittaker 1972, Van der Maarel 1969, Spatz and Siegmund 1973). Stanek (1973) reported no significant difference between the results of a classification using Braun-Blanquet's synthesis table method and a numerical classification procedure done on a computer.

Recently, animal ecologists have rediscovered the community and many of the numerical methods used by phytosociologists are now used in the study of animal communities and niche relationships.

REFERENCES

Austin, M. P. 1976. On Nonlinear Species Response Models in Ordination. *Vegetatio* **33**:33–41.

Beals, E. W. 1973. Ordination: Mathematical Elegance and Ecological Naiveté. *J. Ecol.* **61**:23–36.

Crovello, T. J. 1970. Analysis of Character Variation in Ecology and Systematics. *Ann. Rev. Ecol. Syst.* **1**:55–98.

Curtis, J. T. 1959. *The Vegetation of Wisconsin: An Ordination of Plant Communities.* Madison: University of Wisconsin Press, 657 pp.

Curtis, J. T., and R. P. McIntosh. 1950. The Interrelations of Certain Analytic and Synthetic Phytosociological Characters. *Ecology* **31**:476–496.

Goodall, D. W. 1952. Quantitative Aspects of Plant Distribution. *Biol Rev.* **27**:194–245.

Goodall, D. W. 1953. Objective Methods for the Classification of Vegetation. I. The Use of Positive Interspecific Correlation. *Aust. J. Bot.* **1**:39–63.

Goodall, D. W. 1970. Statistical Plant Ecology. *Ann. Rev. Ecol. Syst.* **1**:99–124.

Goodall, D. W. 1973a. Sample Similarity and Species Correlation. In *Ordination and Classification of Vegetation,* ed. R. H. Whittaker, pp. 107–156. The Hague, The Netherlands: W. Junk Publ.

Goodall, D. W. 1973b. Numerical Classification. In *Ordination and Classification of Vegetation,* ed. R. H. Whittaker, pp. 577–615. The Hague, The Netherlands: W. Junk Publ.

Greig-Smith, P. 1964. *Quantitative Plant Ecology,* 2nd ed. London: Butterworths, 256 pp.

Groenewoud, H. van. 1965. Ordination and Classification of Swiss and Canadian Coniferous Forests by Various Biometric and Other Methods. *Ber. Geobot. Inst. ETH, Stift. Rubel, Zurich* **36**:28–102.

Janssen, J. G. M. 1975. A Simple Clustering Procedure for Preliminary Classification of Very Large Sets of Phytosociological Relevés. *Vegetatio* **30**:67–71.

Kessell, S. R., and R. H. Whittaker. 1976. Comparisons of Three Ordination Techniques. *Vegetatio* **32**:21–29.

Kulczynski, St. 1928. Die Pflanzenassoziationen der Pieninen. *Bull. Int. Acad. Pol. Sci. Lett., Cl. Sci. Math. Nat.* Ser. B. 1972 (Suppl. 2):57–203.

LaFrance, C. R. 1972. Sampling and Ordination Characteristics of Computer Simulated Individualistic Communities. *Ecology* 53:387–397.

Lambert, J. M., and M. B. Dale. 1964. The Use of Statistics in Phytosociology. *Adv. Ecol. Res.* 2:59–99.

Lieth, H., and G. William Moore. 1971. Computerized Clustering of Species in Phytosociological Tables and Its Utilization for Fieldwork. In *Spatial Patterns and Statistical Distributions,* eds. G. P. Patil, E. C. Pielou, and W. E. Waters, pp. 403–422. University Park, Pa.: Pennsylvania State University Press.

Maarel, E. van der. 1969. On the Use of Ordination Models in Phytosociology. *Vegetatio* 19:21–46.

McIntosh, R. P. 1957. The York Woods: A Case History of Forest Succession in Southern Wisconsin. *Ecology* 38:29–37.

McIntosh, R. P. 1958. Plant Communities. *Science* 128:115–120.

McIntosh, R. P. 1967. The Continuum Concept of Vegetation. *Bot. Rev.* 33:130–187.

McIntosh, R. P. 1973. Matrix and Plexus Techniques. In *Ordination and Classification of Vegetation,* ed. R. H. Whittaker, pp. 157–191. The Hague, The Netherlands: W. Junk Publ.

McIntosh, R. P. 1976. Ecology Since 1900. In *Issues and Ideas in America,* eds. B. J. Taylor and T. J. White, pp. 353–372. Norman: University of Oklahoma Press.

Moore, J., S. J. P. Fitzsimons, E. Lamb, and J. White. 1970. A Comparison and Evaluation of Some Phytosociological Techniques. *Vegetatio* 20:1–20.

Noy-Meir, J. 1974. Catenation: Quantitative Methods for the Definition of Coenoclines. *Vegetatio* 29:89–99.

Orloci, L. 1975. *Multivariate Analysis in Vegetation Research.* The Hague, The Netherlands: W. Junk Publ., 276 pp.

Peet, R. K., and O. L. Loucks. 1977. A Gradient Analysis of Southern Wisconsin Forests. *Ecology* 58:485–499.

Ponyatovskaya, V. M. 1961. On Two Trends in Phytocenology. *Vegetatio* 10:373–385.

Sobolev, L. N., and V. D. Utekhin. 1973. In *Ordination and Classification of Vegetation,* ed. R. H. Whittaker, pp. 75–104. The Hague, The Netherlands: W. Junk Publ.

Spatz, G., and J. Siegmund. 1973. Eine Methode zur tabellarischen Ordination, Klassification und okologischen Auswertung von pflanzensoziologischen Bestandsanfnahmen. *Vegetatio* 28:1–17.

Stanek, W. 1973. A Comparison of Braun-Blanquet's Method with Sum-of-Squares Agglomeration for Vegetation Classification. *Vegetatio* 27:323–345.

Whittaker, R. H. 1953. A Consideration of Climax Theory: The Climax as a Population and Pattern. *Ecol. Monogr.* 23:41–78.

Whittaker, R. H. 1956. Vegetation of the Great Smoky Mountains. *Ecol. Monogr.* 26:1–80.

Whittaker, R. H. 1962. Classification of Natural Communities. *Bot. Rev.* 28:1–239.

Whittaker, R. H. 1967. Gradient Analysis of Vegetation. *Biol. Rev.* **42**:207–264.

Whittaker, R. H. 1972. Convergences of Ordination and Classification. In *Grundfragen und Methoden in der Pflanzensoziologie*, ed. R. Tüxen, pp. 39–57. The Hague, The Netherlands: W. Junk Publ.

Whittaker, R. H. 1973. Approaches to Classifying Vegetation. In *Ordination and Classification of Vegetation*, ed. R. H. Whittaker, pp. 323–354. The Hague, The Netherlands: W. Junk Publ.

Williams, W. T., and J. M. Lambert. 1959. Multivariate Methods in Plant Ecology. I. Association Analysis in Plant Communities. *Ecol.* **47**:83–101.

12

Copyright © 1948 by The Royal Danish Academy of Sciences and Letters
Reprinted from *K. Dansk. Vidensk. Selsk. Biol. Skrift* 5(4):3–16, 34 (1948)

A METHOD OF ESTABLISHING GROUPS OF EQUAL AMPLITUDE IN PLANT SOCIOLOGY BASED ON SIMILARITY OF SPECIES CONTENT

AND

ITS APPLICATION TO ANALYSES OF THE VEGETATION ON DANISH COMMONS

BY

THORVALD SØRENSEN

This publication is available from The Royal Danish Academy of Sciences and Letters.

1. Introduction and Survey of Problems Involved.

It is well known that certain combinations of species occur more frequently in nature than others, and it is natural that plant sociology has to a material extent focussed its attention on such combinations of species as are more or less favoured by nature itself. These combinations have come to figure as a permanent skeleton in the systematics of plant sociology as the "pure" units. Partly through abstraction and partly through generalisation, generally accepted concepts, sociations, associations, etc., have been arrived at, according to which rank has been assigned to them.

When making analyses of vegetations the requirement is generally made that "pure" populations be selected, i. e. the most frequently occurring combinations of species that are recognizable with more or less certainty. Such a procedure undeniably facilitates arrangement of the material, which, by the way, may be considered an important reason for adopting it. It cannot be denied, however, that the less frequent combinations of species, too, constitute an exponent of the vegetation of a district, and of the reaction of the species mutually and to life conditions. Only one demand may justly be made on the nature of the vegetation in the limited area under investigation, namely that it be homogeneous with as much approximation to that mathematical concept as nature can offer, and as at any rate becomes apparent to the plant sociologist during the analysis itself in the field.

If an attempt is made to procure, with the demand for homogeneity as sole assumption, and as far as is possible, a comprehensive and exhaustive material of population analyses within a certain area, difficulties arise afterwards, during the arrangement of the material. The various types of vegetation often are so insensibly merged as to form a sliding scale.

Danish plant sociologists, who have most often based their work on RAUNKIÆR's method (determination of frequency), (RAUNKIÆR 1909, 1913, 1934), have shown little interest in a definitive arrangement of the material analyzed into groups corresponding to the sociations of the Uppsala School (DU RIEZ 1930) or the associations of the Zurich School (BRAUN-BLANQUET 1928). Even in cases where a division of this nature might have been possible, Danish plant sociologists have been content with placing populations in a line with the dominating (high frequency) species in a sloping line showing their gliding reaction towards some ecological factor, such as humidity or p_H.

A strictly objective and numerically supported development of this sloping line principle was made recently by TUOMIKOSKI (1942), who arranged the species in correlative sequence. Instead of RAUNKIÆR's frequency percentages he employed degree of coverage as expressed in percentages of the total area, drawing the border-lines between groups where the sloping line displays a greater or lesser degree of discontinuity.

When attempting to classify analyses of vegetations from two geographically separate areas within one system—even when selected "pure" populations are present in both cases—we meet with the same difficulties as when classifying "impure" material from one area only: groups overlap, or we find a continuous, sliding variation. Even with, in some measure, identical floristic content in the two areas, borderlines between the individual species, and so also the borderlines of the units of vegetation, will not conform in the two areas.

These difficulties give rise to two questions: 1) How great is the mutual similarity between populations within a group recognized as a sociation or an association respec-tively, and 2) how great a similarity to already established groups must a population exhibit to permit it to be classified as belonging to one such group? The physiognomical-ecological method of the Uppsala School teaches nothing about this, nor does the floristic method of the Zurich School, just to mention the best known leading sociolo-gical schools. Nor does TUOMIKOSKI take decided ground on the issue, even though his numerical method should provide an indication in this direction.

However, we find a suggestion of the problems outlined above in NORDHAGEN's (1934) outstanding attempt at combining the Uppsala Sociations with Zurich Associa-tions. Where indicative species are partially or totally absent he employs the total floristic similarity between each pair of sociations, placing doubtful sociations where similarity of species is greatest as expressed by JACCARD's coefficient de communauté $\dfrac{c}{a+b-c}$ (JACCARD 1901 a, 1901 b, 1908), where a and b mean the respective numbers of species in the two sociations (A and B) with a percentage of constancy ≥ 50, and c the number of species occurring in both sociations with the aforesaid degree of constancy.

A numerical establishment of difference or similarity between associations already defined according to other points of view has already been made by KULCZYŃSKI (1928) on a basis of total content of species and degree of constancy, the latter expressed in a scale with five steps. The floristic similarity between two associations is expressed here by the Verwandtschaftskoeffizient $V = \dfrac{\left(\dfrac{c}{a} + \dfrac{c}{b}\right) \cdot 100}{2}$. The calculation of V is illustrated most easily by stating KULCZYŃSKI's example (l. c. p. 181):

Species	Ass. A Degree of Constancy	Ass. B Degree of Constancy	Common
N	5	3	3
O	2	4	2
P	1	—	—
Q	2	2	2
R	—	3	—
S	—	1	—
Sum	$a = 10$	$b = 13$	$c = 7$

$$V = \left(\frac{7}{10} + \frac{7}{13}\right) \cdot \frac{100}{2} = 61.92.$$

With V as a basis, associations are classified in floristic groups. KULCZYNSKI, therefore, does not use V for marking the bounds of associations but applies it to making an estimate of the influence of various independent factors upon the floristic composition of associations: adaphic factors, neighbour effect, the mutual competition of species, succession.

Similar methods have subsequently been employed by RENKONEN (1938) and AGRELL (1941, 1945) in dealing with invertebrate communities. However, RENKONEN does not employ constancy, but the relative percentage of dominance. AGRELL (1941, p. 70 et seq.) employs both dominance numbers (quantitative properties) and constancy numbers (qualitative properties). According to AGRELL the two methods produce essentially consistent results. However, AGRELL's calculation formulae, which are different in principle in the two cases, seem, strictly speaking, hardly to permit of a direct comparison.

In conjunction with the works quoted in the foregoing it should only be pointed out here that AGRELL (1944) has indicated "An objective method for characterization of animal and plant communities", based on the quantitative and qualitative characteristics of communities. However, his method provides no objective basis for marking the bounds between communities, which, on the other hand, evidently was not the author's intention (cfr. the author's remark to that effect l. c. p. 14).

One feature still seems to be common to hitherto known statistical methods of treatment: that a statistical treatment is made of already defined groups of populations arranged according to subjective points of view. Hence these methods provide no satisfactory answer to the questions outlined above concerning the actual amplitude of the units. So long as the basis of an objective appreciation of the interdependence of a number of populations that have been included in one group, does not exist, any comparison between individual groups made on a statistical basis must in reality become deceptive to a certain extent.

2. Description of a Method for Classification of Population Analyses in Groups of Equal Amplitude Based on Similarity of Species Content.

During my endeavours to arrive at a consistent grouping of an extensive material of polulation analyses from North East Greenland I have developed a method of classification which seems to realize satisfactory results. The method is an objective one, it being based solely on the similarity of individual populations as expressed by the community of species.

The method is based on the following fundamental principles:

The plant sociological unit is the population analyzed. The unit employed in the statistic treatment is one species in one population, respectively each species in each population. The similarity between two populations is expressed by the quotient of similarity, $qs = \dfrac{c}{\left(\dfrac{a+b}{2}\right)} = \dfrac{2c}{a+b}$, where a means the number of species within the one population, A; b the number of species within the other population, B; c the number of species common to the two populations.

In principle, of course, it may be discussed[1] whether this formula is preferable to JACCARD-NORDHAGEN's $\dfrac{c}{a+b-c}$, or to KULCZYNSKI's $\dfrac{\dfrac{c}{a}+\dfrac{c}{b}}{2}$. The former formula might perhaps seem the more logical one, inasmuch as it expresses the ratio between the species common to the two populations and all (different) species occurring in them. Actually, however, this means that we consider the two populations as one unit, which is confusing seeing that two separate populations were the starting point. When making a graphical representation of $\dfrac{c}{a+b-c}$ we have another difficulty to cope with, namely that both numerator and denominator change simultaneously—in other words that the number of species for two given populations decreases with increasing community of species. It is not possible to represent this function as a straight line. This difficulty is not involved by the two other formulas. If we assume that $a = b$, which, however, will rarely happen in practice, KULCZYNSKI's formula and mine will produce identical results. In cases where a and b differ, KULCZYNSKI's formula will produce the higher values, relatively higher the greater the difference.

It is appropriate at this stage to look a bit more closely at the formula $\dfrac{c}{\left(\dfrac{a+b}{2}\right)}$ and its application. For this purpose we shall use a theoretical example (cfr. table 1).

[1] I should like to give my thanks to Professor K. GRAM, Ph. D., for instructive discussion of, and good advice on, the formulae discussed here.

Let us figure to ourselves two populations, A and B, comprising the species N — T. Presence of a species is indicated by 1, absence by —. By substitution in the formula $\dfrac{2c}{a+b}$ we have $\dfrac{2 \cdot 4}{5+6} = \dfrac{8}{11} = 0.73 = qs.$ (1)

Table 1.

Determination of Quotient of Similarity between two individual populations, A and B.

	A	B	
N.............	1	1	1
O.............	1	1	1
P.............	1	1	1
Q.............	1	1	1
R.............	—	1	0
S.............	1	—	0
T.............	—	1	0
	$a = 5$	$b = 6$	$c = 4$

$$qs = \frac{2c}{a+b} = \frac{2 \cdot 4}{5+6} = \frac{8}{11} = 0.73.$$

The numerator, c, expresses the number of cases in which the individual species occur in both populations. With regard to the denominator it should be noted: Theoretically, each species in population A possesses 1 chance of coinciding with the same species in population B, and likewise each species in B has 1 chance of finding a partner in population A. Thus $a + b$ expresses the sum of the chances of the individual species of coincidence in the two populations separately. Every time 1 chance is made use of by a species in the one population the partner in the other population loses its individual chance, or, in other words, the two partners must share it. This means that $\dfrac{a+b}{2}$ expresses the sum of all theoretically realizable chances of coincidence of species in the two populations under comparison. Thus $\dfrac{c}{\left(\dfrac{a+b}{2}\right)} = \dfrac{2c}{a+b}$ expresses the number of actually realized coincidences divided by the number of possibilities theoretically inherent in the two populations of coincidence of species. This must be a perfectly valid expression of the degree of similarity of species between two populations. $QS = \dfrac{2c}{a+b} \cdot 100$, which expression will for practical reasons be used in the following instead of qs, then represents the percentagewise similarity of species between two populations in proportion to the similarity that is theoretically possible according to the number of species present.

When making a comparison between two individual populations the coincidence —or non-coincidence—of the individual species denotes a case of either-or. It will be obvious, however, that the chance of coincidence in a comparison of groups of

2*

populations, each group comprising several individual populations, must be graduated. I have tried to illustrate this by an example (cfr. Table 2).

Table 2.
Determination of Quotient of Similarity between two groups of populations, A, comprising populations No. 1, 2, and 3, and B, comprising populations No. 4 and 5.

Population No.	A				B			
	1	2	3		4	5		
N	—	- -	----	0	1	1	2	0
O	—	—	—	0	1	1	2	0
P	1	1	---	2	1	1	2	4
Q	1	—	—	1	1	1	2	2
R	1	1	1	3	—	1	1	3
S	1	1	1	3	1	—	1	3
T	—	1	1	2	—	1	1	2
U	1	1	1	3	—	—	0	0
				$a = \dfrac{14}{3}$			$b = \dfrac{11}{2}$	$c = \dfrac{14}{6}$

$$ qs = \frac{2\,c}{a+b} = \frac{2 \cdot {}^{14}/_6}{{}^{14}/_3 + {}^{11}/_2} = \frac{2 \cdot 14}{2 \cdot 14 + 3 \cdot 11} = \frac{28}{61} = 0.46. $$

In this example A comprises 3 populations marked 1, 2, 3; B 2 populations marked 4, 5. Compared with the example given above, one species in one population then represents $\cdot{}^1/_3$ in A, in B $^1/_2$. A species occurring in all 3 populations in A and in both populations in B will then have a chance of coincidence of $^3/_3 \cdot {}^2/_2 = {}^6/_6$. For instance, a species occurring in 2 of the A populations and in 1 of the B populations will have a chance of $^2/_3 \cdot {}^1/_2 = {}^2/_6$. In the example, the sum of coincidences realized is $^{14}/_6 = c$. a is the sum of individual species of the individual populations in A divided by the number of populations, expressed in figures $^{14}/_3$; b, therefore, $^{11}/_2$. qs for the two groups of populations thus is $\dfrac{2 \cdot {}^{14}/_6}{{}^{14}/_3 + {}^{11}/_2} = \dfrac{2 \cdot 14}{2 \cdot 14 + 3 \cdot 11} = \dfrac{28}{61} = 0.46$ (2), or, expressed percentagewise: $QS = 46$. In other words: In practice we have for c: the sum of the products of the numerical presence numbers for all coinciding species; for a: the sum of all individual species in all individual populations in A multiplied with the number of populations in B, and, likewise, for b: the sum of all individual species in all B-populations multiplied with the number of populations in A.

By way of illustration of the procedure of an objective division into groups of a heterogeneous (unclassified) material I shall now go through a simple fictitious example. We imagine the groups A and B in table 2 put into one group, in other words that the 5 populations are unclassified. We calculate the quotients of similarity for the populations in twos: 1,2: $QS = 80$; 1,3: $QS = 67$; 1,4: $QS = 60$; 1,5: $QS = 55$; 2,3: $QS = 89$, and so forth, and arrange the values in a square (cfr. Fig. 1) with the population numbers along the sides of the square. The values are noted both horizontally and vertically (the figure is symmetrical around the diagonal). The columns are added together and divided by the number of values (here 4). We have now an expression of the average *QS* of each population with all the others. It follows from this that populations 1 and 2 occupy a central position in the material, while population 4 is the "poorest". The average of the bottom series of figures provides a summary expression of the mutual similarity existing within the material, the *QS* of the entire

1	2	3	4	5	
	80	67	60	55	1
80		89	40	55	2
67	89		22	40	3
60	40	22		73	4
55	55	40	73		5
262	264	218	195	223	
66	66	55	49	56	

Average: 58

a	b	c	
	39	74	a = 2,3
39		58	b = 4,5
74	58		c = 1
113	97	132	
57	49	66	

Average: 57

A	B	
		49	A = a,c
	49		B = b

Fig. 1.   Fig. 2.   Fig. 3.

material considered as a group. This quantity is in itself of no value if the variation of the individual values in the square is great.

There now remains the task of arranging the material in groups. For this purpose we select a limit value for *QS*, below which no individual value within a group may fall. The position of this limit is chosen arbitrarily and must depend on how high values are at all represented in the material. To illustrate the importance of the position of the limit I shall now go through three cases in which the limit is placed, respectively, at $QS > 70$, $QS > 60$, $QS > 50$.

I. $QS > 70$: In the square (Fig. 1) all values above 70 are marked, and the possible combinations constructed on a basis of the *QS* of each individual population as compared with each of the other populations, the material (the square) being gone through, series after series:

Population 1 can only be combined with population 2 and hence gives the combination 1,2, $QS = 80$.

Population 2 can be combined with populations 1 and 3. From the first series it follows that the combination 1,3 is inapplicable ($QS = 67$, which is below the minimum limit), for which reason 1 is left out. There remains the combination 2,3, $QS = 89$.

Population 3 can only be combined with 2 and likewise gives the combination 2,3.

Population 4 can only be combined with 5 and therefore gives the combination 4,5, $QS = 73$.

Population 5 can only be combined with 4; this combination consequently is identical with the preceding one.

From a restricted material, as illustrated by the example selected, it follows directly which population or populations must be left out in each series. In the case of an extensive material, however, it may become necessary to construct the complete square and then strike out the longitudinal and transversal series containing the inapplicable value or values. After this, the series containing the greatest number of inapplicable values are crossed out in turn. This procedure also facilitates the determination of the average QS of the final combination (cfr. Fig. 1).

All combinations possible are arranged in tabular form, identical combinations of course only being put down once, and combinations merely representing a part of another, more comprehensive combination, are left out.

The practical arrangement is as follows:

Populations	1	2	3	4	5	Average QS of Combination
Combination I ....,....	80	80	. .	. .	. .	80
Combination II	. .	89	89	. .	. .	89
Combination III.........	. .	. .	. .	73	73	73

As some of the combinations generally overlap, it is necessary to provide a system for making the final grouping. As guide therefor the following requirements must be complied with in sequence as set forth below:

1) The groups must in turn be made as comprehensive as possible, that is, the combination comprising the greatest number of populations must be preferred.

2) There must be as few groups as possible, or, in other words, the greatest possible number of the other combinations must be rendered superfluous through the consecutive establishment of groups. This will mean that out of two or more combinations that are equal with regard to 1), the one is to be preferred that "covers" the greatest number of places in the vertical columns of the table.

3) The groups must in turn become as good as possible. This means that out of two or more combinations that are equal with regard to 1) and 2), the one having the greatest QS is to be preferred.

That the requirement of QS is placed last is due to the fact that already through establishment of a minimum limit for QS provision was made that any possible combination is good enough in itself.

The first group selected in this way must, within the selected interval of similarity, be considered as the central group around which the rest occupy positions as secondary groups. These latter are selected in turn from the remaining material according to the same principles.

Theoretically, of course, a grouping after the principles outlined here may be rendered impossible if two or more combinations must be considered equal, even after trial in the three consecutive instances. In practice, at any rate, this will occur very rarely.

In our simple demonstration the result becomes the following grouping:

> Group 2,3 QS = 89, hereafter designated as a
> Group 4,5 QS = 73, hereafter designated as b
> Group 1   , hereafter designated as c.

Next the primary groups (a, b, and c) must be arranged in secondary groups. In establishing the secondary groups we demand a certain minimum value of QS for each pair of primary groups entered. As arbitrary minimum value we chose QS > 60.

A second square is constructed for the primary groups (Fig. 2), similar to the original one for the individual populations, and QS for the groups are inserted. These QS values are determined as the average value of all QS values of the populations of each established group. For instance, the QS of a,b = (40 + 55 + 22 + 40)/4 = 39.

The establishment of secondary groups is now made on the same principle as that used in the first grouping. Combinations possible in this case are:

Primary groups	a	b	c	Average QS of combination
Combination I	74	74	74	

The result is:

> Group a,c QS = 74, hereafter designated as A
> Group b   hereafter designated as B.

For the sake of completeness the operation is continued with construction of the following square (Fig. 3), which constitutes the last phase. We have obtained a division of the material into two groups, the similarity of which is expressed through the quotient of similarity QS = 49

II. QS > 60. Combinations possible in this case are (cfr. Fig. 1):

Populations	1	2	3	4	5	Average QS of combination
Combination I	74	85	78	..	..	79
Combination II	..	..	..	73	73	73

243

which results directly in group 1,2,3 $QS = 79$, hereafter designated as a
group 4,5 $QS = 73$, hereafter designated as b.

The final result has already been reached here as illustrated in figures 4 and 5. QS for a,b is found to be 48.

III. QS > 50. Combinations possible are the following:

Populations	1	2	3	4	5	Average QS of Combination
Combination I	74	85	78	..	..	79
Combination II	58	..	..	67	64	63
Combination III	68	68	..	..	55	64

According to the rules set forth above, combination III must be considered a central group, since it "covers" 7 places, while neither of the two other ones "covers" more than 6, and we get this result:

Group 1,2,5 $QS = 64$, hereafter designated as a
Group 3   hereafter designated as b
Group 4   hereafter designated as c.

After this, the establishment of groups of the second order is made with regard to the minimum limit of $QS > 40$.

Combinations possible for secondary groups are:

Primary Groups	a	b	c	Average QS of Combination
Combination I	65	65	..	65
Combination II ...	58	..	58	58

which gives the groups of the third order:

Group a,b $QS = 65$, hereafter designated as A
Group c   hereafter designated as B.

Here, the QS of A,B is 35.

The gradual comprehension into groups of the material on hand in the three cases is presented in schematic form in figures 4 and 5. With regard to the linear sequence of the groups and the populations respectively it should be pointed out that it has been based on the mutual QS of the groups and the populations respectively. For instance, in case *I* it will be seen (cfr. Fig. 2) that c must be placed in the middle inasmuch as c displays a greater similarity to a and b than the mutual similarity of a and b, and, similarly, for the populations: (cfr. Fig. 1) 1 displays a greater similarity to 2 and 4 than, respectively, to 3 and 5. Hence the linear sequence must become 3, 2, 1, 4, 5.

The two cases first treated, *I*, minimum limit $QS > 70$, and, *II*, minimum limit $QS > 60$, produce an identical grouping already at the limit of $QS > 60$, *I*, however, producing a finer division at the "top" end. *III*, on the other hand, produces

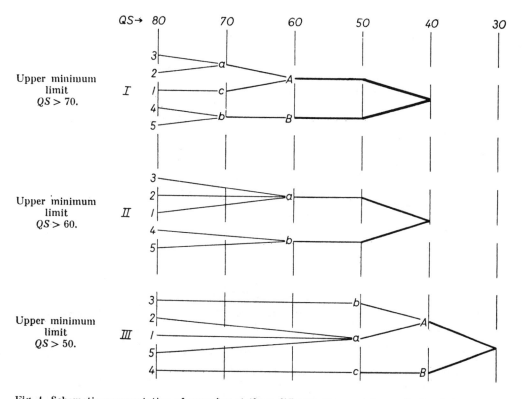

Fig. 4. Schematic representation of grouping at three different upper minimum limits. Only stepwise limits are given.

a rather meaningless grouping, which, at any rate, follows directly from Fig. 5. Now, a glance at Table 2 will at once show that *I* and *II* will give the most natural grouping, and from the combinations possible in *III* it will likevise be seen that combination I would have been preferable. This might be interpreted to mean that the principles applied to the grouping described on p. 10 are wrong. This is not the case, however. The poor result in *III* is due to the fact that the distance between two consecutive groupings was not made sufficiently small (that combination III was in reality not good enough). As a matter of fact, in this case we covered the interval between $QS = 89$ and $QS = 50$ at one stretch. The consequence is that the result obtained in this case was no better than that which would have been obtained by an immediate isolation, of population 4 as being the poorest, as is apparent from Fig. 1. In fact, in *III* we have placed population 4 against all the other populations as a separate unit.

A strictly schematic use of the method on the principles outlined above can be made only by selecting not too great distances between the consecutive groupings.

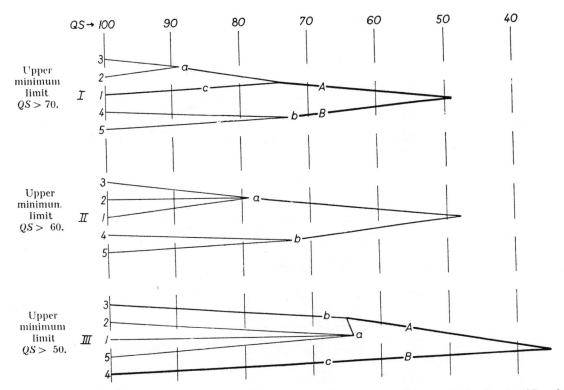

Fig. 5. Schematic representation of grouping at three different upper minimum limits. On the *QS* scale, exact expression of the degree of connexion (average *QS*) within the individual groups is illustrated through the points of confluence of the units.

The limits in the case described were chosen arbitrarily with intervals of 10 as expressed in terms of *QS*. The smaller intervals one selects the better is the guarantee of uniformity within the groups established, and the better is the guarantee that the final seriate arrangement of populations and groups (see Fig. 5) becomes absolutely unassailable. The width of the intervals must be left to individual judgment, as is depends on the use it is desired to make of the result, and on the amount of work that can be expended on the arrangement of the material.

In certain cases, of course, a consideration of the combinations possible will be sufficient to provide an indication as to where it is reasonable to place the limits. If an extensive material is on hand, however, it will often be possible to make such a variety of combinations that the surveyableness is lost, which absolutely necessitates a procedure based on definite principles. The concrete material submitted in the following section will serve as a demonstration that a treatment made in strict accordance with the rules given can lead to a useful result. I wish to point out, moreover, that after training in the use of the method, an ordinary office staff will be able to make the grouping and the seriate arrangement of the groups quite mechanically from the first phase to the last and even to the least details.

However, in the case of an extensive material the calculation of QS for all pairs of populations will be a time-consuming task. For n populations, the number of individual values of QS becomes $n \cdot (n-1)/2$. In practice, therefore, it may be a good plan first to arrange the material roughly into preliminary groups and then to check the mutual interdependence within each individual group. This can be done in two ways: 1) The groups are treated according to (1) (see p. 7), that is, as in the example described above (Fig. 1). The primary groups must be made approximately equally "good"; in other words: they must respect the same minimum limit. 2) The second procedure assumes that the rough grouping of the material was made with relative care, and that we desist in advance from a strict compliance with the requirement of a minimum limit for the individual values within the square (cfr. e. g. Fig. 1). It is then possible to avoid the inconvenience of calculating each QS and so reduce the task to calculating, within the individual groups, the QS of each population against the sum of the others according to (2) (see p. 8). This reduces the number of separate calculations from $n \cdot (n-1)/2$ to n. A practical arrangement for calculating these QS values is shown in Table 3 for the same material as that dealt with in the foregoing, from which it follows that the QS values of the individual population against the sum of the others practically do not differ from the mean values found by the square method (cfr. the bottom series of figures in Fig. 1). The poorest populations of the group will then produce the lowest QS values and may be removed or transferred to another group. The width of variation in the QS values thus found reflects the variation in the material. When establishing co-ordinated groups by this method it is necessary to observe that the variation within each group be slight, also that the groups be approximately at the same QS level. The observance of any definite minimum value is of minor interest in this case.

A drawback of this method is that after removal of the "poor" populations one has no longer any concise expression of the similarity within the group, whereas the square method very easily provides an expression of the uniformity of the improved group after rejection of the inapplicable series in the square. The method is less exclusive than the one first described but may be useful for establishing the "quality" of groups arranged discretionally.

An extensive raw material of population analyses will as a rule be so heterogeneous that it does not pay to calculate each QS of all the populations. Hence the calculation of QS for the various pairs of groups of the first order must be made direct, which is

Table 3.

Table for determination of the Quotient of Similarity between the individual populations in a discretionarily composed group (Average designated the *QS* of the group) on comparison of each population with the sum of the others.

Population No.	1	2	3	4	5		1	2	3	4	5
N....................	—	—	—	1	1	2	—	—	—	2	2
O.....................	—	—	—	1	1	2	—	—	—	2	2
P....................	1	1	—	1	1	4	4	4	—	4	4
Q....................	1	—	—	1	1	3	3	—	—	3	3
R....................	1	1	1	—	1	4	4	4	4	—	4
S....................	1	1	1	1	—	4	4	4	4	4	—
T....................	—	1	1	—	1	3	—	3	3	—	3
U....................	1	1	1	—	—	3	3	3	3	—	—
	5	5	4	5	6	25	18	18	14	15	18

Example: Population No. 1 (A) against Populations No. 2—5 (B). (Cfr. Table 2, p. 8).

$$c = 1 \cdot (4 - 1)/4$$
$$+ 1 \cdot (3 - 1)/4$$
$$+ 1 \cdot (4 - 1)/4$$
$$+ 1 \cdot (4 - 1)/4$$
$$+ 1 \cdot (3 - 1)/4$$

$$c = 1 \cdot (18 - 5)/4 = \frac{13}{4}, \qquad a = \frac{5}{1}, \qquad b = \frac{25 - 5}{4} = \frac{20}{4}.$$

$$qs = \frac{2 \cdot c}{a + b} = \frac{2 \cdot {}^{13}/_4}{{}^5/_1 + {}^{20}/_4} = \frac{2 \cdot 13}{20 + 20} = \frac{26}{40} = 0.65; \quad QS = 65.$$

QS 1 against 2, 3, 4, 5: 65
QS 2 against 1, 3, 4, 5: 65
QS 3 against 1, 2, 4, 5: 54
QS 4 against 1, 2, 3, 5: 50
QS 5 against 1, 2, 3, 4: 56

Average: 58

From the example the following general mode of calculation is derived:

c denotes the presence sum of the entire group for species present in the individual population (A)[1] less the presence sum of the individual population.

a denotes the presence sum of the individual population multiplied by the number of populations of the entire group less 1 (i. e. number of B populations).

b denotes the presence sum of the entire group less the presence sum of the individual population.

done according to (2). The groups are arranged in random sequence in only one table with the number of populations noted for each group. The presence number (constancy number) is put against each species, the sum of the presence numbers being placed at the bottom of each column. Then a direct comparison is made between each pair of groups (cfr. Table 2). The result is arranged in a square and treated according to the method outlined above, until the entire material comes within one group.

[1] Right columns in the Table.

[*Editor's Note:* In the original, material follows this excerpt.]

6. Literature Quoted.

AGRELL, IVAR (1941): Zur Ökologie der Collembolen. Untersuchungen im schwedischen Lappland. — Opuscula Entomologica Suppl. III. Lund.
— (1944): An objective method for characterization of animal and plant communities. — Kungl. Fysiogr. Sällsk. i Lund Förhandl. 15, Nr. 9, Lund.
— (1945): The Collemboles in nests of warm-blooded animals with a method for sociological analysis. — Lund Univ. Årsskrift, N. F. Avd. 2, 41, Nr. 10, Lund.
BRAUN-BLANQUET, J. (1928): Pflanzensoziologie. Grundzüge der Vegetationskunde. Berlin, Verlag Julius Springer.
BRAUN-BLANQUET, J. & M. MOOR (1938): Verband des Bromion erecti. — Prodr. d. Pflanzengesellschaften Fasz. 5.
DU RIEZ, G. E. (1924): Studien über die Vegetation der Alpen, mit derjenigen Skandinaviens verglichen. — Veröffent. Geobot. Inst. Rübel, I. Zürich.
— (1930): Vegetationsforschung auf soziationsanalytischer Grundlage. — Abderhalden: Handb. d. biol. Arbeitsmethoden, Abt. XI, Teil 5. Berlin und Wien.
GRÖNTVED, JOHS. (1927): Formationsstatistiske Undersøgelser paa nogle danske Overdrev. — Bot. Tidsskrift 40, p. 1—71, København.
GUÐJÓNSSON, GUÐNI (1941): Om Aphanes arvensis L. og A. microcarpa (Boiss. et Reut.) Rothm. og deres Udbredelse i Danmark. — Bot. Tidsskrift 45, p. 352—370. København.
IVERSEN, JOHS. (1936): Biologische Pflanzentypen als Hilfsmittel in der Vegetationsforschung. — Kopenhagen, Verl. Munksgaard.
JACCARD, PAUL (1901 a): Distribution de la Flore Alpine dans le Bassin des Dranses et dans quelques régions voisines. — Bull. Soc. Vaudoise des Sc. Nat. 37, p. 241—272. Lausanne.
— (1901 b): Étude comparative de la distribution florale dans une portion des Alpes et du Jura. — Ibid. p. 547—479.
— (1908): Nouvelles recherches sur la distribution florale. — Ibid. 44, p. 223—270.
KULCZYŃSKI, ST. (1928): Die Pflanzenassoziationen der Pieninen. — Bull. Internat. de l'Acad. Polonaise des Sc. et des Lettres Classe d. Sc. Math. et Nat., Ser. B. No. Suppl. II, 1927. p. 57—203. Cracovie.
MOOR, M. (1937): Ordnung der Isoetetales (Zwergbinsengesellschaften). — Prodr. d. Pflanzengesellschaften Fasz. 4. Leiden.
NORDHAGEN, ROLF (1940): Studien über die maritime Vegetationen Norwegens. I. Die Pflanzengesellschaften der Tangwälle. — Bergens Museums Årbok 1939—40, Naturv. Rekke Nr. 2.
— (1943): Sikilsdalen og Norges Fjellbeiter. En plantesociologisk Monografi. — Bergens Museums Skrifter Nr. 22, Bergen.
RAUNKIÆR, C. (1909): Formationsundersøgelser og Formationsstatistik. — Bot. Tidsskrift 30, p. 20—132. — København.
— (1913): Formationsstatistiske Undersogelser paa Skagens Odde. — Ibid. 33, p. 229—243.
— (1934): The Life Forms of Plants and Statistical Plant Geography being the Collected Papers of C. Raunkiær. — Oxford, The Clarendon Press.
RENKONEN, OLAVI (1938): Statistisch-Ökologische Untersuchungen über die terrestrische Käferwelt der finnischen Bruchmoore. — Ann. Zool. Soc. Zool.-Bot. Fennicae Vanamo 6. No. 1. — Helsinki.
TUOMIKOSKI, R. (1942): Untersuchungen über die Untervegetation der Bruchmoore in Ostfinnland. I. Zur Methodik der pflanzensoziologischen Systematik. — Ann. Bot. Soc. Zool.-Bot. Fennicae Vanamo 17, No. 1. Helsinki.
TÜXEN, REINHOLD (1937): Die Pflanzengesellschaften Nordwestdeutschlands. — Mitt. d. Floristisch-soziologischen Arbeitsgemeinschaft in Niedersachsen 3, p. 1—170. Hannover.

Indleveret til Selskabet den 12. Februar 1947.
Færdig fra Trykkeriet den 15. April 1948.

13

Reprinted from *Northwest Sci.* **25**(1):17–19, 24–31 (1951)

A Criticism of the Plant Association and Climatic Climax Concepts

ROBERT H. WHITTAKER
Department of Zoology
The State College of Washington

T HREE OF THE MOST SIGNIFICANT concepts in our present view of vegetation are the association as a unit of vegetation into which species are organized, succession of plants toward a climax community, and the climatic climax of a region toward which all successions converge. These are fairly old ideas in American ecology now, considering that they were among the early concepts developed in the field. It may be noted that they were developed from field observations interpreted without critical scientific controls, by methods which may have been the best in their time but were far short of techniques which can be applied now. Doubts of the validity of these concepts in their original form have been expressed by various authors since 1926 (Gleason, 1926, 1939; Cain, 1947; Mason, 1947; Egler, 1947). It would seem appropriate that they should receive further testing in the field.

The first concept, the association, is the traditional unit of synecology, through which natural communities were studied, into which actual stands were grouped. It has always been assumed that it was a valid unit, and that it had boundaries so that it could be treated as a scientific unit. The association as a unit can be tested against reality in the field in clear and unequivocal ways. Species cannot be organized into valid units by coincidence alone, there must be organizing forces of sufficient strength to impose unity. If such organization exists, it must appear in similarity of distributions—not for all species, but for most—since it is hardly conceivable that truly dependent and interdependent species would be independently distributed. We can test organization into units as it appears in distribution patterns in the field. The logic of the test may be simply stated: if species populations are significantly organized into associations, that organization must appear significantly in distributions.

DIFFERENT TECHNIQUES USED

Attempting to test the validity of the association with conventional methods which assume its validity would be a dubious procedure, but suitable techniques can be designed. Some of these techniques, which may be described as gradient analysis of vegetation, are designed to follow the distribution of plant populations along environmental gradients without regard to supposed associations. The familiar field transect may be used and can most conveniently be applied to the vegetation of old mountain ranges with mature topography, where gradients are largely unbroken by topographic discontinuities and communities of primary succession. One may then take samples, counts of plant populations, at uniform and fairly frequent intervals along continuous gradients, either up a mountainside at 100- or 200-foot intervals through the zones or at regular intervals along a moisture gradient from the floor of a valley out to a dry southwest slope. When distributions of plant populations are traced along the gradient through the samples, it may then be seen to what extent they are organized into zones or associations with distinct boundaries. There are some limitations in the method. For one thing, if a transect crosses a discontinuity or abrupt change in environment, an abrupt change in vegetation, with a number of species having their limits there, will almost necessarily appear. It is probably possible to support one's point either for or against the association if the right place is chosen for a single transect. To make a fair decision it is necessary to take a series of transects chosen without prejudice so far as possible, in order to see whether discontinuities between associations always appear and whether, if they do, they are always in the same place relative to the gradient and the communities involved. A series of field transects were made in the Great Smoky Mountains and seemed to demonstrate convincingly the continuity of plant populations through most of the traditional associations (Whittaker, 1948).

A modification of the field transect to give a more generalized picture based on a larger mass of data may be called the composite or synthetic transect. Samples are taken of all stands encountered in a study area that are not clearly successional. Stands are sampled without regard to whether they fit into associations; they are taken at random, so far as possible, without prejudice, to represent every vegetation type within the limits used by more than one sam-

251

ple. One may use such samples for elevation transects—as for valley bottoms or dry southwest slopes at all elevations—or for moisture transects. For the latter in the Great Smoky Mountains, 50 or 60 samples were taken within each 1000-foot elevation belt, arranged in series from most mesic to most xeric, and grouped into 12 or 13 "stations" along the gradient. Arrangement in series can be on the basis of the topographic position of the site, of the composition of the vegetation itself, as the gradation from mesic to xeric is known from field transects, or of actual soil-moisture determinations. The first two of these have been used successfully; direct determination of moisture for the number of samples used should be feasible in some cases. Some difficulties of the approach may be evident. It involves the assumption that the very complex gradients of elevation and moisture-balance can, in mature topography at least, be treated successfully as single gradients. And taking a valid random sample of vegetation while excluding seral stands is by no means as easy and as safely objective as one might wish. By these means, however, a general picture of the relation of plant populations to gradients may be obtained (Tables 1, 2, and 3).

SPECIES NOT IN ASSOCIATION UNITS

The picture which emerges is that of a series of population curves along the gradient (Fig. 1). It may be noted that the curves are of binomial form, with tapered tails, and flow continuously into one another, and that no two curves are alike. It is difficult to interpret these curves except on the basis that each species has its own mode or center of population away from which it declines gradually, its mode and dispersion corresponding to its own capacity for maintaining a population against the gradients of environment and competition. Four of the traditional associations (left to right: mixed mesophytic, oak-hickory, oak-chestnut, and pines) are represented along this gradient, and they are not separated by boundaries of any kind, nor do the species fall into four corresponding groups.

So far as our test of the association is concerned, it may be observed that on this and the other transects species were not organized into association units. To this evidence from the Smokies may be added the mass of biogeographical evidence of maps of distribution (e.g., Munns, 1938), and even published tables of frequencies in associations (e.g., Braun, 1940; Brown, 1941; Dansereau, 1943; Fautin, 1946), which indicate how species change frequency or abundance along gradients between communities but are not really organized

[Editor's Note: Tables 1 and 2 have been omitted owing to limitations of space.]

Table 3.—Transect of the moisture gradient at low elevations (below 2500) feet in the Great Smoky Mountains to show relation of binomial distributions of all stems (one-half inch and larger at breast height) to canopy stems. Figures are percentages of stand

	1.* Deep cove forest		2. Cove forest		3. Cove forest transition		4. Oak-hickory forest		5. Oak-chestnut forest		6. Oak-chestnut heath		7. Pine forest	
	Stems %	Can. %	Stems %	Can. %	Stems %	Can. %	Stems %	Can. %	Stems %	Can. %	Stems %	Can. %	Stems %	Can. %
Carya cordiformis (Wang.) K. Koch	0.1	—	1.7	1	—	—	—	—	—	—	—	—	—	—
Cladrastis lutea (Michx. f.) K. Koch	0.1	11	0.6	—	—	—	—	—	—	—	—	—	—	—
Betula alleghaniensis Britton.	4.5	16	4.4	5	0.1	4	—	—	—	—	—	—	—	—
Aesculus octandra Marsh.	17.2	1	3.7	—	0.8	5	—	—	—	—	—	—	—	—
Fagus grandifolia Ehrh.	0.4	15	1.3	3	2.9	2	0.1	—	—	—	—	—	—	—
Tilia heterophylla Vent.	13.1	5	5.8	11	1.1	1	0.3	—	—	—	—	—	—	—
Fraxinus americana L.	1.2	—	0.2	1	1.0	1	1.2	1	—	—	—	—	—	—
Acer saccharum Marsh.	4.0	22	5.4	7	1.0	8	0.5	7	0.2	—	—	—	—	—
Liriodendron tulipifera L.	0.3	1	4.9	7	3.1	3	2.4	—	1.3	—	—	—	—	—
Magnolia acuminata L.	0.2	—	0.8	3	0.7	—	0.3	—	0.2	—	—	—	—	—
Halesia monticola (Rehder.) Sarg.	50.3	12	10.6	3	2.5	·1	0.6	—	1.7	1	—	—	—	—
Ilex opaca Ait.	0.1	—	0.5	—	0.8	—	0.3	—	—	—	0.2	—	—	—
Quercus borealis Michx. f.	0.2	1	2.5	3	2.7	4	7.0	30	4.3	3	2.4	1	0.2	—
Tsuga canadensis (L.) Carr.	6.6	11	18.5	47	15.1	37	2.9	1	0.7	—	0.6	—	0.1	—
Cornus florida L.	0.8	—	8.7	—	11.8	—	14.3	—	7.3	—	3.3	—	—	—
Carpinus caroliniana Walt.	—	—	0.5	—	0.7	—	—	—	—	—	—	—	—	—
Acer rubrum L.	0.8	3	4.6	4	20.2	8	26.0	4	29.1	4	27.5	4	5.0	1
Magnolia tripetala L.	—	—	0.5	—	0.7	—	—	—	—	—	—	—	—	—
Magnolia fraseri Walt.	—	—	2.1	3	2.9	2	—	—	—	—	—	—	—	—
Betula lenta L.	—	—	10.2	—	4.2	1	0.2	—	2.1	—	—	—	—	—
Acer pensylvanicum L.	—	—	2.0	—	2.9	—	0.3	—	0.3	—	0.9	—	—	—
Hamamelis virginiana L.	—	—	1.9	—	5.4	—	1.7	—	0.2	—	0.3	—	—	—
Carya glabra (Mill.) Sweet	—	—	0.5	—	3.0	—	4.9	11	2.6	5	3.3	—	—	—
Carya ovalis (Wang.) Sarg.	—	—	0.2	—	—	—	0.8	14	1.6	3	1.1	1	0.2	—
Nyssa sylvatica Marsh.	—	—	0.7	1	1.5	—	1.8	1	2.6	—	2.4	—	2.1	1
Alnus rugosa (Du Roi.) Spreng	—	—	—	—	0.5	—	—	—	—	—	—	—	—	—
Clethra acuminata Michx.	—	—	—	—	0.3	—	—	—	—	—	—	—	—	—
Quercus falcata Michx.	—	—	—	—	0.2	—	—	—	—	—	—	—	—	—
Ilex montana T. and G.	—	—	—	—	0.1	—	—	—	—	—	—	—	—	—
Liquidambar styraciflua L.	—	—	—	—	0.1	—	—	—	—	—	—	—	—	—

Table 3.—(Continued)

	1.* Deep cove forest		2. Cove forest		3. Cove forest transition		4. Oak-hickory forest		5. Oak-chestnut forest		6. Oak-chestnut heath		7. Pine forest	
	Stems	Can.	Stems	Can.	Stems	Can.	Stems	Can.	Stems	Can.	Stems	Can.	Stems	Can.
	%	%	%	%	%	%	%	%	%	%	%	%	%	%
Salix nigra Marsh.	—	—	—	—	0.1	—	0.1	—	0.1	—	—	—	—	—
Amelanchier canadensis (L.) Medic.	—	—	—	—	0.4	..	—	—	—	—	—	—	—	—
Pyrularia pubera Michx.	—	—	—	—	0.1	—	—	—	—	—	0.2	—	6.5	—
Oxydendrum arboreum (L.) DC.	—	—	—	—	5.1	1	8.4	6	11.3	—	12.7	—	0.3	—
Carya alba (L.) K. Koch	—	—	—	—	1.1	—	6.9	—	1.4	—	0.2	—	0.1	—
Robinia pseudoacacia L.	—	—	—	—	0.2	—	0.5	—	1.4	—	3.5	6	0.5	1
Quercus alba L.	—	—	—	—	2.2	3	5.7	10	1.3	1	7.1	8	2.8	3
Pinus strobus L.	—	—	—	—	1.7	7	2.4	2	2.8	—	17.0	31	2.6	9
Quercus montana Willd.	—	—	—	—	1.7	2	5.2	15	18.8	44	9.3	43	0.8	4
Castanea dentata (Marsh.) Borkh. (dead)	—	—	—	—	1.5	8	0.6	4	5.2	33	0.8	1	1.2	1
Quercus velutina Lam.	—	—	—	—	0.3	—	2.4	4	2.4	3	3.5	3	7.2	13
Quercus coccinea Muench.	—	—	—	—	0.1	—	0.1	—	0.1	—	—	—	—	—
Cercis canadensis L.	—	—	—	—	—	—	0.9	—	—	—	—	—	—	—
Juglans nigra L.	—	—	—	—	—	—	0.3	—	—	—	—	—	—	—
Ulmus fulva Michx.	—	—	—	—	—	—	0.1	—	—	—	—	—	—	—
Sassafras albidum (Nutt.) Nees.	—	—	—	—	—	—	0.7	—	1.5	1	0.5	1	—	—
Pinus virginiana Mill.	—	—	—	—	—	—	—	—	0.6	—	1.8	—	41.2	40
Pinus rigida Mill.	—	—	—	—	—	—	—	—	0.5	—	1.5	—	26.7	26
Quercus marilandica Muench.	—	—	—	—	—	—	—	—	—	—	—	—	2.5	—
Total stems	907	103	937	115	1035	110	1391	120	870	100	663	72	962	108
Stands in sample	1		5		5		8		7		6		14	
Canopy limit in inches (largest 10 per cent)	14		10		8		8		7		8		8	

* These larger samples are along the same gradient from moist cove forests to dry pine forests as Table 1; the station numbers of the two transects do not correspond.

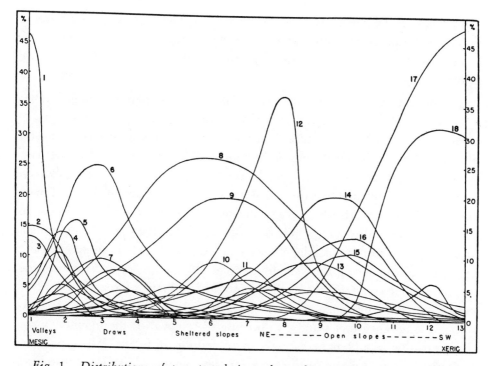

Fig. 1.—*Distributions of tree populations along the moisture gradient, illustrating the binomial form and the continuity of populations through associations (mixed meso-phytic, oak-hickory, oak-chestnut, and pine, from left to right). The curves are smoothed from the data of a composite transect (Table 1) for low elevations, 1500 to 2500 feet, in the Great Smoky Mountains; values of the ordinate are percentage of stems in the stand.* Major species indicated: *1,* Halesia monticola; *2,* Aesculus octandra; *3,* Tilia hetero-phylla; *4,* Betula alleghaniensis; *5,* Liriodendron tulipifera; *6,* Tsuga canadensis; *7,* Betula lenta; *8,* Acer rubrum; *9,* Cornus florida; *10,* Carya alba; *11,* Hamamelis virginiana; *12,* Quercus montana; *13,* Quercus alba; *14,* Oxydendrum arboreum; *15,* Pinus strobus; *16,* Quercus coccinea; *17,* Pinus virginiana; *18,* Pinus rigida.

into distinct communities. The communities we know as zones may have a boundary of sorts in one direction at least, and communities of special habitats may seem sharply defined; but these are special cases. So far as the general picture, on which general concepts may be based, is concerned, the conception of the association as a fundamental unit does not agree with the evidence and must be abandoned. This criticism of the association may be applied, with perhaps some modification, to the union, formation, zone, and other units. It does not mean that some of these are not useful or even unavoidable for many types of ecological research, and criticism of their use as arbitrary groupings for some purposes is not intended. In many problems of vegetation study, however, associations or other groupings are not needed and should be avoided as misleading.

The succession concept is more secure, although, as is well enough known, a sere is a kind of variable and irregular continuum and not an orderly series of distinct steps. The conclusion, however, that all seres in an area converge ultimately on a climatic climax is as much subject to doubt as is the association. We can observe that a tendency toward convergence exists, but we cannot therefore assume that this convergence can be extrapolated to a climatic climax. It is more difficult to make unequivocal tests of the climatic climax concept against reality than to test the association, but with certain preliminary qualifications a test is possible. We cannot very well include in succession, a biological phenomenon, the very long-range physiographic processes in order to produce our convergence. In the number of millions of years necessary to reduce a mountain range to base level the climate would have changed through long-range cycles and also because of removal of the mountains, and the vegetation would have evolved to something different. We cannot use perfectly flat country as a justification for the climatic climax since featureless topography is exceptional. Most of the earth's surface has some relief; if the climatix climax is not to be a generalization from a special case it must fit regions with a fair degree of topographic diversity. The climatic climax concept can be tested, then, by asking whether it can be demonstrated that on mature topography the vegetation of all site types is actually converging toward the same canopy and undergrowth and whether, in old mountain ranges and in the tropics, where vegetation has probably had time to develop as far as it is going to on existing sites under present climate, such convergence has occurred.

If we turn again to such old mountain ranges as the Great Smoky Mountains, we do not find the convergence in question. We find a variety of vegetation corresponding to the variety of site conditions. The vegetation has not converged in these mountains, and stand-tables from all parts of the gradient give no support to the idea that convergence is now occurring. The various oak types have the largest extent at lower elevations, but the mixed mesophytic forests are not changing, and those who have studied the pines at the xeric extreme seem to agree that they are climax for their sites (Whittaker, 1948; Braun, 1936; Cain, 1937). The vegetation in sites not recently burned appears to be stable, in balance with environment, except as death of the chestnuts has changed the composition of some stands. Succession does not really lead to a regional climax type; it leads to some kind of climax suited to the specific conditions of the site where succession occurs. Since there is always a variety of moisture conditions in mature topography, succession, on a time-scale which can reasonably concern ecologists, can only lead to a variety of climax types. The fact that a given growth-form may prevail regionally and correspond to climate does not imply unity of climax for different sites.

Some types are described as pre- and postclimax. If the vegetation is not actually changing in one direction or another under existing climate, the prefixes have not the meaning they imply; all types are equally climax. In fact, for rooted plants, climatic moisture factors act only through the specific moisture conditions of the site, and climax is relative to site. We may judge, then, that the climatic climax is at best a special case. If it is not unequivocally false, it is at least an unwarranted extropolation which does not agree sufficiently with the evidence and which should be replaced with a more reasonable interpretation. The alternative would seem to be the study of actual climax patterning and the description of regional prevailing climaxes when it can be determined that a given type actually prevails on a majority of the sites in an area. It may be recognized that in many areas there may be no prevailing climax.

MOSAIC CHART FOR CLIMAX PATTERN

Better methods of interpreting vegetation patterns than those based on the climatic climax can be developed. In the Great Smoky Mountains the transects were first used to trace vegetation change along gradients of moisture and elevation. Then a device called the mosaic chart was used, a chart with elevation and moisture, the latter determined here by topographic location of the site, as axes. The random samples of vegetation, about 300 of them, were then classified into arbitrary "types" and plotted against the axes of the chart. When lines were drawn around the types, the vegetation pattern shown in Figure 2 appeared. It may be emphasized that the division into types and the position of the lines separating them are arbitrary; the only partial discontinuities existing in the field are those separating the gray beech forests and grassy balds from the rest of the pattern. The picture is not of one or a few distinct climaxes, but of a climax pattern which mostly changes continuously from site to site. Climax vegetation here is a complex continuum of plant populations. It is

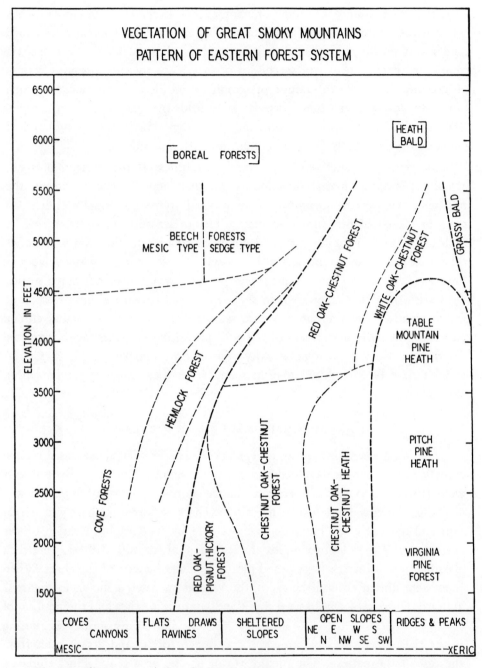

Fig. 2.—Climax pattern for the Eastern Forest System in the Great Smoky Mountains, mosaic chart based on 300 site-samples.

suggested that the climax pattern is a more concrete and realistic view of vegetation than the climatic climax abstraction. It is far closer to being a picture of what is actually out in the field.

The association and climatic climax may be regarded as the preliminary approximations of early ecologists in their attempt to describe the reality of vegetation. They are projective concepts, Aristotelian in orientation as most ecological thinking has been; and the classifications and abstractions of the observer were projected into nature and seen there as part of reality. It is often true that our interpretation of reality is more to be compared with the image of a projection machine than that of a camera unless we have means of testing what part of the picture is cast from outside in and what from inside out. The gradient analysis techniques described here are one means of testing reality by bypassing the association and climatic climax in order to work up to our picture of natural communities from more certain phenomena—gradients and the distribution of populations along them. The techniques described here only touch the surface of possible applications of the basic approach. The discarding of early concepts is no loss to ecology if methods not based on them are at hand and if interpretations like the association as an arbitrary grouping and the climax pattern can be drawn from them to give a clearer picture. This picture, which has been emerging since Gleason's paper in 1926, should serve us well enough for the present even though, as our knowledge increases and methods improve, the adjustment of concept to reality should be a continuing process.

Literature Cited

Braun, E. L. 1936. The vegetation of Pine Mountain, Kentucky. Am. Midl. Nat., 16:517–65.

———. 1940. An ecological transect of Black Mountain, Kentucky. Ecol. Monog., 10:193–241.

Brown, D. M. 1941. The vegetation of Roan Mountain: a phytosociological and successional study. Ecol. Monog., 11:61–98.

Cain, S. A. 1947. Characteristics of natural areas and factors in their development. Ecol. Monog., 17:185–200.

———., et al. 1937. A preliminary guide to the Greenbrier-Brushy Mountain nature trail, the Great Smoky Mountains National Park. Unpubl. ms., University of Tenn. 29 pp.

Dansereau, Pierre. 1943. L'Érablière Laurentienne. I. Valeur d'indice des espèces. Contrib. de l'Institut Botanique de l'Universite de Montreal, 45:66–93.

Egler, F. E. 1947. Arid southeast Oahu vegetation, Hawaii. Ecol. Monog., 17:383–435.

Fautin, Reed W. 1946. Biotic communities of the northern desert shrub biome in western Utah. Ecol. Monog., 16:251–310.

Gleason, H. A. 1926. The individualistic concept of the plant association. Bull. Torrey Bot. Club, 53:7–26.

———. 1939. The individualistic concept of the plant association. Am. Midl. Nat., 21:92–110.

Mason, H. L. 1947. Evolution of certain floristic associations in western North America. Ecol. Monog., 17:201–10.

Munns, E. N. The distribution of important forest trees of the United States. U. S. Dept. Agric., Misc. Publ. 287.

Whittaker, R. H. 1948. A vegetation analysis of the Great Smoky Mountains. Thesis, Ph.D., University of Illinois.

14

Reprinted from *Ecology* 32(3):476–496 (1951)

AN UPLAND FOREST CONTINUUM IN THE PRAIRIE-FOREST BORDER REGION OF WISCONSIN [1]

J. T. Curtis

University of Wisconsin, Madison

and

R. P. McIntosh

Middlebury College, Middlebury, Vermont

Introduction

It has been said that the desirable order of ecological research in a given region is first, the study of communities, second, the study of the individual species, and last, the study of the habitat (Yapp 1922). Sociological investigations of constituent species and their numerical relations should precede autecological studies if the latter are to have maximum meaning. This sequence should lead to a greater efficiency in research, in that an initial knowledge of any natural groupings of species may enable the autecology of the groups to be studied simultaneously, with a resultant saving in time and travel expenses. The grossly inadequate state of autecological knowledge of even our most common species indicates that the current haphazard method of attack is faulty; if aggregations of species occur together under similar environmental conditions in nature, they could be studied in a systematic program. A long-term investigation of the communities of Wisconsin has been underway since 1946 with a view to determining the existence and the floristic nature of such aggregations (Curtis and Greene 1949). The present paper deals with the upland hardwood forests of the prairie-forest border region of the state and their interrelations with each other and with certain physical factors of the environment. Deep appreciation is expressed to Professor P. B. Whitford, who

[1] This work was supported in part by the Research Committee of the Graduate School of the University of Wisconsin from funds supplied by the Wisconsin Alumni Research Foundation.

generously placed his files on forest composition at our disposal; to Professors Grant Cottam, H. C. Greene, N. C. Fassett, R. S. Muckenhirn, M. L. Partch, and S. A. Wilde for valued advice and consultation; to Messrs. Orlin Anderson, and R. S. Brown for indispensable aid in the collection of field data; to Miss Margaret Gilbert for aid in the computation; to Messrs. A. M. Fuller, Neil Harrington and Fred Wilson for help in the location of stands; and to the many members of the Soil Conservation Service, the Regional Foresters of the Wisconsin Conservation Department and the advisers of the Junior Wisconsin Academy of Sciences, who so willingly imparted their knowledge of local conditions in their regions. The authors are also indebted to Professor K. J. Arnold, of the University Computing Service, for aid in the statistical phases of the work.

The taxonomic nomenclature used in this paper follows that of Gray's Manual of Botany, 8th edition, 1950.

Literature Review

The earliest references to the vegetation of southwestern Wisconsin were made by the explorers and pioneer travelers in the years following 1698. For the most part, they contained only general descriptions, with remarks about unusual species. The first account by a competent botanist was that of Increase A. Lapham in 1852, but he dealt primarily with the distribution of individual species. The first description of plant communities as such was given by T. C. Chamberlain in 1877. This eminent geologist, who was later to exert such great influence on H. C. Cowles at the University of Chicago, was at that time the State Geologist of Wisconsin. He distinguished three forest communities in southern Wisconsin, but made clear that no abrupt line of demarcation existed between them, with repeated statements illustrating "a gradual transition from one to the other" and "an almost imperceptible merging of one into the other." His three communities were the "oak group" including *Quercus alba* (white oak), *Q. macrocarpa* (bur oak), *Q. rubra* (red oak), and *Q. velutina* (black oak), with *Populus tremuloides* (aspen), *Carya ovata* (shagbark hickory), *Prunus serotina* (black cherry) and *Pyrus ioensis* (wild crab apple) as lesser members; the "oak and maple group" in which *Acer saccharum* (sugar maple) was most conspicuous but *Quercus alba* and *Q. rubra* were still common, with *Acer rubrum* (red maple), *Ulmus rubra* (slippery elm), and *Tilia americana* (basswood) in small numbers; and the "maple group" with *A. saccharum* dominant, but with *Tilia americana, Ulmus rubra,* and *Ostrya virginiana* (ironwood) as characteristic and abundant associates. He comments about the "oak and maple group" as follows: "This group is not characterized by the exclusive presence of any prominent plant but by a distinctive association of plants common to several groups."

Bruncken (1900) also described three forest associations, but only two were found on uplands—the "hemi-xerophytic oak-hickory" and the "mesophytic basswood-maple." Additional observational studies on the forests of the region were reported by Pammel (1904), Marks (1942) and Stout (1944), but the first account in which quantitative data on actual composition were presented was that of Wagner (1947), who studied a stand in Jefferson County which fits into Chamberlain's oak and maple group. Cottam (1949) investigated an oak woods in Dane County by quantitative methods and reported in detail on the historical development of oak openings

and the changes which have taken place in them since settlement. He attributed major importance to fire in the maintenance of oak openings and to cessation of fire as the cause of the subsequent development of the present closed canopied stands.

The most comprehensive paper on the forests of southern Wisconsin was by Whitford (1951). Although this study was primarily concerned with the relation between the clone size of herbs and the age of the forest stands, considerable attention was devoted to the establishment of a method for classifying the stands. In the course of the investigation, Whitford studied a considerable number of stands and presented detailed data on tree, shrub, and herb composition. He arranged his series of 28 stands into three groups on the basis of shade tolerance of the tree species. He assigned each species to one of three tolerance classes, and then determined the relative fraction of all trees in a stand which belonged to each of these classes, on a summation basis. This method resulted in a sequence of stands from those composed entirely of intolerant species to those made up mostly of tolerant species. Whitford divided his stands into three groups—"oak-hickory, intermediate, and maple-basswood," but the division was made for pragmatic reasons and was not based on any internally apparent separation of the data into three groups.

Studies in Illinois, Iowa and Minnesota indicate that a series of forests similar to those in Wisconsin are found in at least portions of these states. No quantitative investigations in northern Illinois have come to our attention, but the general descriptions of Fuller and Strasbaugh (1919), and DeForest (1921) indicate a series of forest types from those dominated by *Quercus* and *Carya* to those dominated by *Acer saccharum*. In Iowa, the studies of Clark (1926) and of Aikman and Smelser (1938) show a succession from *Quercus macrocarpa* and *Q. velutina* in the initial stages through

Quercus rubra-Tilia americana stands to the terminal forests of *Acer saccharum* and *Tilia americana*. Daubenmire (1936) reported on the structure of the *Acer-Tilia* community and on its relation to the surrounding *Quercus* forests in the "Big Woods" of Minnesota.

While the results of the various investigations in these four states are by no means in complete agreement, there does appear to be a general uniformity of opinion as to the topographic distribution of the forests, with some type of *Quercus* forest on the most xeric forested sites and an *Acer* forest in the most mesic habitats. The disagreement seems to lie in the determination of the number of discrete communities that may be recognized, in the floristic nature of the boundaries between them, and in their exact composition. Many of the early workers emphasized the lack of sharp lines of demarcation, while later workers, who could study only the more completely dissected and isolated stands resulting from the highly agricultural land use of the present time, were more inclined to describe discrete communities of several types.

Although the flora of Wisconsin has many diverse floral elements, it was early pointed out that there are two major floristic provinces, separated by a diagonal line from northwest to southeast (Cheney 1894). The southwestern portion of the state contains the prairies and hardwood forests, while the northeastern part contains both hardwood and coniferous forests. The boundary line between them, actually a zone or band, is comparable to the tension zone proposed by Griggs (1914). Its position can be determined within fairly narrow limits by the method proposed by Clements (1905: 187) "The limiting line or ecotone of a . . . province is a composite obtained from the limits of principal species and checked by the limits of species typical of the contiguous vegetations."

Applying this to Wisconsin, a number of trees including *Quercus velutina*, *Carya ovata*, and *Juglans nigra* reach their northern boundaries in a zone which also includes the southern limits of such trees as *Betula lutea*, *Abies balsamea*, and *Populus balsamifera*. This same zone marks the northward or southward extent of a great many herbaceous species, both native and exotic; it also separates closely related varieties of single species, as in the case of *Acer saccharum saccharum* and its subspecies *A. s. nigrum* (Desmarais 1948) and *Brachyelytrum erectum erectum* and *B. e. septentrionale* (Salamun 1950). The floristic province southwest of the zone is here termed the prairie-forest province, while that to the northeast is the northern hardwoods province (roughly equivalent to the Lake Forest of Weaver and Clements 1938). It is recognized that these names may be of local descriptive value only and are considered to be tentative. Perhaps the most characteristic feature of the floristic provinces from the standpoint of vegetation is the occurrence of oak woods in the early forest succession south of the zone in contrast to pine woods in the same successional position north of the zone. The terminal forests in both provinces are dominated by sugar maple. Within the narrow zone itself various elements of the two provinces occur in puzzling mixtures. Their elucidation must await completion of studies on both sides of the zone.

These floristic provinces are similar to the natural areas of Cain (1947) and the floristic areas of Raup (1947) and Egler (1948). They are characterized by the presence of a uniform flora such that "all stands have equal chances of containing all the same species" (Cain 1947). It is only within such limitations that plant associations can possibly achieve objective reality. One of the aims of the present paper is to determine whether discrete communities with definite structure and definable boundaries actually exist within such a natural floristic province, or whether plant associations are "metaphysical approximations in a field where there are unlimited variables, combinations and permutations" (Cain 1947) even within the narrow limits set by the province.

DESCRIPTION OF AREA

Physiography

The south and west borders of the area included in this study (the Wisconsin-Illinois boundary and the Mississippi River) were prescribed by various non-ecological but nevertheless compelling reasons. The remaining boundary was based on the biologically more valid presence of the tension zone discussed above which extends across Wisconsin from northwest to southeast (Fig. 1). The major physiographic differentiation of the region is its separation into glaciated and non-glaciated areas. The Driftless Area in the western portion of the study area is a region of mature drainage patterns and rugged, almost mountainous, topography as contrasted with the glaciated eastern portion which possesses immature drainage and a more or less rolling topography. The glaciated areas consisted primarily of Wisconsin drift with lesser areas of older (Illinoian)

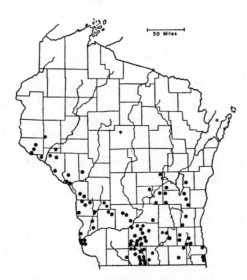

FIG. 1. Map of Wisconsin showing location of stands studied. The cross near the center of the map indicates 90° W. Long., 45° N. Lat.

drift. The underlying bedrock is composed predominantly of old sedimentary rocks—Cambrian sandstones and Lower Magnesian, Ordovician and Silurian limestones.

Climate

The area under consideration lies almost wholly in the transition zone between climatic regions I and IV of Borchert (1950). As such, its climate is intermediate between that of the Northeastern forest region of snowy winters and reliable summer rains and that of the central Prairie region with relatively dry winters and frequent summer droughts. Partaking of both forest and grassland climates, it supports both prairie and forest and their intermediate expression, the oak savanna. Excellent summaries of the ecologically significant climatic variables for the general region are mapped by Borchert (1950).

Soil

The forest soils of the area belong to the grey-brown podzoic soil group and are largely of the Fayette-Dubuque series in the Driftless Area and the Miami-Bellefontaine series in the eastern glaciated area. The topsoils in both series are usually composed of loess or a mixture of loess and residual soil. Profile characteristics are similar throughout the study area. In genera, there is a well-marked dark A_1 layer varying from one-half inch to as many as ten inches in depth, although one to four inches is more typical. The A_2 is not well defined in terms of color but may be detected by textural differences; it is usually one-half to two inches in depth. The brownish B layer generally shows a well-defined nut structure.

General vegetation

The southwestern half of Wisconsin lies in the broad ecotonal belt between grassland and summergreen forest. In presettlement times its vegetation consisted of scattered prairies, oak openings of savanna-like aspect, and broadleaf deciduous forests of varying density and composition. The present-day stands of natural vegetation are mostly derived from these three communities, although other vegetation types are also present in the form of relics (McIntosh 1950), The deciduous forests occur on a range of topographic sites, from very dry through mesic to very wet. The driest sites, on cliffs, sandy soils and southwest slopes are characteristically covered with scattered stands of small oaks (*Quercus velutina, Q. macrocarpa*) or with red cedar (*Juniperus virginiana*) glades. As the moisture conditions improve, the forests become more dense, with a greater number of important tree species. The most mesic conditions commonly support sugar maple, basswood or other species of similar requirements. Wet soil forests are largely restricted to the flood plains of streams and rivers but are found to a limited extent also on the shores of glacial lakes in the eastern portion of the area. Their most characteristic trees are *Ulmus americana, Acer saccharinum, Fraxinus pensylvanica lanceolata* and *Quercus bicolor*.

Most of the original prairie has been placed under cultivation but small remnants still occur along railroads and highways and occasionally elsewhere. Prairies of several sorts occurred on the same range of topographic types as the deciduous forests (Curtis and Greene 1949). Prairies formed a part of the presettlement vegetation of all counties included in the present study, although the area covered varied from less than one per cent in Richland County to over 50 per cent in LaFayette County. It is estimated that 20 per cent of the total area was in prairie when the first settlement by Europeans began in the 1830's.

METHODS

The plan of the investigation used in this work is an adaptation of the mass collection method used by taxonomists (Anderson 1941; Fassett 1941). It is our conviction that variation in floristic composition is one of the most important characteristics that may be determined in the study of any vegetation, and this may be understood only after a great many examples of the type have been analyzed. It is important that the initial ideas of the type limits be sufficiently broad to preclude the dangers of subjective selection of only those stands which fit a preconceived notion as to what a particular community should be. The only criteria used for the selection of stands in the present study were: (1) that they be natural forests (*i.e.*, not artificially planted) of adequate size, fifteen acres being the minimum, with forty acres or more preferred; (2) that they be free from disturbances in the forms of fire, grazing or excessive cutting; (3) that they be on upland land forms on which run-off waters never accumulate. If a stand fulfilled these conditions, it was studied, regardless of what species of trees might be present. It might be expected that homogeneity would have been a required characteristic, but, as will be shown below, it was found that this could be determined statistically from the data at a later time. If a stand was then found not to be homogeneous, its data were eliminated from the compilations.

A few of the stands were known to us before the study was begun, but the majority were found in response to a circular letter sent out to state and federal forestry agencies, county agents, Soil Conservation Service personnel, and others whose work brings them into contact with woodlands. All of the stands discovered in this way were checked in the field, and, if they met the criteria of selection, they were incorporated in the study. In this way, an essentially random selection of 95 stands in 29 counties was obtained without prejudice as to kind. The geographical distribution of the samples (Fig. 1) was believed adequate to give an accurate picture of the forests of the area.

Field methods

Each stand was sampled by use of the random pairs method (Cottam and Curtis 1949). Forty pairs of trees were used for the determination of frequency, density and dominance. Presence was recorded for all vascular species on specially prepared forms which contained the names of the 200 plants most commonly found in upland forests of the region. Frequency of herbs, shrubs and tree seedlings was obtained by 20 quadrats, each ¼ milacre in size. Density was not determined for these forms.

Saplings over 1″ DBH were counted in a transect between the two trees of each pair. Since the distance between the trees was variable, the transects resulted in different total areas in each stand and thus prevented the use of frequency as a measure. Density, therefore, is the only type of information available for the saplings. It appears desirable to devise some other methods of sampling saplings in future studies, for in many cases the sample derived by the method described above was inadequate.

As will be shown in detail elsewhere, the random pairs method, using 40 pairs of trees per stand, gives results whose accuracy compares very well with that obtained by the quadrat method. For example, a composite vegetational index (see below) by random pairs for three stands on which adequate quadrat data were also available showed an average deviation of only 2.9% from the index calculated from the quadrat figures. Duplicate analyses by the random pairs method made in three stands in two consecutive years by two sets of investigators showed an average trial to trial variation in the index of only 1.34%. This compares with an average deviation of 2.11% in five sets of quadrats made in as many years in one of the same stands. The random pairs method may therefore be assumed to give reproducible results of sufficient precision for the purposes at hand.

We knew of no way by which the homogeneity of stands could be determined objectively in the field, so it became necessary to devise a method of checking this after the data were collected. Since 40 separate pairs of trees are measured and recorded in sequence in the field, it is possible to segregate the data from successive groups of 10 pairs, each representing a different areal portion of the stand. Using these four groups as separate samples, the Chi-square test for homogeneity (Snedecor 1946) may be used to determine whether the actual occurrences of the major trees in any group deviate significantly from the occurrences expected on the basis of uniform distribution. The requirement of an expected value of at least five individuals in each sample makes this test applicable only to those tree species represented by at least 20 trees in 40 points. These species are the ones which determine the character of the stand in any case and if they are uniformly distributed, deviations on the part of lesser tree species are not likely to be significant. When this technique was applied to the current data it was found that in 95 cases only five Chi-square values exceeded the expected value at the five per cent level. In other words, five of 95 differed significantly from a homogeneous distribution. This approximates very closely the expected number (5 out of 100) of Chi-square figures which would exceed the expected by chance. On this basis it seemed reasonable to accept these as results of chance and reason for not rejecting the stands. This technique may be applied in the field in the future; if a stand is found not to be homogeneous further study will be in order.

Soil samples were collected separately from the A_0, A_1, and B horizons in most stands. This was accomplished by the use of three small pits located in representative portions of the wood, with the three subsamples pooled for later analysis.

Treatment of data

The data were tabulated and recorded on standard data sheets. Measures of number, size, and distribution for each tree species were calculated from the figures provided by the random pairs method for each stand. Relative (per cent) density and relative (per cent) dominance were determined in the ordinary manner, but frequency was calculated as percentage sum of frequency or relative frequency (Raunkiaer 1934, Curtis and McIntosh 1950), rather than the more usual simple frequency. The purpose of this change was to place frequency on the same mathematical basis as density and dominance, so that a summation index (density plus frequency plus dominance), with a constant value of 300 for all species in a stand could be prepared. The relative importance of a tree species is more clearly expressed by this method than by the DFD (density-frequency-dominance) index in its original form (Cottam 1948; Whitford 1949, Stearns 1951) where the sum was indeterminate. The index is calculated by adding the separate values of relative density, frequency and dominance for each species. Its magnitude is an excellent indication of the vegetational importance of a species within a stand, since it is sensitive to such variables as apparent contagion or exceptional basal area. This summation index of a particular species within a stand will be referred to herein as its *importance value* in order to differentiate it from the DFD index used by earlier workers. Not only does it differ because of its use of relative frequency, but because frequency as determined by the random pairs method is not strictly comparable to frequency measured by quadrat sampling. Quadrat frequency values bear a mathematical relationship to density values, with a fixed minimum ratio and a variable maximum depending chiefly on the degree of randomness or contagion exhibited by the species (Curtis and Mc-

Intosh 1950). Random pair frequency numbers, however, are more limited, in that they may range from equal to the density, down to one half the density, but they never can be smaller than one half the density values. As a result, frequency numbers usually are close to densities in absolute magnitude. The importance value, therefore, is weighted toward density, in that the number of trees present exerts a greater effect on the index than does their size. This seems proper in a floristic index.

Soil acidity and water-retaining capacity were determined in the laboratory on air-dried samples which were ground with a mortar and pestle and sieved through a 2 mm screen. Acidity was determined by the Helige-Truog method, while water-retaining capacity was measured by the Hilgard method. Analyses for available phosphorus, potassium, magnesium, and for organic matter were made by the Helige-Truog and Wilde methods at the State Soils Laboratory on the University campus. Exchangeable calcium was determined by volumetric analysis, also at the State Soils Laboratory. Nutrient ion concentrations are expressed as pounds per acre for the measured depth and weight of the particular horizon sampled. In the interests of brevity, only the data for the A_1 horizon are presented in this paper.

RESULTS

One of the difficulties inherent in the "mass collection" method of ecological study is the problem of handling the large quantities of diverse data that accumulate during the course of the investigation. It soon became apparent in the current work that the first requirement was a scheme for organizing or arranging the stands so that suitable comparisons of data could be made. To this end, an attempt was made to classify the stands into groups which had similar structure on the basis of tree composition. However, detailed examination of the data revealed no clear cut distinctions which would readily separate the stands into such groups without the exercise of a considerable amount of subjective judgment.

It is possible to classify the stands objectively on the basis of the leading dominant trees in each. If the single leading dominant is used for this purpose it would be necessary to recognize 9 types in the 95 stands. The use of the first two tree species in order would necessitate recognizing 30 types; of the first three species, 75 types; of the first four, 95 types. In other words, no two stands of the 95 examined had the same arrangement of their first four most important tree species. The number of units recognized is thus determined entirely by the basis of selection and in no way represents an approximation of any natural groups. Accordingly, we attempted to devise a method whereby each species present in the stand might be taken into account in the classification of the stand. In this way, any tendency of the stands to fall into distinct groups would have maximum opportunity for expression.

A study of the importance values for all species soon showed that only a few species ever achieved high levels of importance. The majority of tree species never occurred as important elements in any of the stands. The ability of a tree species to achieve a given importance value, or what may be termed its importance potential, is the result of various hereditary physiological attributes whose nature is but poorly defined at present. Some species, like *Quercus alba*, reach major dominance in many forest stands covering a wide geographic area. Others, like *Gymnocladus dioica*, apparently never are important members of the forest community in any part of their range. Four tree species—*Quercus velutina, Q. alba, Q. rubra* and *Acer saccharum*—have an importance potential markedly greater than that of any other trees of southern Wisconsin (Table I). Since they are the major tree components

TABLE I. *List of major tree species found in 95 upland hardwood stands of southern Wisconsin with data illustrating importance potential*

Species	No. of stands of occurrence	Average importance value	Maximum importance value
Quercus rubra	80	90.4	228
Acer saccharum	45	82.8	201
Quercus alba	86	64.1	202
Quercus velutina	40	62.8	206
Tilia americana	63	39.6	179
Ulmus rubra	53	33.4	140
Quercus macrocarpa	25	24.8	170
Fraxinus americana	28	20.0	94
Prunus serotina	56	17.3	114
Ostrya virginiana	48	16.2	42
Carya ovata	39	15.7	61
Acer rubrum	18	13.3	58
Juglans nigra	21	12.5	30
Carya cordiformis	43	12.2	44
Juglans cinerea	23	12.1	53

of the upland forests of the region, it was decided to study their interrelations as a possible initial basis for community delimitation.

These four species were the leading dominants in 80 of the 95 stands studied. When these stands were placed in groups with the same leading dominant, it was found that seven stands had *Quercus velutina* as the major tree, 18 had *Q. alba*, 34 had *Q. rubra*, and 21 had *Acer saccharum*. The percentage occurrence (constancy, based on a sample of 80 trees in 40 random pairs) of each of the four species and their average importance values were calculated for each of these groups. The results are shown in Table II. The stands in which *Q. velutina* was the leading dominant had a greater average amount of *Q. alba* than of either of the other two tree species, and *Q. alba* was present in a much greater fraction of the stands. In stands dominated by *Acer saccharum*, *Q. rubra* was characteristically second in importance. On the other hand, stands dominated by either *Q. alba* or *Q. rubra* each had two species of about equal rank in second place. Under these circumstances, the most logical arrangement of stands is that indicated in the table, which shows a decreasing order of *Q. velutina* and an increasing order of *A. saccharum*, with *Q. alba* and *Q. rubra* in intermediate positions.

TABLE II. *Average importance value (IV) and constancy % of trees in stands with given species as the leading dominant*

(For species with highest importance potential only—80 stands)

Species	Leading dominant in stand			
	Q. velutina	*Q. alba*	*Q. rubra*	*A. saccharum*
Q. velutina				
Average IV	165.1	39.6	13.6	0
Constancy %	100.0	72.3	38.3	0
Q. alba				
Average IV	69.9	126.8	52.7	13.7
Constancy %	100.0	100.0	97.1	66.7
Q. rubra				
Average IV	3.6	39.2	152.3	37.2
Constancy %	25.0	94.5	100.0	76.3
A. saccharum				
Average IV	0	0.8	11.7	127.0
Constancy %	0	5.6	29.4	100.0

TABLE III. *Average importance value (IV) and constancy % of trees in stands with given species as the leading dominant*

(Eleven species of intermediate importance potential—80 stands)

Species		Leading dominant in stand			
		Q. velutina	*Q. alba*	*Q. rubra*	*A. saccharum*
Q. macrocarpa	IV	15.6	3.5	4.2	0.1
	C %	50.0	38.9	20.6	4.8
Prunus serotina	IV	21.4	21.8	5.9	1.4
	C %	87.5	89.0	64.8	19.0
Carya ovata	IV	0.3	8.8	5.2	5.9
	C %	12.5	61.2	38.3	33.3
Juglans nigra	IV	1.5	1.2	2.2	1.9
	C %	12.5	11.1	20.6	23.8
Acer rubrum	IV	3.9	2.3	2.4	1.0
	C %	12.5	33.3	23.5	4.8
Juglans cinerea	IV	0	2.7	1.7	4.8
	C %	0	11.1	20.6	47.6
Fraxinus americana	IV	0	1.9	5.1	7.6
	C %	0	11.1	20.6	42.8
Ulmus rubra	IV	4.6	7.7	8.3	32.5
	C %	25.0	27.8	53.3	85.7
Tilia americana	IV	0.3	5.9	19.0	33.0
	C %	12.5	16.7	73.5	100.0
Carya cordiformis	IV	2.5	5.8	4.1	8.2
	C %	12.5	33.3	41.2	66.7
Ostrya virginiana	IV	0	2.4	5.5	16.2
	C %	0	22.2	41.2	95.3

TABLE IV. *Average frequency and constancy % of understory species in stands with given tree species as the leading dominant (80 stands)*

Species	Leading dominant in stand			
	Q. velutina	O. alba	Q. rubra	A. saccharum
Cornus foemina				
Average frequency %	20.0	23.9	16.2	0.0
Constancy %	66.6	100.0	67.8	12.5
Sanguinaria canadensis				
Average frequency %	0	7.5	7.8	14.0
Constancy %	0	35.7	46.5	50.0
Laportea canadensis				
Average frequency %	0	0	0.2	9.4
Constancy %	0	0	3.6	37.5

This ranking of the stands seemed to be in agreement with the general descriptions of the forest series reported for the region and thus gave promise of further fruitful investigation. Accordingly, the same method (based on groups with *Q. velutina*, *Q. alba*, *Q. rubra* and *A. saccharum* as leading dominants) was used to calculate average constancy percentages and importance values for a series of other tree species of intermediate importance potential. The results as shown in Table III have been arranged in a natural order in that those first in the list (*Q. macrocarpa*, *Prunus serotina*, *Carya ovata*) seem to be allied to *Q. velutina* and *Q. alba*, while those placed last (*Carya cordiformis*, *Tilia americana*, *Ostrya virginiana*) are clearly associated with *Acer saccharum*.

Average frequency and constancy

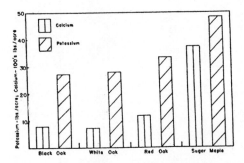

FIG. 2. Available potassium and calcium in the A_1 soil horizon of stands in which the indicated species are the leading dominants.

values for several herbs and shrubs were determined for the same groups of stands (Table IV); they showed the same types of distribution, with optimum values in a particular group for each species. Measurements of certain soil characters were also averaged by these groups (Fig. 2) and they too showed similar trends.

The alignment of lesser species with the four major dominant trees shown in Tables III and IV is not very precise, since it does not separate those species which may be associated with the same major dominant. To make further separation possible, the stands were classed into eight groups, based on the two leading dominants in order of their importance values. Thus the stands which had *Q. alba* as the main dominant were split into two subgroups—those with *Q. velutina* as the second most important tree and those with *Q. rubra* as the second tree. Constancy percentages and average importance values were calculated for all species in each subgroup. Some of the results are given in Figure 3. They are arranged with the sequence of the groups based on the information in Table II and with the individual species arranged in order from those most closely associated with *Q. velutina* to those most allied to *Acer saccharum* as shown by the position of the stands in which they reach optimum importance. On the basis of this information, the

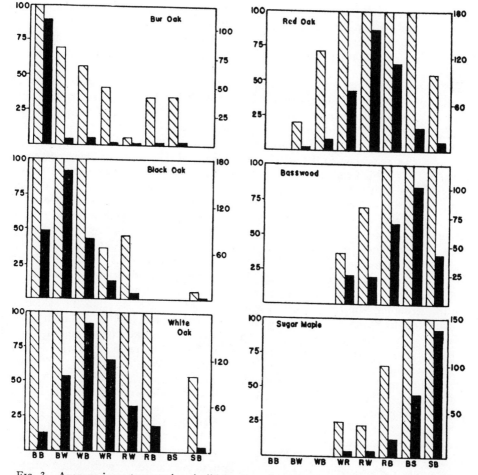

Fig. 3. Average importance value (solid bars) and constancy % (hatched bars) for major species in stands grouped according to the two leading dominants. The scales on the left ordinates are constancy percentages, those on the right ordinates are average importance values. The group of stands are indicated by the paired letters below the graphs, where BB represents the stands in which Bur Oak (*Q. macrocarpa*) is the leading dominant and Black Oak (*Q. velutina*) is the second leading dominant; BW represents stands with Black Oak as first dominant and White Oak (*Q. alba*) as second dominant; WB with White Oak leading and Black Oak second; WR with White Oak leading and Red Oak (*Q. rubra*) second; RW with Red Oak first and White Oak second; RB with Red Oak first and Basswood (*Tilia americana*) second; BS with Basswood first and Sugar Maple (*Acer saccharum*) second; and finally SB with Sugar Maple the leading dominant and Basswood second.

species can be arranged in the following order, from pioneer to climax species: *Quercus macrocarpa* → *Q. velutina* → *Prunus serotina* → *Q. alba* → *Carya ovata* → *Juglans nigra* → *Q. rubra* → *Acer rubrum* → *Fraxinus americana* → *Tilia americana* → *Juglans cinerea* → *Carya cordiformis* → *Ulmus rubra* → *Ostrya virginiana* → *Acer saccharum*.

The method of leading dominants thus served to arrange the stands into groups which showed a reasonable phytosociological pattern but it could not demonstrate the existence of discrete communities since it was based on an artificial, even though objective, choice of unifying species. In addition, it suffered from the vagaries resulting from the calculation of averages from differing numbers of stands in each group or subgroup. It seemed desirable, therefore, to treat the stands individually, rather than to average them in groups.

To facilitate handling this large amount of data, the moveable strip method was devised. A strip of opaque white celluloid, ¼ inch wide and 10 inches long, was prepared for each stand. The importance value of each tree species in the stand was shown on this strip by means of colored marks, located on the basis of the scale of importance values from 0 to 300. By this method the kinds and relative importance of the tree species in any stand could be determined at a glance, and by placing the strips side by side in a suitable frame, comparison of many stands could be accomplished by inspection. Natural groupings of tree species might be expected to become apparent, if any existed.

The most enlightening arrangement of stands was made on the basis of the information gained from the leading dominant method. The strips were placed so that the importance values of *Q. velutina, Q. alba, Q. rubra* and *A. saccharum* most nearly approached the order of these species which was indicated in Figure 3. This subjective procedure could not be very precise, but a good approximation was achieved (Fig. 4) as shown more clearly by Figure 5, where the same data are averaged by successive groups of

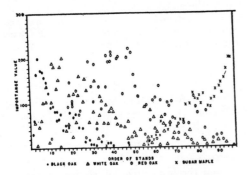

Fig. 4. Importance values of four major species in individual stands arranged in an order most closely approaching that shown in Fig. 3.

five stands, smoothed by the formula: $(a + 2b + c)/4$.

The other species were not used in establishing this original rank, but they were found to be distributed in characteristic positions when their importance values in each stand were arranged in the order established above. Figure 6 shows distributions of *Q. macrocarpa, Prunus serotina, Tilia americana* and *Ostrya virginiana*. *Quercus macrocarpa* and *Prunus serotina,* on the basis of their distribution relative to the major species, are seen to be more nearly akin to the pioneer species, while *Ostrya virginiana* and *Tilia americana* are seen to approach the climax species, thus substantiating the results of Figure 3. The remainder of the species are distributed similarly when plotted in this way.

The relative position in which each species approaches its optimum development can be fixed by means of diagrams similar to Figures 5 and 6 where sufficient data are available. The breadth of any of the figures and the absolute position of the mode would be changed if more stands were added to the series. However, the relative position remains essentially the same once a sufficient number of points is available. At the end of the first year's study, with data available for only 60 stands, the curves for the major species were well established in the same relative positions as given by the final data on 95 stands.

A study of all the figures shows that each species has a minimum, an optimum and a maximum point for development, with the points located by reference to similar states for other species. The only exceptions are *Q. macrocarpa* and *Acer saccharum,* which lack one of the cardinal points due to their location at opposite ends of the series within the limits of the study. It is to be emphasized that the optimum points for each species are purely relative. The strip method can show that *Juglans nigra* reaches its best development in forest stands which are intermediate between stands dominated by *Q. alba* and others dominated by *Q. rubra* but it cannot give a precise measure of the difference. Nevertheless, the arrangement of stands provided by this method shows that the tree species found in the upland hardwoods of southern Wisconsin form a continuous series, with overlapping ranges but distinct conditions for optimum development of each species. No discrete groupings of species are apparent; rather, the entire assemblage forms a continuum.

Climax adaptation numbers

The above results indicate that each tree species reaches its optimum development in stands whose position is fixed in a definite relationship to the other tree species. The sequence of the species in this pattern is such that pioneer species are at one end and climax species at the other. Pioneer and climax are here used in their adjectival sense and refer to the entire syndrome of specific physiological attributes which enables the one to live in the high light, variable moisture and immature soil conditions of initial stands and the other in the low light, medium moisture and mature soil conditions of the terminal forests.

The results of this study, as well as those of other investigations in this area, indicate that *Acer saccharum* is the tree species best equipped to persist in the terminal forests of the area. All other species are less efficient in this respect, and their relative degree of "climaxness"

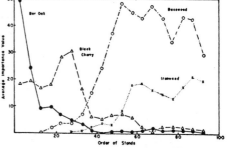

FIG. 6. Importance values of four lesser species, arranged in the order given in Fig. 4 and averaged by successive groups of 5 stands.

may be evaluated by the spatial relations of their optimum development curves as illustrated in Figures 5 and 6. Thus *Tilia americana*, whose highest importance values are reached in stands near the *A. saccharum* end of the pattern, is better adapted to terminal forest conditions than *Q. rubra*, with a peak closer to the middle of the series, while *Q. macrocarpa*, which is most important in stands remote from those dominated by *A. saccharum*, is the least well adapted.

Using *Acer saccharum* as a standard, all other species can be rated on the basis of the distance of their optimum stands from those dominated by *Acer*. The rating was accomplished by arbitrarily assigning a relative value of ten to *A. saccharum* and lesser numbers to the other species according to their position, as shown in Table V. Those species designated by an asterisk are tentatively placed with an approximate value inasmuch as they occur only rarely in the samples and thus their position is not clearly established by the few points available.

The choice of suitable words to designate the two ecologically distinct ends of the series is a troublesome one. The word "climax" for the terminal end is especially doubtful, since it already has a plethora of meanings and probably would not be defined the same by any two practicing ecologists. It has been suggested that the term "mesophily," as employed by Dansereau (1943), be used in the

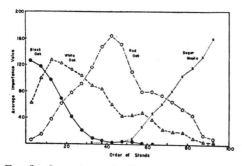

FIG. 5. Same data as in Fig. 4 averaged by successive groups of 5 stands.

TABLE V. *Scientific and common names of tree species found in stands studied, with the climax adaptation numbers of each*

Scientific name	Common name	Climax adaptation number
Quercus macrocarpa Michx.	Bur oak	1.0
Populus tremuloides Michx.	Trembling aspen	1.0
Acer negundo L.	Boxelder	1.0
Populus grandidentata Michx.	Large-tooth aspen	1.5
Quercus velutina Lam.	Black oak	2.0
Carya ovata (Mill.) K. Koch	Shagbark hickory	3.5
Prunus serotina Ehrh.	Black cherry	3.5
Quercus alba L.	White oak	4.0
Juglans nigra L.	Black walnut	5.0
Quercus rubra L.	Red oak	6.0
Juglans cinerea L.	Butternut	7.0
Ulmus thomasi Sarg.	Rock elm	7.0
Acer rubrum L.	Red maple	7.0
Fraxinus americana L.	White ash	7.5
Gymnocladus dioica (L.) Koch	Kentucky coffee tree	7.5
Tilia americana L.	Basswood	8.0
Ulmus rubra Muhl.	Slippery elm	8.0
Carpinus caroliniana Walt.	Blue beech	8.0
Celtis occidentalis L.	Hackberry	8.0
Carya cordiformis (Wang) K. Koch	Yellowbud hickory	8.5
Ostrya virginiana (Mill.) K. Koch.	Ironwood	9.0
Acer saccharum Marsh	Sugar maple	10.0

* Climax adaptation number of these species is tentative, because of their low frequency of occurrence in this study.

present case instead of climax. It is true that the climax plants here discussed are extreme mesophytes, in the sense that they achieve optimum development only in habitats of medium and relatively constant conditions of both atmospheric and edaphic moisture. However, it is equally true and probably of greater significance that these climax plants are shade-loving plants (sciophytes) able to grow and reproduce in conditions of very low light intensity. For the most part they are also plants of high nutrient demands, with tendencies toward partial heterotrophism. They are therefore not median with respect to light or nutrition and the use of the word "mesophily" might carry unfortunate connotations. It would involve an extension of meaning as great as in the case of "climax." What is perhaps needed is a word to encompass all of those adaptive physiological and morphological characters that differentiate plants of initial stands from plants of terminal stands. In the absence of such a word, we have deemed it best here to use the phrase "climax adaptation" as a tentative description of these characters, and to call the relative values assigned to each species "climax adaptation numbers." High numbers indicate good adaptation to all of the environmental factors present in terminal stands while low numbers indicate adaptation to the conditions of initial stands. This admittedly unsatisfactory compromise at least has the merit that it does not unduly emphasize one factor in an obviously multiple factor situation even though it does further burden the already overworked word climax. "Succession number" has been suggested, but it is inadequate, since an intermediate stage in the series may be reached by retrogressive as well as by successional processes.

A comparison of these climax adaptation numbers with the shade tolerance ratings of foresters indicates that they are roughly parallel. This is not unexpected, since the characteristics that contribute to great shade tolerance are largely the same as those that make for persistence in terminal stands. The method used here in establishing the climax adaptation numbers may offer a supplement to the usual procedures for the subjective determination of shade tolerance. It offers greater possibility of refinement than the coarse, qualitative estimates generally used. It must be recognized, however, that the application of this type of rating is only feasible within the bounds of a single floristic province, since shade tolerance as measured by the adaptation number of a species may vary in other areas with different component floras.

Vegetational continuum index

To establish a basis for evaluating forest stands in terms of their total tree composition it was decided to use the climax adaptation number of each tree species as a means of weighting its importance value as determined from the sampling data. The adaptation number of each species (as given in Table V) was multiplied by its importance value in a particular stand. The products were added and the resultant weighted total used as a basis of placing the stand in its proper relation to other stands. The maximum possible range of these weighted numbers is 2,700 units from 300 to 3,000. A stand composed entirely of *Q. macrocarpa* or a mixture of *Q. macrocarpa* and *Populus tremuloides* would have a sum of 300; a stand composed solely of *A. saccharum* would have a sum of 3,000. A stand consisting of mixtures of species whose climax adaptation numbers lie between one an ten would have a sum somewhere between these extremes. This range of

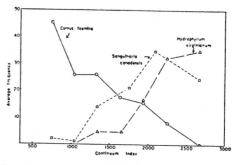

FIG. 8. Average frequency values of three herbs, arranged in order of continuum index.

weighted numbers is termed the *vegetational continuum index*. A continuum index value was calculated for each stand in the study. When the celluloid strips for each stand were arranged in the numerical sequence given by these index values (Fig. 7) it was found that the distributional pattern for each of the tree species retained the same relative position as before.

The vegetational continuum index is actually a means of utilizing all of the tree species in a stand to express the position of that stand on a gradient. Major importance in assessing the position of a stand falls upon those species which have a high importance value but each of the other tree species contributes in some degree.

Tree reproduction is distributed on the vegetational continuum index in a pattern similar to that of the mature trees. The reproduction should possibly be used to further weight the position of a stand in the vegetational continuum index, but, as indicated earlier, the sampling method used in this study gave inadequate data on saplings in most cases and therefore such weighting was not attempted.

Herb and shrub species which played no part in developing the vegetational continuum index were found to be distributed on it in patterns similar to those of the tree species. The distributions of certain shrub and herb species is illus-

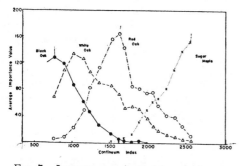

FIG. 7. Importance values of four major species arranged in numerical order of continuum index and averaged by successive groups of 5 stands.

trated in Figure 8. They are found to form curves with characteristic minimum, optimum and maximum points for development. These curves locate the lesser species relative to the tree species and to each other. Some herbaceous species, unlike most trees, are found to cover the entire range of stands, but it is nevertheless possible to predict with reasonable accuracy the important herb species to be found in stands in any given range of the vegetational continuum index.

Soil factors also may be related to the forest cover by means of the vegetational continuum index. The relation of exchangeable calcium and of soil reaction to the index is shown in Figures 9 and 10. The values in these graphs were

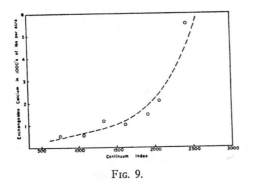

FIG. 9.

obtained by averaging successive groups of three stands for which data were available. Similar correlations were obtained with the water retaining capacity and the organic matter content.

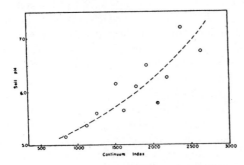

FIG. 10. Acidity of the A_1 layer in stands arranged on the continuum index.

DISCUSSION

Three possibilities as to the nature of the upland hardwood forests of this area may be postulated. The first of these is that the forests represent an amorphous collection of plants in which no units or patterns are discernable. In other words, they are chance aggregates according to the viewpoint of Mason (1947) and are dependent solely on a "coincidence of tolerance" between plants and environment. The presence of a pattern definable in terms of tree composition and the demonstration that trees and other plants are not found together in chance mixtures but in a rather definite pattern would indicate that this interpretation is not tenable in the present case.

Secondly, one could maintain that these forests represent several discrete communities, distinguishable from each other by boundaries which are reasonably distinct in terms of the measurements available to the plant ecologist. If this were true, then determination of the relative sociological importance of the trees in each stand should show that certain groups of tree species reach their optimum development in the same stands, while other groups occur together in other stands. If the boundaries between communities are actually distinct, then the species of one group should never occur as important members of another group. The results of this study indicate that such a discrete grouping of species does not occur in Wisconsin. Rather, each species reaches a high level of importance in only a few stands, with no groups of two or more species showing optimum development in the same stands.

It is thus necessary to accept the third possibility, namely that the upland hardwoods of southwestern Wisconsin represent a continuum in which no clearly defined subdivisions are discernable but in which a definite gradient is exhibited. If this is truly the case then the entire physiognomic upland hardwood forest is to be interpreted as an entity (or portion of an entity) without discrete subdivisions. The failure of any one or any group of the environmental factors measured to indicate any separation substantiates the conclusion of unity based on the tree data. The presence in certain of these environmental factors of some indication of a gradient or trend related to the tree gradient is further indication that these forests represent a continuous cline from initial stages composed of pioneer species to terminal stages composed of climax species.

This conclusion would appear to substantiate the individualistic association hypothesis of Gleason (1926) and its emendation by Cain (1947). It differs, however, in that *"unlimited* variables, combinations and permutations" (Cain) do not occur. All things are not possible, only some, and these possible permutations seem to follow a pathway prescribed by the physiological potentialities of the available flora within the limits of the existent physical environment as modifiable by that flora. Within a given floristic province, the amplitude of environmental tolerance of each species is more limited than the full range of environment in that province. Furthermore, although the tolerance range of each species may overlap that of many other species, the range for optimum development of each species is different from that of all other species. There are, therefore, no groups of tree species which regularly occur together and only together, except for accidental duplications of narrow environmental ranges. Rather, tree species occur in a continuously shifting series of combinations with a definite sequence or pattern, the resultant of a limited floristic complement acting on, and acted upon by, a limited range of physical environmental potentialities. Such a gradient of communities is here called a vegetational continuum, to differentiate it clearly from the controversial concept of a plant association. As applied to forests, it is similar in scope to the "sylvan continuum" of Darling (1949).

The geographical limits of this continuum outside of Wisconsin are not known at present, but it appears that southeastern Minnesota, northeastern Iowa and northern Illinois are definitely included. There are indications that the upland forests on the southern border of the prairie peninsula are similar also, especially those in Missouri, Nebraska (Aikman 1926), Arkansas (Arend and Julander 1948) and Kentucky. This is perhaps to be expected on the basis of climatic similarities (Borchert 1950). The continuum throughout is intimately associated with the prairie-forest border. Its major trees are better adapted to xeric conditions, particularly those associated with periodic severe summer droughts, than are most other species of the deciduous forest. It has borne the brunt of repeated prairie advances in post-Wisconsin time and possibly in earlier interglacial periods as well (Lane 1941, McComb and Loomis 1944). As a result, it is probably simpler in both floristic composition and developmental pattern than the forest aggregations farther east and south. It is recognized that this simplicity was an important factor in the recognition of the continuum pattern and that similar patterns may be more difficult to ascertain in other regions of greater floristic variety.

If the vegetation of a particular physiognomy on a particular topography in a particular floristic province (*i.e.,* deciduous hardwood forests on the uplands of southwestern Wisconsin) forms a continuum without identifiable individual segments, as is the contention of the present paper, then certain far-reaching implications are apparent. Without at this time entering the perennial argument concerning the relative scope of the word "association" as used by American and European workers, it is obvious that both groups usually apply it to discrete communities which are thought to be recognizable as distinct entities in the field. It is on this characteristic that the whole series of "synthetic" characters of plant communities was erected by Rübel (1922), Braun-Blanquet (1927) and other European workers. Such well-established community measures as presence, constancy, and fidelity become of greatly restricted value when applied to stands which form a continuum. For example, in the present case, a number of herbs are known which may be found in almost every initial stand in which the leading dominant is *Quercus velutina* and thus would have a high degree of presence or constancy in those stands. All of them, however, also occur in lesser amounts and in a smaller percentage of the stands which are dominated by *Q. alba*. This diminution continues in the *Q. rubra* stands and most of the species become very rare or absent in the terminal *Acer saccharum* forest. A similar situation in reverse exists for a group of species most common in the *Acer* stands. The absolute value of presence or constancy, therefore, is determined almost solely by the amplitude allowed by the investigator in his choice of stands to be studied. If a large portion of the continuum is used, then all species except the most ubiquitous and least demanding have very low values of presence and constancy. These ubiquitous species are commonly found in nearby non-related assemblages and hence are of very low value as community indicators in any case. As the portion of the continuum used becomes restricted in scope, then the presence and constancy values increase in magnitude. As we have seen, however, scarcely any two stands exist which have precisely the same complement of tree species, to say nothing of the more numerous herbs, and hence, presence and constancy values rarely achieve truly high levels.

It appears probable from preliminary observations that the lowland forests (which receive at least temporary accumulations of run-off water) of this area will eventually prove to be part of the same continuum herein discussed. If this should indeed be the case, then the

synthetic characters would be still further reduced in usefulness, since their arithmetic values would be of very low magnitude. Presence and constancy may continue to maintain their useful roles in phytosociology if each investigator accompanies them with a detailed account of the limits set by him on the range of communities included within his study. When used in this way, the synthetic characters may be very valuable, particularly in comparative studies, as between grazed and non-grazed woodlots or the like.

It is hoped that the concept of a vegetational continuum and the methods for its analysis as developed in this paper will prove to be widely applicable to the forest stands of other regions. Whether or not all floristic provinces have similar continuous gradients rather than recognizably discrete communities is not known at present, although a few studies based on the analysis of many stands in a single region (Hanson and Whitman 1938; Whittaker 1948; Allen 1950; Potzger 1950) clearly indicate that similar continua are present. Dyksterhuis (1949) also shows a continuously varying composition of grassland under the impact of overgrazing which is essentially similar to the forest continuum here described.

In those areas in which its construction is possible, the vegetational continuum index offers a new approach to ecological studies of many types. It may be used to describe local deviations in tree composition such as occur when pioneer species enter spot openings in a mature forest resulting from a restricted fire or other disturbance and such other situations as are included in the "gap phase" concept of Watt (1947). Autecological studies of an herbaceous species may be conducted in stands which are known to be optimum for the development of that species and comparative life history studies of groups of species may be made readily in stands known to be similar on the basis of trees and soils. The index provides a more detailed framework for the study of the relation of random and contagious distributions of herbaceous species in relation to age of occupancy as discussed by Whitford (1949). Mammal, bird, insect or other animal populations, their fluctuations and interrelationships, are all inseparably interwoven with plant communities. One of the major difficulties heretofore encountered in studies of natural animal populations has been the lack of an adequate means of delimitation of the nature of the plant segment of the habitat complex. The vegetational continuum index affords a more accurate means of describing this habitat factor and in addition immediately places a stand in its proper relation to other stands. As further information is collected on the correlations between the physical factors of the environment and the index and on the relations of the microbiota or other special groups of organisms to the index, it will be possible to summarize a very considerable amount of information by means of an index number. There is no thought here that the index will serve as the final touchstone of phytosociologists, nor that the complex of physical and biological agents of importance in a community can possibly be reduced to a single number. The index rather should serve as a framework upon which a great amount of independently derived information may be arranged and interrelated.

SUMMARY

1. The upland hardwood forests of the prairie-forest floristic province of southwestern Wisconsin were studied by means of a random sampling of undisturbed stands distributed through 29 counties covering an area 160 miles by 230 miles. Data on tree, shrub, and herb composition were obtained in 95 stands by the random pairs method of sampling. Soil acidity, nutrient concentration and moisture constants were also determined for these stands.

2. The relative ecological importance of each tree species in each stand was expressed by a summation index of the relative frequency, relative density and relative dominance herein called the *importance value*. This value has a constant total of 300 for each stand.

3. Four tree species (*Quercus velutina, Q. alba, Q. rubra* and *Acer saccharum*) attained an average importance value in all stands much higher than that of any other species. When the stands were arranged into groups each dominated by one of these four species, it was found that each of the lesser tree species was associated with one of the groups. Further breakdown of the stands into subgroups according to the second most important species, resulted in a more accurate alignment of all species into a series beginning with the pioneer species, *Quercus macrocarpa, Populus grandidentata,* and *Quercus velutina,* and ending with the climax species *Ostrya virginiana* and *Acer saccharum.* Each species reached optimum development at a given point along this series, but no groups of species were found which achieved their peak in the same stands. Thus no discrete communities could be detected on the basis of the leading dominants in the stands.

4. Use of a strip method for visual comparison of the importance values of all species in each stand resulted in a ranking or order of stands which agreed with the order obtained from the leading dominants. The initial stands in the rank were composed of shade-intolerant, non-mesic, pioneer species, while the terminal stands contained the very shade-tolerant, mesic climax species. Upon the assumption that *Acer saccharum* was the most successful species in the terminal forests, it was possible to arrange all other tree species in a descending order according to the position of the stands in which they reached optimum development, relative to the stands containing *Acer saccharum.* Arbitrary numbers from one to ten, called *climax adaptation*

numbers, were assigned to the species to indicate the degree to which they approach the climax *Acer* in autecological characteristics. For the more important trees of the region, the adaptation numbers are as follows: *Quercus macrocarpa* —1.0, *Quercus velutina*—2.0, *Carya ovata*—2.5, *Prunus serotina*—3.5, *Quercus alba*—4.0,*Quercus rubra*—6.0, *Ulmus rubra*—7.0, *Tilia americana*—7.5, *Ostrya virginiana*—8.0, and *Acer saccharum*—10.0.

5. The climax adaptation numbers were used as a means of weighting the importance values of each species in a stand, as determined by field sampling. The summation of these weighted numbers resulted in an index which served to locate the stand along a gradient. All species present contributed to the index, with the greatest effect exerted by the dominant species. On this basis, the 95 stands studied were distributed uniformly between an index value of 632 and a value of 2,650 (out of a possible range from 300 to 3,000). No distinct groups of stands were apparent—rather the entire series of communities formed a continuum in which a definite gradient was exhibited from initial stages composed of pioneer species to terminal stages composed of climax species. Such a gradient of communities is here called a *vegetational continuum.*

6. Herb and shrub species were found to be distributed in typical patterns when their frequencies were plotted against the continuum index. Each of these understory plants was restricted to a particular range on the index—each reached optimum developments only within a particular group of closely related stands. Various measurements of soil properties showed a good correlation with the index. This was especially true of the concentration of available calcium, the water retaining capacity and the organic matter content of the A_1 soil layer.

7. It is suggested that the vegetational continuum index may be a useful tool for the coordination of much independently derived information on animal populations, microbiota, and physical factors of the environment.

LITERATURE CITED

Aikman, J. M. 1926. Distribution and structure of the deciduous forest in eastern Nebraska. Univ. Nebraska Studies **26**: 1–75.

Aikman, J. M., and A. W. Smelser. 1938. The structure and environment of forest communities in central Iowa. Ecology **19**: 141–170.

Allan, P. F. 1950. Ecological basis for land use planning in Gulf Coast marshlands. Jour. Soil and Water Conserv. **5**: 57–62, 85.

Anderson, Edgar. 1941. The technique and use of mass collections in plant taxonomy. Ann. Missouri Bot. Garden **28**: 287–292.

Arend, J. L., and O. Julander. 1948. Oak sites in the Arkansas Ozarks. Arkansas Agr. Exp. Sta. Bull. **484**: 1–42.

Borchert, J. R. 1950. The climate of the central North American grassland. Ann. Assoc. Amer. Geog. **40**: 1–39.

Braun-Blanquet, J. 1928. Pflanzensoziologie. Julius Springer. Berlin.

Bruncken, E. 1900. On the forest conditions in the vicinity of Milwaukee. Bull. Wisconsin Nat. Hist. Soc. **1**: 179–184.

Cain, S. A. 1947. Characteristics of natural areas and factors in their development. Ecological Monog. **17**: 185–200.

Chamberlain, T. C. 1877. Native vegetation. Geology of Wisconsin **2**: 176–187.

Cheney, L. S. 1894. Is forest culture in Wisconsin desirable? Trans. Wisconsin State Hist. Soc. **24**: 163–170.

Clark, O. R. 1926. An ecological comparison of two types of woodland. Proc. Iowa Acad. Sci. **33**: 131–134.

Clements, F. E. 1905. Research methods in ecology. Univ. Publ. Co., Lincoln, Neb.

Cottam, G. 1949. The phytosociology of an oak woods in southwestern Wisconsin. Ecology **30**: 271–287.

Cottam, G., and J. T. Curtis. 1949. A method for making rapid surveys of woodlands by means of pairs of randomly selected individuals. Ecology **30**: 101–104.

Curtis, J. T., and H. C. Greene. 1949. A study of relic Wisconsin prairies by the species-presence method. Ecology **30**: 83–92.

Curtis, J. T., and R. P. McIntosh. 1950. The interrelations of certain analytic and synthetic phytosociological characters. Ecology **31**: 434–455.

Dansereau, P. 1943. L'erabliere Laurentienne. Can. Jour. Res. **21**: 66–93.

Darling, F. F. 1949. History of the Scottish forests. Scottish Geog. Mag. **65**: 132–137.

Daubenmire, R. F. 1936. The "Big Woods"

of Minnesota: Its structure, and relation to climate, fire and soils. Ecological Monog. 6: 235–268.

DeForest, Howard. 1921. The plant ecology of the Rock River woodlands of Ogle Co., Illinois. Ill. State Acad. Sci. Trans. 14: 152–193.

Desmarais, Yves. 1948. Dynamics of leaf variation in the sugar maples. Ph.D. thesis. Univ. of Wisconsin.

Dyksterhuis, E. J. 1949. Condition and management of range land based on quantitative ecology. Jour. Range Manage. 2: 104–115.

Eggler, W. A. 1938. The maple-basswood forest type in Washburn County, Wisconsin. Ecology 19: 243–263.

Egler, F. E. 1947. Arid southeast Oahu vegetation, Hawaii. Ecological Monog. 17: 383–435.

Fassett, N. C. 1941. Mass collections: *Rubus odoratus* and *R. parviflorus*. Ann. Missouri Bot. Gard. 28: 299–374.

Fuller, G. D., and P. D. Strausbaugh. 1919. On the forests of LaSalle Co., Ill. Ill. State Acad. Sci. Trans. 12: 246–272.

Gleason, H. A. 1926. The individualistic concept of the plant association. Bull. Torrey Bot. Club 53: 7–26.

Griggs, R. F. 1914. Observations on the behavior of species at the edge of their range. Bull. Torrey Bot. Club 41: 25–49.

Hanson, H. C., and W. Whitman. 1938. Characteristics of major grassland types in western North Dakota. Ecological Monog. 8: 57–114.

Lane, G. H. 1941. Pollen analysis of interglacial peats of Iowa. Iowa Geol. Survey 37: 233–262.

Lapham, I. A. 1852. Woods of Wisconsin. Trans. Wisconsin State Agr. Soc. 2: 419–434.

Marks, J. B. 1942. Land use and plant succession in Coon Valley, Wisconsin. Ecological Monog. 12: 113–134.

Mason, H. L. 1947. Evolution of certain floristic associations in western North America. Ecological Monog. 17: 201–210.

McComb, A. L., and W. E. Loomis. 1944. Subclimax prairie. Bull. Torrey Bot. Club 71: 46–76.

McIntosh, R. P. 1950. Pine stands in south-western Wisconsin. Trans. Wis. Acad. of Sci., Arts, Letters. 40: 243–258.

Pammel, L. H. 1904. Forestry conditions in western Wisconsin. Forestry & Irrig. 10: 421–426.

Potzger, J. E. 1950. Forest types in the Versailles State Park area, Indiana. Amer. Midland Nat. 43: 729–746.

Raunkiaer, C. 1934. The life forms of plant and statistical plant geography. Oxford, Clarendon Press. London.

Raup, H. M. 1947. Some natural floristic areas in boreal America. Ecological Monog. 17: 221–234.

Rübel, E. 1922. Geobotanische Untersuchungsmethoden. Berlin.

Salamun, P. J. 1950. The interpretation of variation in populations of *Spiraea tomentosa, Brachyelytrum erectum, Osmorhiza claytoni,* and *Osmorhiza longistylis.* Ph.D. thesis. Univ. of Wisconsin.

Snedecor, G. W. 1946. Statistical methods. Fourth Ed. Iowa State College Press. Ames.

Stearns, F. 1951. The composition of the sugar maple-hemlock-yellow birch association in northern Wisconsin. Ecology 32: 245–265.

Stout, A. B. 1944. The bur oak openings of southern Wisconsin. Trans. Wisconsin Acad. 36: 141–161.

Wagner, R. D. 1947. The ecological composition of an oak-maple subclimax forest in Jefferson County, Wisconsin. M.S. thesis. Univ. of Wisconsin.

Watt, A. S. 1947. Pattern and process in the plant community. Jour. Ecology 35: 1–22.

Weaver, J. E., and F. E. Clements. 1938. Plant Ecology. McGraw-Hill. New York.

Whitford, P. B. 1951. Estimation of the ages of hardwood stands in the prairie-forest border region. Ecology 32: 143–146.

——. 1949. Distribution of woodland plants in relation to succession and clonal growth. Ecology 30: 199–208.

Whittaker, R. H. 1948. A vegetation analysis of the Great Smoky Mountains. Ph.D. thesis. Univ. of Illinois.

Yapp, R. H. 1922. The concept of habitat. Jour. Ecol. 10: 1–17.

15

Reprinted from *Australian J. Bot.* 2(3):304–324 (1954)

OBJECTIVE METHODS FOR THE CLASSIFICATION OF VEGETATION

III. AN ESSAY IN THE USE OF FACTOR ANALYSIS*

By D. W. GOODALL†

(*Manuscript received April 5, 1954*)

Summary

The possibilities of using the statistical technique of factor analysis in describing variations in plant communities are explored. This method enables the variations to be treated as continuous, instead of resulting in a separation of the stands studied into a limited number of discrete associations or other synecological categories. It further provides a means for testing whether such separation can be objectively justified. It may often facilitate the recognition of the complexes of environmental factors which mainly determine differences in vegetation, and provides a means of estimating the relative value of the various species as indicators of these environmental complexes.

In the present paper, the "principal axes" technique of factor analysis is applied to the analysis of data for percentage cover for 14 species in the Victorian Mallee. It is shown that their distribution, in so far as it does not depend on factors peculiar to individual species, can be represented in terms of at most five orthogonal "factors". The two most important "factors" are interpreted in terms of catenary changes in the vegetation. Other less common species not included in the analysis show high correlations with these "factors". In units of 1.28 ha there is no evidence that more than one continuously varying population is represented in the area; but in units of 25 sq. m. the majority of quadrat records fall into one or other of two principal categories, representing the valley and ridge communities. The potential value of factor analysis in plant sociology, and difficulties in its application to this field, are discussed.

I. INTRODUCTION

Whenever the classification of vegetation, taking into account the presence and quantity of more than one or two species, is considered, ideas of interspecific correlation are always involved tacitly, if not explicitly. In the work of the Zurich-Montpellier school, for instance (see Goodall 1952), they remain tacit. They became explicit in such investigations as those of Kulczynski (1927) and Sørensen (1948); Tuomikoski (1942) attempted actually to use product-moment correlations to designate axes in a continuum, in which vegetational records could be placed in order according to their specific composition. The present series of papers continues this line of thought, and endeavours to make use of statistical tools better than those available to the earlier investigators.

---

* A paper embodying some of the material in the present publication was read at the International Biometric Conference in Bellagio, Italy, on September 4, 1953.

† At the time the field observations were made, the author was on the staff of the University of Melbourne. His present address is Department of Agricultural Botany, University of Reading, England.

In the first paper of this series (Goodall 1953a) reasons were given for regarding the lack of correlations between the quantities of species present in sample areas as evidence of the homogeneity of the area sampled, and a method was described by which a heterogeneous collection of quadrat data could be divided into groups answering to this criterion of homogeneity. It was indicated that this treatment was regarded as no more than an approximation; that it divided into discrete groups what should probably be regarded as regions in a single continuum, without any natural boundaries. It is the purpose of the present paper to conduct a preliminary exploration of a method which enables the features of this postulated continuum to be studied. Until this has been done, it does not seem possible to test the appropriateness of these two methods of treatment respectively.

The first and simplest hypothesis to be tested is that the collection of data is homogeneous, and that the variations in it can be regarded as arising from random causes only, there being no correlation between the quantities of the various species present; let us call this hypothesis A. If hypothesis A has to be rejected, the simplest hypothesis which may still be tenable would appear to be that the data arise from a single population, the non-random variations found in which can be ascribed to correlations among the species; this will be called hypothesis B. Such correlations will usually depend on some overriding environmental factors which are not uniform over the area studied; but the existence of such factors is not hypothesis B, and to test it directly would require environmental as well as vegetational data. Almost equal in simplicity to hypothesis B is the supposition (hypothesis C) that the observational data sample a number of discrete populations, within each of which the variables are subject only to random variation; this is the hypothesis tacitly accepted as the background to most systems for the classification of vegetation. The way to test hypothesis C lies through hypothesis B. The simplest forms of these hypotheses — namely, that there is a single population showing a single set of interdependent correlations among the species, or that there are just two separate populations without internal interspecific correlations — will be called respectively hypotheses B_1 and C_1.

Considered geometrically, the records for each sample area of the quantities of each species, after standardization, may be regarded as specifying a point in a coordinate system of as many axes as there are species. If a single homogeneous population is concerned, the points for the various sample areas will form a single hyper-spherical cluster. If hypothesis B_1 is justified, the points will again fall into a single cluster, but the shape of this cluster will be that of a hyper-ellipsoid rather than a hyper-sphere, one axis being elongated while the others remain equal. If, on the other hand, hypothesis C_1 is true, the points will form two separate hyper-spherical clusters. Now if hypothesis C_1 is true, but hypothesis B_1 is provisionally accepted, the process of identifying the vector corresponding to the long axis of the postulated hyper-ellipsoid will in fact give the vector joining the centres of the two hyper-spherical clusters. If then the distribution along this vector of the projections of the points be studied, they will be found to be distributed bimodally, and this bimodality provides evidence supporting hypothesis C_1 against hypothesis B_1. This argument can, of course, be extended to the more general hypotheses B and C.

If a hypothesis of separate populations of the general form of hypothesis C is accepted, discriminant analysis (Goodall 1953b) forms the appropriate method of treatment of the observational data, particularly where questions such as the allocation of intermediate points are being considered. But if factorial hypotheses such as B are being tested — if, that is, a hypothesis of separate populations has not yet been proved — the methods of factor analysis are appropriate.

Factor analysis is a technique developed in the first instance — primarily by Spearman (1927) — for the treatment of psychological data. In the testing of intelligence and ability, the psychologist is confronted with a mass of data consisting of the scores of a number of individuals in each of a battery of tests; from these data he wishes to draw conclusions as to the general intelligence of each individual, and as to any particular abilities independent of general intelligence which have affected the results of the tests. The methods developed by Spearman and by the psychologists and statisticians who studied the question afterwards enabled the original mass of data, too extensive to be readily grasped as it stood, to be simplified so that the collection of scores for each individual could be represented by a single figure expressing "general intelligence", or a small group representing particular abilities. When these general factors had been removed from the test scores, the residuum of variation depended only on factors peculiar to each test.

The analogy between this situation in psychology and that with which the plant sociologist is confronted is close. For intelligence tests, one reads plant species; for individuals under test, one reads sample areas; for test scores, the quantity of the species present; for general intelligence or particular abilities, one may substitute xerophytism, salinity tolerance, nutritional requirements, and the like. And, just as the general intelligence shown by the test scores of a child may depend on his genetic constitution, the diseases he has suffered, and his nutrition, so the general level of xerophytism of the plants growing in a sample area will depend on its soil and subsoil structure, rainfall, slope, evaporation, and so forth.

Various techniques have been used for factor analysis (see Holzinger and Harman 1941). Of these techniques, that first described by Hotelling (1933) and termed the method of "principal axes" lends itself best to tests of significance, although the computing work is somewhat heavier than with other techniques. The appropriate significance tests have been discussed by Bartlett (1950), and the method of "principal axes", with these tests, has been used in the present work.

TABLE 1

CORRELATIONS BETWEEN PERCENTAGE COVER (AFTER ANGULAR TRANSFORMATION) FOR 14 SPECIES OF PLANTS IN THE VICTORIAN MALLEE

	B. uniflora	C. pseudomicro-phyllum (prostrate form)	C. pseudomicro-phyllum (erect form)	D. semiannularis	D. bursariifolia	E. calycogona	E. dumosa	E. oleosa	M. uncinata	S. variabilis	T. irritans	V. triloba	W. rigida	Z. apiculatum
Bassia uniflora (R.Br.) F. Muell.	1.000	0.174	0.095	−0.510	−0.603	0.328	−0.471	0.601	−0.431	0.254	−0.628	0.032	0.213	0.683
Chenopodium pseudomicrophyllum Aellen (prostrate form)	0.174	1.000	0.271	−0.129	−0.335	0.307	0.119	0.047	−0.270	0.393	−0.157	0.400	0.120	0.025
C. pseudomicrophyllum (erect form)	0.095	0.271	1.000	0.309	0.058	0.208	0.255	0.197	−0.181	0.616	−0.166	0.060	0.090	−0.009
Danthonia semiannularis R.Br.	−0.510	−0.129	0.309	1.000	0.449	−0.115	0.295	−0.382	0.377	0.265	0.270	0.198	−0.263	−0.450
Dodonaea bursariifolia Behr.	−0.603	−0.335	0.058	0.449	1.000	−0.325	0.456	−0.419	0.327	−0.074	0.547	−0.295	−0.298	−0.508
Eucalyptus calycogona Turcz.	0.328	0.307	0.208	−0.115	−0.325	1.000	0.072	0.059	−0.654	0.366	−0.220	0.001	0.400	0.327
E. dumosa A. Cunn.	−0.471	0.119	0.255	0.295	0.456	0.072	1.000	−0.512	−0.044	−0.036	0.469	0.199	−0.186	−0.542
E. oleosa F. Muell.	0.601	0.047	0.197	−0.382	−0.419	0.059	−0.512	1.000	−0.499	0.194	−0.688	−0.056	0.224	0.586
Melaleuca uncinata R.Br.	−0.431	−0.270	−0.181	0.377	0.327	−0.654	−0.044	−0.499	1.000	−0.183	0.412	0.154	−0.478	−0.427
Stipa variabilis Hughes	0.254	0.393	0.616	0.265	−0.074	0.366	−0.036	0.194	−0.183	1.000	−0.372	0.239	0.128	0.163
Triodia irritans R.Br.	−0.628	−0.157	−0.166	0.270	0.547	−0.220	0.469	−0.688	0.412	−0.372	1.000	−0.200	−0.206	−0.399
Vittadinia triloba D.C.	0.032	0.400	0.060	0.198	−0.295	0.001	0.199	−0.056	0.154	0.239	−0.200	1.000	−0.151	−0.013
Westringia rigida R.Br.	0.213	0.120	0.090	−0.263	−0.298	0.400	−0.186	0.224	−0.478	0.128	−0.206	−0.151	1.000	0.285
Zygophyllum apiculatum F. Muell.	0.683	0.025	−0.009	−0.450	−0.508	0.327	−0.542	0.586	−0.427	0.163	−0.399	−0.013	0.285	1.000

For $P = 0.05$, $r = 0.349$.

II. The Observational Data and Their Analysis

(a) The Original Data

The collection of data to which factor analysis was first applied was that from the Victorian Mallee, which has already been used in other connexions (Goodall 1953*a*, 1953*b*). Information was available on the percentage cover by each species of flowering plant in each of 256 quadrats of 25 sq. m. area distributed by restricted randomization within an area of 640 m square occupied by virgin mallee scrub vegetation. Of the 57 species recorded, the 14 most frequent (all, that is, recorded in more than 20 quadrats), including the two distinct forms of *Chenopodium pseudomicrophyllum*, were selected for study.

Table 2

JOINT DISTRIBUTION OF PLANTS IN THE VICTORIAN MALLEE: FACTOR COEFFICIENTS

Species	Factor *A*	Factor *B*	Factor *C*	Factor *D*	Factor *E*
Bassia uniflora	0.838	—0.102	0.142	—0.023	—0.104
Chenopodium pseudomicrophyllum (prostrate form)	0.309	0.574	0.086	0.494	—0.122
C. pseudomicrophyllum (erect form)	0.150	0.726	0.022	—0.466	—0.175
Danthonia semiannularis	—0.556	0.480	0.224	—0.333	0.323
Dodonaea bursariifolia	—0.740	0.077	—0.217	—0.414	—0.159
Eucalyptus calycogona	0.502	0.432	—0.502	0.156	0.231
Eucalyptus dumosa	—0.534	0.529	—0.335	0.232	—0.386
Eucalyptus oleosa	0.759	—0.152	0.176	—0.347	—0.307
Melaleuca uncinata	—0.674	—0.206	0.553	—0.020	0.259
Stipa variabilis	0.343	0.723	0.209	—0.327	0.190
Triodia irritans	—0.760	—0.057	—0.342	0.149	0.088
Vittadinia triloba	0.045	0.431	0.590	0.520	0.052
Westringia rigida	0.479	0.056	—0.509	0.004	0.437
Zygophyllum apiculatum	0.768	—0.223	0.017	—0.077	0.117

To reduce the volume of the data, and the number of zero values, the quadrat records were first combined in groups of eight. Since the area studied had first been divided into an 8 by 8 grid, this means that the units for the factor analysis consisted of rectangles 160 by 80 m, in each of which values for eight quadrats were available; the means of these groups of eight values were the data used in the analysis.

(b) The Correlation Matrix and its Analysis

The percentage cover data were first subjected to angular transformation; for each pair of species, the product-moment correlation coefficients between these transformed data were then computed. The resulting correlation matrix is shown in Table 1. It will be noted that, of the 91 correlations, 32 reach the 5 per cent. level of significance. The initial test of the determinant of the whole matrix (Bartlett 1950), as might be expected, showed significance at an extremely high level, with $\chi^2 = 238.4$ ($n = 91, P < 10^{-15}$), so that the hypothesis of homogeneity was definitely unacceptable.

A first factor, corresponding to the longest axis of the postulated hyper-ellipsoid, was accordingly extracted; this will be termed factor A. It accounted for a substantial part of the correlations observed, for the corresponding latent root was 4.760 (as against a mean of unity for all 14 latent roots of the matrix). The factor coefficients* for the 14 species are shown in the second column of Table 2. It will be noted that the species with the highest positive coefficients for factor A are *Bassia uniflora*, *Zygophyllum apiculatum*, and *Eucalyptus oleosa*, while high negative coefficients occur with *Triodia irritans*, *Dodonaea bursariifolia*, and *Melaleuca uncinata*. It is thus clear that factor A may be regarded as representing adaptation to poorer drainage and the other conditions that go with it, for the first group of species is found in the hollows that run from east to west across the area, while the other group is found on the intervening ridges. This is illustrated in Figure 1(*b*), where the values of factor A for each of the quadrat groups are shown in their appropriate positions on a plan of the area; it may be compared with the contour map in Figure 1(*a*).

A test of the claims of hypotheses B_1 and C_1 is now possible. In Figure 2 the frequency distribution of the values of factor A for different quadrat groups has been plotted as a histogram. There seems from this diagram no *prima facie* evidence of the existence of two discrete types of vegetation in the area, and a test of deviations from normality (though, of course, insensitive with only 32 observations) gave a value for χ^2 of 1.75 (with 3 degrees of freedom, the distribution being divided for this purpose at the pentiles). Hypothesis C_1 may thus be rejected.

The existence of this common factor A by no means accounted for the whole of the interspecific correlations; when the determinant of the residual matrix after its elimination was tested, it still gave a significant value of χ^2 (see Table 3). A second factor (factor B) was accordingly extracted.

The coefficients of this second factor are shown in the third column of Table 2. Here, the highest coefficients are those for *Stipa variabilis* and *Chenopodium pseudomicrophyllum*, and the lowest those for *Zygophyllum apiculatum* and *Melaleuca uncinata*. This is much more difficult to interpret in terms of adaptational features than factor A, for there seems little in common between *Z. apiculatum* and *M. uncinata* — and, in fact, they initially showed a strong negative correlation. However, if dot diagrams are prepared to show the relationship between the quantities of the individual species and the values of factor A, the significance of factor

---

* These factor coefficients give the direction ratios of the vector corresponding to the longest axis of the hypothetical hyper-ellipsoid; their actual magnitudes are such as to provide the coordinates of a point on the vector distant from the origin by an amount equal to the standard deviation of the factor.

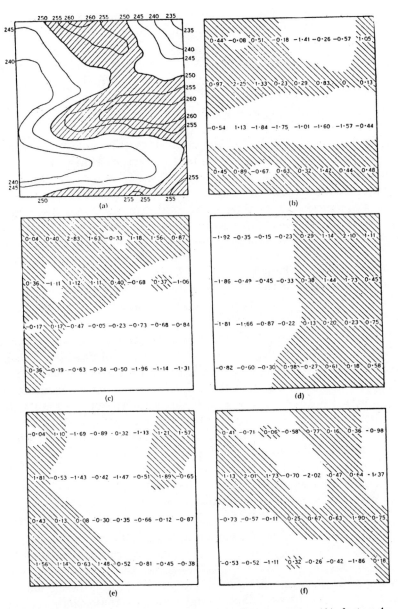

Fig. 1.—Maps of area under study: (*a*), Contour map; (*b*) factor *A*;
(*c*) factor *B*; (*d*) factor *C*; (*e*) factor *D*; (*f*) factor *E*. In (*a*) the
area more than 250 ft above sea-level has been shaded; in (*b*) to (*f*)
the areas shaded are those with positive values for the factor in
question.

B becomes clear. This has been done in Figure 3. Whereas *C. pseudomicrophyllum* and *Stipa variabilis* show a marked maximum towards the
centre of the range of factor *A*, the relationships in *M. uncinata* and

Z. apiculatum are practically linear, decreasing and increasing respectively with increasing values of factor *A*. Thus factor *B* simply represents a non-linear component of the relationship of the distribution of the species to factor *A*. To put it in adaptational terms, if factor *A* represents

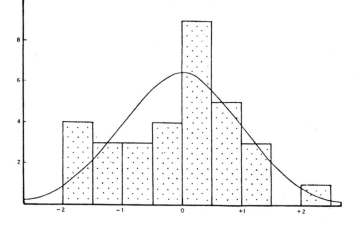

Fig. 2.—Frequency polygon (with fitted normal curve) for values of factor *A*, groups of eight quadrats.

adaptation to conditions of restricted drainage (species with large negative coefficients being poorly adapted to conditions of restricted drainage, but well adapted to free drainage), then species best adapted to intermediate drainage conditions but growing poorly under either extreme of restricted or free drainage will have coefficients close to zero for factor *A*, but high positive coefficients for factor *B*.

TABLE 3

LATENT ROOTS OF CORRELATION MATRIX CORRESPONDING TO SIGNIFICANT FACTORS

Factor	Latent Root	χ^2	n	P
A	4.760	238.4	91	$<10^{-15}$
B	2.399	153.0	78	$<10^{-6}$
C	1.595	114.3	66	0.0005
D	1.348	88.6	55	0.002
E	0.801	63.9	45	0.03
Sum for all others	3.097	29.9	36	0.75

In view of this interpretation of factor *B*, it is not surprising that the spatial distribution of positive and negative values of this factor is irregular (Fig. 1(*c*)), showing little correlation with any topographical features. Factor *B* also provides no evidence that the population is multiple; Figure 4 shows, in the form of a dot diagram, the joint distribution of values for factors *A* and *B* — and no signs of clumping can be seen.

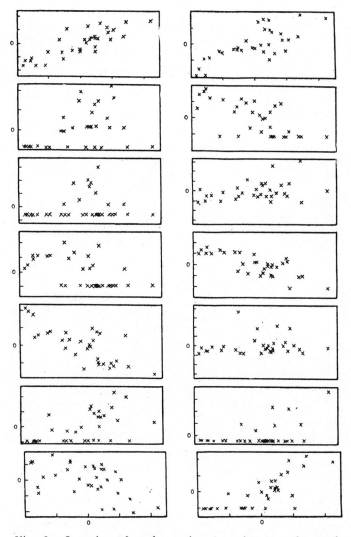

Fig. 3.—Quantity of each species (angular transform of percentage cover) corresponding to different values of factor A:

Bassia uniflora

Chenopodium pseudomicrophyllum
  (prostrate form)

C. pseudomicrophyllum
  (erect form)

Danthonia semiannularis

Dodonaea bursariifolia

Eucalyptus calycogona

E. dumosa

E. oleosa

Melaleuca uncinata

Stipa variabilis

Triodia irritans

Vittadinia triloba

Westringia rigida

Zygophyllum apiculatum

The determinant of the residual matrix after the elimination of factor B was still significant (see Table 3), and a third factor (C) was accordingly extracted. The coefficients may be seen in Table 2. Here the largest positive values are for *Vittadinia triloba* and *Melaleuca uncinata*, the largest negative ones *Westringia rigida* and *Eucalyptus calycogona*. It is again difficult to interpret this factor in terms of similarities between the species mentioned, and the distribution diagrams of Figure 3 do not suggest an explanation on the lines of that for factor *B*, based on a

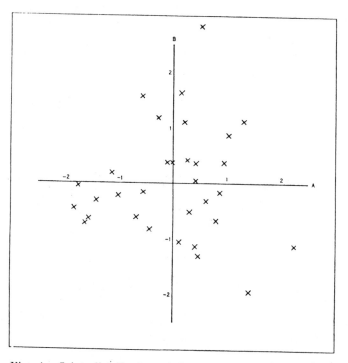

Fig. 4.—Joint distribution of factors *A* and *B*, groups of eight quadrats.

different set of coincident "humps". The spatial distribution of values of factor *C*, however (Fig. 1(*d*)), shows a remarkably simple pattern, all the positive values being grouped together on the east side of the area studied. On a larger scale, this would almost suggest mutual replacement by two distinct floristic elements, but in the present case this explanation would seem too far-fetched. It is conceivable, however, that a mosaic variation on a scale even larger than that giving the vegetational differences between ridges and hollows is superimposed on it, and that the area studied includes portions of two elements of this mosaic.

The joint distribution of values of factors *A*, *B*, and *C* cannot well be represented in two dimensions, but the joint distributions of factor *C* with factors *A* and *B* are shown separately in Figure 5(*a*), (*b*). Again, there is no clear indication that more than one population is present.

The determinant of the residual matrix of correlations remained significant (see Table 3), and a fourth factor (*D*) was accordingly extracted. Once again, the factor coefficients are shown in Table 2; the largest positive coefficients are those for *Vittadinia triloba* and the prostrate form of *Chenopodium pseudomicrophyllum*, the largest negative ones those for the erect form of *C. pseudomicrophyllum* and *Dodonaea bursariifolia*. Apart from the interesting distinction between the two forms of *C. pseudomicrophyllum*, it is difficult to make biological sense of this factor. Its distribution pattern in the field (Fig. 1(*e*)) is fairly simple, though not obviously related to features of relief or to the other

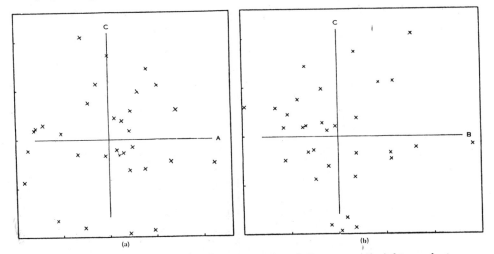

(a)   (b)

Fig. 5.—(*a*) Joint distribution of factors *A* and *C*, groups of eight quadrats.
(*b*) Joint distribution of factors *B* and *C*, groups of eight quadrats.

associated environmental factors. Once again, the dot diagrams (Figs. 6(*a*), (*b*), (*c*)) indicating its joint distribution with the three factors already extracted show no clear evidence of clumping such as one would expect if the population were multiple.

The determinant of the residual matrix had still not been reduced to non-significance by the elimination of these four factors. It was, however, sufficiently close to the usual 5 per cent. significance level to make it doubtful whether a fifth factor should be extracted. However, a factor *E* was computed; the coefficients are shown in the final column of Table 2 and the distribution pattern in the field in Figure 1(*f*). After the removal of this factor, there is no reasonable doubt that factorization is complete.

(c) Relevance of the Factor Analysis to Smaller Sample Areas

It was initially stated that, for convenience of treatment and for better approximation to normality, quadrats were combined in neighbouring groups of eight, so that the units in the analysis were areas measuring 80 m in the east-west direction by 160 m from north to south, each of which had been sampled by eight 25-sq. m. quadrats under restricted randomization. The units of area were deliberately made longer from north to south since the most obvious vegetational differences were in this direction, and it was feared that otherwise these most obvious differences might overshadow and obscure any others.

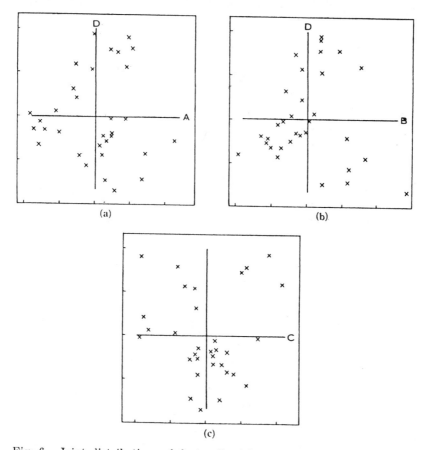

Fig. 6.—Joint distributions of factor *D* with (*a*) factor *A*, (*b*) factor *B*, and (*c*) factor *C*, groups of eight quadrats.

Since, in most types of vegetation, variations on several scales are superimposed one upon the other, it was unlikely that the relative importance of these different types of variation would be the same for individual quadrats as for the large rectangular areas; consequently, one would not expect the results of factor analysis applied to data for individual quadrats to be exactly the same as for the groups of eight quadrats. But, if the larger-scale variations are important, they will be reflected in the factors extracted in either case, and the smaller-scale variation will simply increase the number of significant factors.

It is accordingly of interest to evaluate for individual quadrats the factors extracted for the larger rectangular areas. This has been done for factors A, B, and C. In Figure 7 the distribution of these values is shown for each factor separately. It is noteworthy that, for factor A, the distribution is clearly bimodal (for deviations from normality, $\chi^2 = 64.74$; $n = 10$; $P << 0.001$) and suggests that individual quadrats may fall into two distinct populations, the ranges of values for factor A

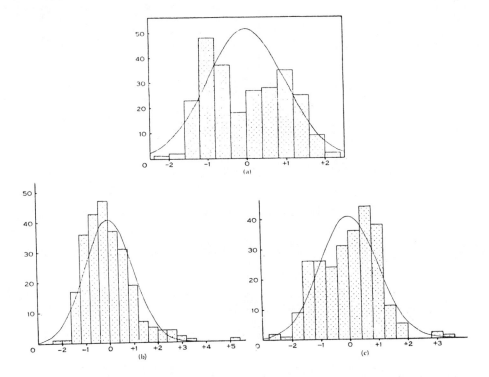

Fig. 7.—Frequency polygon (with fitted normal curve) for standardized values of (a) factor A, (b) factor B, and (c) factor C, individual quadrats.

overlapping somewhat. Factor B is unimodal, but skewed to a significant extent ($\chi^2 = 33.62$; $n = 11$; $P < 0.001$), which is probably connected with the skew distributions of some of the species contributing most strongly to this factor. The distribution of factor C suggests bimodality, though less strongly than in factor A, and in any case is not normal ($\chi^2 = 27.90$; $n = 11$; $P < 0.01$). The joint distribution of factors A and B, and A and C, are shown in Figures 8 and 9. Figure 8 shows again the bimodal distribution commented upon in relation to Figure 7(a); the two principal clusters of points are in fact separated by a band which is almost empty, and the points which occur between them in the one-dimensional representation are here displaced in the positive direction of factor B. Thus the effective separation of the two populations has been increased by taking two factors into account, and they are now connected by a rather attenuated U-shaped band. The botanical interpretation is, of course, clear: the two clusters represent the two ends of the catena; quadrats with intermediate composition are less common, and are characterized by additional species which do not occur, or occupy less ground, at either extreme. Figure 9 shows a clearly marked cluster with positive values for both A and C, and a suggestion of distinction of two clusters with negative values of A, but positive and negative respectively for C.

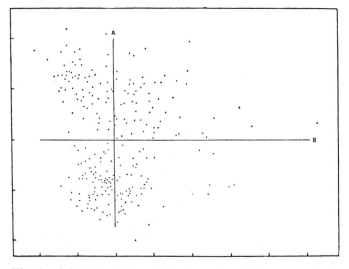

Fig. 8.—Joint distribution of factors A and B, standardized values for individual quadrats.

For factor A, the spatial distribution of the values for individual quadrats has been plotted in Figure 10. Comparing this with Figure 1(b), it is clear that combining data for quadrats in groups of eight has partially obscured the dependence of this factor on the topography of the area. The ridge community on the southern edge, for instance, is clear in Figure 10, but not shown in Figure 1(b).

(d) Relevance of the Factor Analysis to Other Species

Since the factor analysis computations become very unwieldy when the number of variates is large, only 14 of the 57 species recorded were included in the analysis. If, however, the results were applicable to the whole vegetation, one would expect the distribution of the other, less frequent, species to be correlated fairly closely with the values of the factors extracted from the data for the commoner species. That this is so is shown in Table 4. For this purpose, the factor values for individual quadrats have been used; a 2×2 table has been constructed showing the association of the species with high or low values of the factor, and tested by the χ^2 or the exact test, as appropriate. For species occurring in more

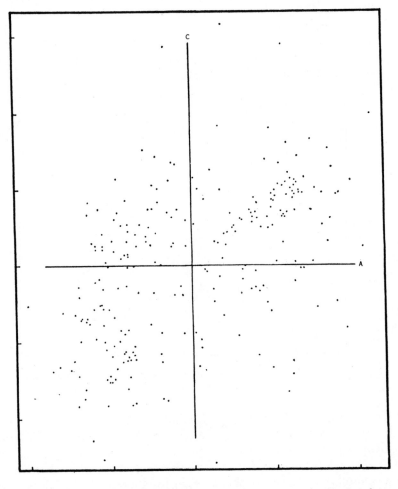

Fig. 9.—Joint distribution of factors A and C, standardized values for individual quadrats.

than nine quadrats the factor values were divided at the median, for those occurring in from three to nine quadrats at the upper or lower quartile; species occurring in one or two quadrats only were not included.

For factor A, no fewer than 12 out of the 29 species included in Table 3 show significant association, and for each factor the number of entries in the table is significantly greater than expectation.

III. Discussion

This first* attempt to apply the methods of factor analysis to the problems of plant sociology has shown that it can display neatly and conveniently many of the relationships subsisting between the distribution of different species; but it has also shown that there are a number of obstacles to be surmounted before it can be regarded as a fully appropriate instrument for this purpose.

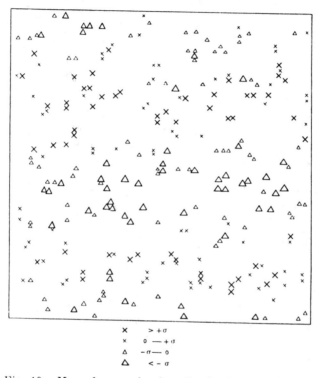

✕	$> +\sigma$
✕	$0 - +\sigma$
△	$-\sigma - 0$
△	$< -\sigma$

Fig. 10.—Map of area, showing distribution of values of factor A for individual quadrats.

The most important difficulty revealed in this study is the non-linear relationship of the distribution of several species to the factors extracted. This does not invalidate the analysis, for it is still true that the number of factors extracted indicates the number of independent terms required adequately to specify for any quadrat the quantities of the various species expected, in so far as this does not depend on features peculiar to the one

* When a paper based on this was read at the Third International Biometric Conference, I was interested to learn that work on similar lines at Wageningen was approaching completion. Data collected for grassland communities by Dr. D. M. de Vries were being analysed by Dr. G. Hamming. It seems probable that the results of this work will be published at about the same time as the present paper.

species. But it means that, though the factors may be statistically orthogonal, they are not biologically independent; the interpretation thus becomes more complicated.

TABLE 4

SIGNIFICANCE OF ASSOCIATION OF LESS FREQUENT SPECIES WITH HIGH OR LOW VALUES OF THE FACTORS

Species	Number of Occurrences	Factor				
		A	B	C	D	E
Acacia microcarpa F. Muell.	4	..	..	..	..	..
A. rigens A. Cunn.	5	..	..	--	..	..
A. sclerophylla Lindl.	4	..	..	..	++	..
Aotus villosa Sm.	3	+	..	..	..	..
Atriplex stipitata Benth.	4	..	..	..	..	++
Baeckia behrii F. Muell.	4	++	..	..	..	..
Bassia parviflora Anders.	6	..	++	..	..	..
Beyeria opaca F. Muell.	5	+	..	+	..	..
Callitris verrucosa R. Br.	18	---	..	---	..	-
Cassia eremophila A. Cunn.	20	+++	..	..	++	+
Dianella revoluta R. Br.	3	+	..	..	..	..
Dodonaea stenozyga F. Muell.	6	..	..	..	..	..
Enchylaena tomentosa R. Br.	10	++	..	..	..	..
Grevillea huegelii Meiss.	19	..	..	..	..	..
Halgania cyanea Lindl.	3	..	..	-	..	..
Kochia pentatropis R. Tate	12	+	..	..	..	..
K. tomentosa (Moq.) F. Muell.	4	..	..	..	..	..
Lepidosperma viscidum R. Br.	7	--	..	--	..	--
Lomandra leucocephala (R. Br.) Ewart	8	..	..	..	..	..
Melaleuca pubescens Schau.	8	--	..	..	..	..
Micromyrtus ciliatus J. M. Black	6	..	..	---	..	..
Myoporum platycarpum R. Br.	5	..	..	..	..	..
Olearia muelleri Benth.	19	+++	..	..	-	..
Pittosporum phillyraeoides D.C.	4	++	..	++	--	..
Santalum murrayanum F. Muell.	7	..	..	..	..	..
Stipa elegantissima Lab.	6	..	..	..	..	..
S. mollis R. Br.	18	..	++	..	..	-
Trichinium seminudum J. M. Black	4	..	..	..	..	..
Vulpia myuros (L.) Gmel.	4	..	..	..	..	..

+++ or --- indicates positive or negative association, with $P < 0.001$;
++ or -- indicates positive or negative association, with $P : 0.001\text{-}0.01$;
+ or - indicates positive or negative association, with $P : 0.01\text{-}0.05$.

The situation does not arise in the present data, but there is the possibility that, in a series of diagrams like those of Figure 3, one species might have a "hump" — perhaps a narrow one — in a unique position shared with no other species. In other words, the presence of that species in quantity would be diagnostic of a narrow range for the factor. This would not appear in the analysis, since no interspecific correlations would reflect it. Consequently the valuable information on the values of the factor which could be afforded by the presence of this species would go unused.

Another possible difficulty concerns the inclusion of data for the less common species. As is shown in Table 3, these species are often highly correlated with one or other factor, and may sometimes be diagnostic of a narrow range of one factor. But the extremely skew distributions, over-loaded with zeros, which they show make one very diffident about including them. Consequently the precision which they might be able to add to the estimation of the factors for different sample areas is lost.

A point needing careful consideration is the relation between the sociological factors extracted in the course of factor analysis, and the environmental factors which play an important part in controlling the distribution of plants. They are not the same, though they may be closely related. The sociological factor is, in fact, simply an element in a description of the composition of the vegetation in a sample area. It relates to the *stand*, not the *biotope*. It may be interpreted as indicating something in common between the species of plants having similar coefficients — similarity in some particular feature of physiology, biology, or perhaps chorology; the value of the factor for a particular sample area may then be regarded as designating the extent to which this feature is represented in the plant population in that sample area. At times, the interpretation of the sociological factor may be closely related to some environmental factor — as, for instance, xerophytism, or salt tolerance; at others, the group of plants whose intercorrelations are responsible for a sociological factor may be a group of parasites and their common host, or a group of potential "nurse-plants" with the species needing their protection. Or, if the study covers a large area, it may be simply a group of species which, for historical reasons, have similar geographical distribution. But even where there is a close relation between the sociological factor and an environmental factor, it would be incorrect to equate the two. This would be equivalent to equating the dryness of a site with the xerophytism of the plants growing there, or the salinity of a site with the salt tolerance of its plant population — presumptive cause, that is, with presumptive effect. The effect may, used circumspectly, provide an excellent indication of the cause, but should be kept strictly distinguished from it.

It is not to be expected, however, that sociological factors — or even the distribution of individual species of plants — can be closely related to any single, simply measurable environmental factor. Whether or not a plant can grow successfully in a given position depends on the whole complex of factors occurring there, including their values at each successive stage of development; interactions (in the statistical sense) between these factors are often of great importance — for instance, rain in July may be favourable to the growth of annuals if there has been none in June, unfavourable otherwise. It is the high correlation between different environmental factors that often suggests a deceptively simple relationship between plant distribution and the environment. There is much to be said for the view that the complexes of environmental factors determining plant distribution can be indicated and measured better indirectly, through the plants themselves, than by direct physical measurements; this is, of course, the idea behind the use of "phytometers" in agricultural meteorology, and the attention devoted to phenology. If this is being done, factor analysis enables the whole vegetation to be used in the most appropriate way for the indication and indirect measurement of environmental complexes, once the complex of environmental factors on which a particular sociological factor depends has been approximately identified, and is bound to be a considerable improvement on the use of individual indicator species.

The population analysed, consisting as it did of groups of eight quadrats randomized within rectangular areas of 1.28 ha, gave no evidence of qualitative heterogeneity — there was no occasion to regard it as separable into two or more discrete communities. When the same factor analysis was applied to the individual quadrat data, however, it was clear that these quadrats fell in the main into two distinct classes; a minority of them contained intermediate quantities of the species mainly distinguishing these two classes, but also included several species missing from the two main classes. That the two classes should be apparent with individual quadrats and not with groups of quadrats presents no difficulty; the two classes occur on different parts of a catena, and the groups of quadrats were large enough, and extended in the right direction, to cover most or all of a catena — in other words, the two classes represent elements in a mosaic on a smaller scale than the areas on which the factor analysis was based. The proportions in which the elements of the small-scale mosaic are represented in the different rectangular areas vary sufficiently to provide a factor — and the most important factor — in the analysis, but the discrete variation in the mosaic has been smoothed out by combining quadrats.

It is difficult to decide what status to ascribe to the two discrete populations observed among the individual quadrat data. Are they to be regarded as separate plant communities or not? The commonsense view is that environmental conditions change continuously along a catena, and that in the absence of evidence of abrupt change the vegetational changes along it should also be treated as a continuum; the fact that the areas available to the ridge and valley vegetation are greater than those for the slope vegetation (leading to two clusters of points, and a thinner scatter in between) should not cause one to regard this continuum as two distinct communities with transitional zones. The plant-sociological purist, however, believing that variations in vegetation should be considered in the first instance independently of variations in environmental factors which might be responsible for them, might well prefer the latter view. At the moment there seems to be no criterion by which to decide between these views. If no clumping of points in the original variate-space, or the simpler derived factor-space, can be detected, one is entitled to rule out the possibility of discrete communities. But the converse does not hold so clearly, and the interpretation of such cases seems largely a matter of taste.

The extent to which factor analysis can be used in plant sociology will be severely limited by the heaviness of the computing work involved — at least until electronic computing facilities become more generally available. If the difficulties mentioned above could be overcome, there seems little doubt that it would be the most satisfactory technique for resolving and reducing to relative simplicity the complexities of plant distribution. But with ordinary computing methods it is bound to be very laborious, and is likely to be used in fundamental investigations into the structure of vegetation rather than in more descriptive work.

Even where not supported by clumping of points in factor-space, division of vegetation into arbitrary classes can well be justified for purposes of practical convenience, provided that its arbitrariness is recognized and that it is not claimed to represent an objective state of things. The most reasonable divisions would probably be obtained by making use of the factor analysis, separating the classes at convenient values (e.g. the means) of the factors. Very similar classifications to those resulting from such arbitrary division of the factor-space would probably be obtained by other methods based on interspecific correlation but not using factor analysis. It might be expected, for instance, that the associations of Braun-Blanquet and his school would represent particular regions of the factor-space which would be obtained if the vegetation on which they had been working were treated in this way, though substantial regions would be regarded as transitional, and left unallotted. Similarly, the classes separated by successive elimination of interspecific correlation (Goodall 1953a, 1953b) in the data used in the present paper might be expected to correspond fairly closely with regions in the factor-space described above.

Factor analysis does not result in a classification of vegetation in the ordinary sense, but in arrangement of the vegetational data in a multi-dimensional series. For such an arrangement, there appears to be no word in English which one can use as an antonym to "classification"; I would like to propose the term "ordination" (cf. Goodall 1954). In the present series of papers, the same data have now been subjected to objective procedures for both classification (Goodall 1953a) and ordination. But these two procedures by no means exhaust the possibilities of objective simplification of vegetational diversity. In subsequent investigations, it is intended to test other methods serving the same ends. In particular, it is hoped to bring forward further evidence bearing on the suitability of classification and ordination respectively as the basic approach in plant sociology. Until other possible methods have been tested, and data for other vegetation types analysed, it would be premature to attempt any comparative appraisal of the methods used so far.

IV. ACKNOWLEDGMENTS

I am grateful to Professor A. S. Boughey, Professor J. S. Turner, and Dr. F. E. Binet for reading and criticizing this paper in draft. My very hearty thanks are also due to Dr. F. E. Binet for the great pains he took to explain to me the techniques of factor analysis.

V. REFERENCES

BARTLETT, M. S. (1950).—Tests of significance in factor analysis. *Brit. J. Psychol. Statist. Sect.* 3: 77-85.

GOODALL, D. W. (1952).—Quantitative aspects of plant distribution. *Biol. Rev.* 27: 194-245.

GOODALL, D. W. (1953a).—Objective methods for the classification of vegetation. I. The use of positive interspecific correlation. *Aust. J. Bot.* 1: 39-63.

GOODALL, D. W. (1953b).—Objective methods for the classification of vegetation. II. Fidelity and indicator value. *Aust. J. Bot.* 1: 434-56.

GOODALL, D. W. (1954).—Vegetational classification and vegetational continua. *Angew. PflSoziol.*, *Wien*. Aichinger Festschrift (in press).

HOLZINGER, K. J., and HARMAN, H. H. (1941).—"Factor Analysis. A Synthesis of Factorial Methods". (University of Chicago Press: Chicago.)

HOTELLING, H. (1933).—Analysis of a complex of statistical variables into principal components. *J. Educ. Psychol.* 24: 417-41, 498-520.

KULCZYNSKI, S. (1927).—Die Pflanzenassoziationen der Pieninen. *Bull. Int. Acad. Cracovie (Acad. Pol. Sci.)* 3B, Suppl. 2: 57-203.

SORENSEN, T (1948).—A method of establishing groups of equal amplitude in plant sociology based on similarity of species content. *Biol. Skr.* 5 (4): 1-34.

SPEARMAN, C. (1927).—"The Abilities of Man: Their Nature and Measurement". (Macmillan: London.)

TUOMIKOSKI, R. (1942).—Untersuchungen über die Untervegetation der Bruchmoore in Ostfinnland. I. Zur Methodik der pflanzensoziologischen Systematik. *Ann (Bot.-Zool.) Soc. Zool.-Bot. Fenn. Vanamo* 17 (1): 1-203.

16

Reprinted from *Ecol. Monogr.* 27:325–334, 337–349 (1957), with permission of the publisher, Duke University Press, Durham, North Carolina

AN ORDINATION OF THE UPLAND FOREST COMMUNITIES OF SOUTHERN WISCONSIN*

J. Roger Bray† and J. T. Curtis

Department of Botany, University of Minnesota, Minneapolis, Minnesota
Department of Botany, University of Wisconsin, Madison, Wisconsin

TABLE OF CONTENTS

INTRODUCTION

A renewed interest in objective and quantitative approaches to the classification of plant communities has led, within the past decade, to an extensive examination of systematic theory and technique. This examination, including the work of Sörenson (1948), Motyka *et al.* (1950), Curtis & McIntosh (1951), Brown & Curtis (1952), Ramensky (1952), Whittaker (1954, 1956), Goodall (1953a, 1954b), deVries (1953), Guinochet (1954, 1955), Webb (1954), Hughes (1954) and Poore (1956) has accompanied theoretic studies in taxonomy [Fisher (1936), Womble (1951), Clifford & Binet (1954), Gregg (1954)] and in statistics (Isaacson 1954). It is a conclusion of many of these studies that nature of unit variation is a major problem in systematics, and that whether this variation is discrete, continuous, or in some other form, there is a need for application of quantitative and statistical methods. In ecologic classification, an increased use of ordinate systems, which has been stimulated by the development of more efficient sampling techniques and the collection of stand data on a large scale, has prompted the proposal of the term "ordination" (Goodall 1953b). Goodall (1954a) has defined ordination as "an arrangement of units in a uni- or multi-dimensional order" as synonymous with "Ordnung," (Ramensky

1930), and as opposed to "a classification in which units are arranged in discrete classes." The present study is an attempt to examine the upland forests of southern Wisconsin in relation to a suspected multidimensional community structure by the use of ordination method and in so doing, to review the theoretic position necessary to such an examination.

LITERATURE REVIEW

The application of quantitative techniques to community classification has been based, in part, upon the assumption that quantitative community composition, as determined by suitable sampling methods, can be a primary basis for the building of ordination systems. This assumption was emphasized by Gleason (1910) in an examination of the relationship between biotic and physical factors. Gleason quoted Spalding (1909), "The establishment of a plant in the place which it occupies is conditioned quite as much by the influence of other plants as by that of the physical environment," and concluded from his own observations, ". . . the differentiation of definite associations is mainly due to the interrelation of the component plants; and the physical environment is as often the result as the cause of vegetation." Further emphasis upon vegetation in itself was made by Clements & Goldsmith (1924), and more recently, by Mitsuderą (1954), who regard the community as an instrument which, if properly examined and manipulated, might be a key to the relation of biotic and physical phenomena. Cain (1944), Goodall (1954a, 1954b), Whittaker (1954) and Williams (1954) have questioned the relevance of considering single physical environmental factors apart from an environmental complex. Insistence upon the study of vegetation on

* This work was supported, in part, by the Research Committee of the Graduate School of the University of Wisconsin, from funds supplied by the Wisconsin Alumni Research Foundation.

The Junior F. Hayden Fund of the University of Minnesota provided partial support of publication costs.

† Present address: Department of Botany, University of Toronto, Toronto, Canada.

its own level has been reiterated in many studies with recent examples in Shreve (1942), Curtis & McIntosh (1951), Brown & Curtis (1952), and Ramensky (Pogrebnjak 1955).

Included in information on community structure are two basic relationships: that of individual species to each other, and that of stands or plots as a whole to each other. These relationships have determined the development of two complementary but separate approaches to the problem of classification.

The species approach stresses the degree of mutual occurrence of a species with other species. This approach received an early quantitative background in the work of Forbes (1907a&b, 1925) who proposed a coefficient of associate occurrence which he applied to the study of bird and fish communities. Further development of the concept of associate occurrence led to the elaboration of indexes of interspecific association (Dice 1948; Cole 1949; Nash 1950) and to the construction of indexes of relative species occurrence (deVries, 1953, 1954; Bray 1956a). Indexes of interspecific association have been applied, in classification, (1) to correlate species with host specificity or with environment (Agrell 1945; Hale 1955), (2) to determine community groupings by the selection of groups of species with high interspecific correlations (Stewart & Keller 1936; Tuomikoski 1942; Sörenson 1948; Goodall 1953a; Hosokawa 1955-1956) and (3) to determine degree of amplitudinal overlap of species as an indication of kind of community variation (Gilbert & Curtis 1953). Indexes of relative species occurrence have been used as the basis for a spatial ordination of the species (deVries 1953, 1954) or for an objective assignment of species adaptation values (Bray 1956a).

The second basic approach, that of correlation among stands as a whole, can be roughly divided into three methods. The first uses information other that that derived from the vegetation in order to establish a primary series of gradients or regimes along which a subsequent vegetation alignment is undertaken (Wiedemann 1929; Vorobyov & Pogrebnjak 1929; Pogrebnjak 1930; Ramensky 1930; Hansen 1930; Whittaker 1956). These regimes do not usually represent direct physical environmental factors, but more often express environmental complexes of interrelated factors, such as soil moisture, or they follow environmental controls, such as elevation, which determine a complex of factors. Ramensky, for example, used previously established soil moisture and soil nutrient regimes to construct a primary stand ordination from which "Functional Averages" were extracted by a series of eliminations of aberrations in the compositional stand data. These averages were considered the median conditions of the biocoenosis and served as bases for final stand orientation. The work of Ramensky included some of the first intergrading bell-shaped species distributions to be demonstrated along vegetational gradients.

The second stand method is the use of objective techniques to show relationships among stands which have previously been classified into discrete units, usually within the Braun-Blanquet system. This use was given an early formulation in the work of Lorenz (1858) who was apparently the first to apply quantitative methods in community classification when he compared various kinds of moors on the basis of "per cent of species similarity." Later techniques, including those of Kulczyński (1929), Motyka et al. (1950), Raabe (1952) and Hanson (1955) use Jaccard's Coefficient of Community or one of its quantitative modifications to show the compositional similarity of units on various hierarchical classification levels.

The third method attempts, from a direct analysis of quantative vegetational data, to demonstrate degree of relationship by the construction of compositional gradients which are independent of environmental or other considerations. The use of the various techniques of factor analysis (Goodall 1954b), of stand weighting devices based on the assignment of species adaptation values (Curtis & McIntosh 1951; Brown & Curtis 1952; Parmalee 1953; Kucera & McDermott 1955; Horikawa & Okutomi 1955), and of attempts to utilize directly indexes of quantitative coefficients of community (Whittaker 1952; Bray 1956a) or indexes of occurrence probability (Kato et al. 1955) are examples of the above approach.

Previous Treatment of the Upland Forest of Wisconsin

A linear ordination of the stands of the upland forest of southern Wisconsin was presented by Curtis & McIntosh (1951). Subsequent studies were made in which soil fungi (Tresner et al. 1954) were arranged along this ordination and in which corticolous cryptogams were related in part to the ordination and in part to host specificity (Hale 1955). Other studies were made of the forest herbs (Gilbert 1952), of autecological characteristics of herbs (Randall 1951), and of the savanna transition into prairie (Bray 1955).

Limitations to a linear presentation became apparent from continued Wisconsin field work. One was the observation of ecological substitution in which two separate species alternated in sharing what appeared to be identical ranges of environmental tolerance (McIntosh 1957). Further reason to suspect the existence of a possible multidimensional structure came from a growing realization of the importance of past history in determining the composition of any stand.

Source of Data

All of the 59 stands of this study were sampled by the same methods. The trees were measured by the random pairs technique (Cottam & Curtis 1949) using 40 points and 80 trees per stand. The characters here used are absolute number of trees per acre and total basal area per acre, both on a species basis. The shrubs and herbaceous plants were sampled by 20 quadrats, each 1 m. sq., laid at every other point. The character used is simple frequency.

The stands employed were selected from the large number available by a stratified random procedure, so devised as to give an equal number of stands from each major geographic portion of the southwestern one-half of Wisconsin. All stands were at least 15A in size, were on upland sites upon which rain water did not accumulate, and were in reasonably undisturbed condition. As actually applied, this last criterion meant that the stands were ungrazed, had not been subject to fire within the recent past, and had never been logged to such an extent as to create large openings in the canopy. In most cases, a few trees had been removed at various intervals in the past, as witnessed by an occasional stump. The limited logging probably created serious errors in the measured amounts of *Juglans nigra,* since this high-value species was deliberately searched for on an intensive scale during World War I. It is believed that the population densities of the remaining species did not vary greatly from those which would normally be produced by natural death and windthrow. A very few stands were totally undisturbed for at least the past 50 yrs.

The sampling methods employed for both the trees and the understory were not, as could be expected, completely free from error. Estimates of sampling error were made by repeated sampling of the same stand, using different investigators in both the same and in different years. Two extensive series of such tests, in a maple woods and an oak woods, showed a standard error of 10.8% for the individual tree species and 7.1% for the understory plants. On this basis a conservative estimate of over-all error in the individual stand measures of about 10% seems reasonable. Obviously, this error would be much less for the most common species and greater for the rare species (Cottam *et al.* 1953).

We are indebted to Dr. Orlin Anderson, Dr. R. T. Brown, Dr. Margaret L. Gilbert, Dr. George H. Ware, Dr. Richard T. Ward, and especially to Dr. R. P. McIntosh for their aid in the collection of the original data. Professor Grant Cottam of the University of Wisconsin and Professor J. W. Tukey of Princeton University were very generous with their time and advice on various problems.

Taxonomic nomenclature in the present paper is after Gleason (1952).

TREATMENT OF THE DATA

NATURE OF THE APPROACH

The ordination approach was selected for the present study in order to provide statements (1) which depict, with a sufficient degree of quantitative exactness, the compositional structure of a community and (2) which might be able to give some initial indication of the over-all patterns of interaction between biologic and physical phenomena. The possibility of using ordination statements to suggest causal reactions is dependent on the concept of physical and biotic factors interacting in a relationship in which each factor is, to some degree, mutually determined

by the others. There is, therefore, as is often noted in ecologic writing, no simple cause and effect relationship between physical phenomena (as primarily causal) and biotic phenomena (as primarily effectual), especially in the more complex environments. There is, rather, instead of a domain which is determined by a small number of independent factors (that is, a system of mechanist causality), a field of interrelated units and events (configurational causality). The ordination of this field is, then, a plotting of the changes in some biotic and/or physical features from area to area within the system, or, in another sense, a mapping of its complexity. Such a mapping indicates, by the relative proximity of different features and their varying spatial patterns, the degree to which the features may participate in a mutually determined complex of factors. With the completion of this mapping, it may then be possible to apply statistical tools which indicate more fully the causal interactions in any one part of the ordination.

Of the two major approaches to ordination study, that of stand or of species orientation, it was decided to use a technique which gave theoretic spatial relations of stands as a first result. With such a framework, the distribution patterns of individual species can be easily studied by directly plotting some measure of their behavior in each stand. A similar plotting can be made for measures of environmental or historic factors in each stand, or for general descriptive features. If the ordination is originally based upon species rather than stand relationships, then the location of relative stand position (and as a consequence, correlation with environmental features) becomes more difficult.

USE OF SCORE SHEETS

If the degree of similarity of stands, one to another, is to be assessed, then some decision must be made as to what criteria are to be used in judging this similarity. There is wide agreement amongst temperate-zone ecologists that community comparisons must be made on a floristic basis and that environmental or other features are not valid for primary comparisons. Unfortunately, this agreement does not extend much beyond the general idea of floristics. Clements and the Anglo-American school generally recommend the use of the dominants as the main criterion of community or stand relationship, while Braun-Blanquet and his adherents use characteristic species of high fidelity, even though these may be small, rare or otherwise dynamically insignificant members of the assemblage. Lippmaa (1939), Daubenmire (1954) and others use synusia, either singly or in combination, for their characterization of community resemblance. In no case has the total flora of a given stand been used, since the determination of all of the bacteria, soil fungi, soil algae, liverworts, mosses, lichens and vascular plants is usually beyond the facilities at the command of ecologists. The principle is well recognized, therefore, that the entire species complement is not needed for meaningful statements about community composition. The disagreement lies in the question of where to draw the line short of the total flora.

In the similarly complex problem of soil classification, it is possible that the degree of resemblance between a series of soil samples might be determined by applying a single test to each sample. Such tests might measure the texture of the soil, its percentage content of sand, or some other simple character. It would then be possible to arrange the group in descending order, as from very sandy through sandy-loam to non-sandy. The sandy loams in the center would have certain features in common and would differ greatly from the two extremes, but they might well include soils which differed widely among themselves. The application of a second test to the samples, such as fertility level, would serve to differentiate soils which were of similar texture, but, even so, the resulting groups might not be homogeneous—they might differ in organic matter content, pH, color, or other characters. A similar problem was recognized by Pirie (1937), "In an earlier part of this essay the transition from living to non-living was compared to the transition from green to yellow or from acid to alkaline. If this comparison were valid, it would be possible to lay down a precise but arbitrary dividing line. But as it has been shown that "life" cannot be defined in terms of one variable as colours can, the comparison is not strictly valid and any arbitrary division would have to be made on the basis of the sum of a number of variables any one of which might be zero." The best approach might therefore be to apply a series of tests, each examining some pertinent or important aspect of soil makeup. A study of the results of the series of tests would give a firm basis for comparison of the original group and would easily pick out the soils most nearly related to each other and least closely related to others. It might be desirable to weight the test results, in order of their importance. The calcium content is more important than the sodium content in most temperate forest soils and its results might be appropriately weighted to show this importance. Statistically, the degree of relationship could be shown by a suitable measure of the correlation between the sets of test values for any two soil samples.

The same procedure can be employed in the floristic analysis of plant communities. A standard series of tests can be applied to each stand and the results for pairs of stands correlated with each other. If this is done for all stands in the series, then similar stands should have high mutual correlation values, but dissimilar stands should show low correlation. If an appropriate measure of correlation is used, then the resulting index should give a linear measure of the difference between any two stands.

One test of the similarity of two stands could be the quantity of *Acer saccharum* that each contains, analogous to a determination of the calcium content of soil samples. A series of such tests, using other dominants, would increase the precision of the comparison. Similarly, a series of herbs and shrubs, chosen to include some which were restricted to the formation under study and others known to be sensitive to the varying conditions present in different stands of the formation, could be added to the list of tests.

In the current studies, measurements of 26 species were used as the tests. Twelve of these species were dominant trees, while the remainder were herbs and shrubs. The trees included all of the species with a presence value above 33% in the 59 stands studied. The herbs and shrubs were chosen at random in blocks along the original continuum gradient from among those species which were neither overly common nor rare and which had shown clear cut distribution patterns in previous studies. Weighting of the 12 dominant tree species was accomplished by using two separate measurements of each species (absolute density per acre, and absolute dominance per acre) as independent tests.

There were, thus, 38 tests employed for each stand: frequency (in 1 m quadrats) of 14 species of herbs and shrubs, density of 12 species of trees, and dominance (in square inches of basal area at breast height) of the same 12 trees. The results of the tests for each stand were recorded in a separate score sheet for that stand. (Of course, many of the test scores on a given sheet were zero, when one or more test species was missing from that stand.) The test scores were in different units, since the original measurements were made in three different classes. This discrepancy was rectified by expressing each score as a percentage of the maximum value attained by that test on any of the sheets. These corrected scores thus indicated, in comparable units, the behavior of each test species in relation to its optimum behavior in the entire series. Since the number of test scores and the sum of these scores varied from stand to stand, the scores for each stand were adjusted to a relative basis. They finally indicated, therefore, the relative amount contributed by each test organism to the combined score for the stand. In one stand, for example, species A may contribute 17.1% of the total score while species B may represent only 1.9%. The adjusted scores on a relative basis appear to offer the best basis for making comparisons between stands (Whittaker 1952).

Index of Similarity

The choice of a suitable index is largely dependent on the choice of an ordination technique. There are, however, several characteristics of available indexes which should be examined. The standard correlation coefficient, "r," incorporates a square transformation which leads to the weighting of the importance of entries with high values. Thus, if a pair of stands have one or two species in common which have high score values, the stands will have a high correlation coefficient regardless of the relative similarity or dissimilarity of their lesser species. It can be shown that high values of "r" cover a wide range in interstand variation and are relatively insensitive in the medium to high areas of stand similarity. There is a high sensitivity in the lower range of coefficient values of "r," but this sensitivity lies in an area where the ecologic differences between stands are not very significant due to the residue of widely plastic species which inhabit many of the stands of any geographic area. Of the available indexes of similarity employed in phytosociology, both Gleason's quantitative modification of Jaccard's Coefficient of Community (Gleason 1920) and Kulczyński's index (1927) can be shown to have a greater ability to differentiate stands within the area of medium to high similarity than has the correlation coefficient. When the sum of score values is relative and equals 100, both Gleason's and Kulczyński's coefficients can be expressed in the terms later used by Motyka *et al.*

(1950) as $C = \dfrac{2w}{a+b}$ where a is the sum of the quantitative measures of the plants in one stand, b is the similar sum for a second stand, and w is the sum of the lesser value for only those species which are in common between the two stands (Oosting 1956). Thus, if two stands by chance had exactly the same scores for exactly the same species, the index would be 1.00, since (a) and (b) would be equal and both would equal (w). If there were no species in common, then the index would be zero. The range from no resemblance to complete identity is appropriately covered by the range from 0 to 1. This index appears to be the best approximation yet available to a linear measure of relationship.

As used in the present study, the index reduces simply to (w), or the sum of the lesser scores for those species which have a score above zero in both stands. This is due to the use of relative scores, such that (a) plus (b) is always 2.00 in every pair

of stands and $\dfrac{2w}{2.00} = w$. In practice, the score sheets

were so arranged that the final adjusted scores were recorded in the last column on the extreme edge of the sheet. One sheet could then be superimposed, in turn, on every other sheet in a slightly offset position, and the lesser values added on a machine for all tests where a positive value was present on both sheets.

When a large number of stands are studied, the calculation of the w index becomes burdensome, since

there are n x $\dfrac{n-1}{2}$ comparisons to be made. Thus,

for 10 stands, 45 comparisons are needed, while for 100 stands, 4950 are required. In such cases, recourse should be had to electronic calculators, using punch cards as score sheets. In the present case, a complete comparison was made by hand for 59 stands, resulting in 1711 values of the w index. These values were arranged by stand number in a matrix. We would be happy to correspond with anyone who is interested in obtaining a copy of the matrix for further work.

THE ORDINATION METHOD

The use of stand data and of a summation of a series of tests of these stands has been outlined as the quantitative basis for the present study. Of the ordination techniques which were reviewed, many depend upon the use of a previous knowledge and sometimes classification of either the vegetation or physical environment. Although this use is not necessarily undesirable in ordination studies, it is apparent that a technique which can extract an ordination directly from the available data would be best suited to the present study. One quantitative and completely objective technique which makes this extraction is factor analysis.

Factor analysis seeks to draw "functional unities" (factors) from an oriented table (i.e. matrix) of correlation coefficients. These coefficients can be calculated, as in Goodall (1954b), from a correlation among species, which are correspondent to tests in factor analysis; or, if the relationship among stands is needed, the use of direct interstand correlation is permissible (Tucker 1956). In either case, the standard techniques of factor analysis are applicable. The extraction of a functional unity is followed, in most factor techniques, by the computation of a new matrix, called the residual matrix, from which the next unity can be obtained. This series of extractions results in a number of linear vectors, called the factor matrix, and an attempt is then made to identify each vector with an underlying cause. Factor analysis is used in areas where no hypotheses are available about the causal nature of the domain, and is based upon the assumption, according to Thurstone (1947), that ". . . a variety of phenomena within a domain are related and that they are determined, at least in part, by a relatively small number of functional unities or factors." When applied to ordination study, however, these functional unities are not, as is emphasized by Goodall (1954b), direct environmental forces but are rather, sociologic factors. A sociologic factor is defined as "an element in the description of the composition of the vegetation," and it may or may not be related to environmental factors.

The application of factor analysis to the present study was carefully considered, but was rejected for the following reasons: (1) the heavy computational load involved in handling 59 stands, (2) the disadvantages, which have been discussed above, in applying "r" (the correlation coefficient) to stand data, and (3) a hesitancy in interpreting factor anaylsis when applied to stand data especially in regard to the difficulties noted by Goodall (1954b), "though the factors may be statistically orthogonal, they are not biologically independent; the interpretation thus becomes more complicated." The construction of a preliminary empiric method with the following criteria therefore seemed desirable: (1) vegetation structure is regarded as a possible key to the nature of the interaction of factors, and, as such, must be studied on its own level, (2) an extraction of an ordination directly from objectively derived data without previous classification would be desirable.

The basis of a technique which might satisfy the above criteria is the same as that for ordination systems in general: the degree of phytosociologic relationship between stands can be used to indicate the distance by which they should be separated within a spatial ordination. The degree of relationship of vegetation units has usually been measured by some estimate of the similarity of stand composition, with a high degree of similarity signifying a close spatial proximity. The technique which will be outlined attempts, therefore, to extract from a matrix of measurements of interstand similarity, a spatial pattern in which the distance between stands is related to their degree of similarity.

Given a matrix of values of distance between points in Euclidean space, it is possible, without a prior knowledge of their location, to reconstruct their spatial placement (Torgerson 1952). This reconstruction depends upon simple techniques in which, in two dimensional space, for example, 3 points not in the same location and not on the same line, are used to locate the other points by their relationship to the 3 reference points. If, for example, there were in a matrix, 4 points, of which A, B, and C were each separated by a distance of 40 units and point D was separated from A and B by 20 units and from C by 34.64 units, then the position of these points could be established, with A, B, and C forming the apicies of an equilateral triangle and D occurring midway on line AB.

When coefficients of community are used, however, as indicators of spatial distance, then exact interstand distances are not available, since the position of a stand in relation to another stand occurs within an area of uncertainty originating in the sampling error made in surveying the stands. Furthermore, it is likely that stands occupy proximate instead of exact theoretic positions in relation to each other. The occurrence of stands within an area of uncertainty results in a matrix of estimated proximate distances. If such proximate distances are available, then the location of stand D in the previous example might be in a different position relative to reference stands A, B, and C as compared to three other reference stands (if exact interstand distances were available, the positions would be the same regardless of the choice of reference stands). The technique to be developed, therefore, is a preliminary attempt to derive an ordination from estimates of proximate interstand distance which gives a single spatial configuration most closely approximating the matrix distances.

This technique depends upon the selection of a pair of reference stands for the determination of stand positions on any one axis. Given proximate interstand distances, the choice of reference stands is of crucial importance. In making this choice, it is evident that reference stands are comparable, in part, to sighting points as used in plane-table surveying

and that those stands which are furthest apart will be more accurate for judging interstand distance than those which are in close proximity. This accuracy is especially desirable because of the area of uncertainty in which each stand fluctuates relative to the positions of neighboring stands. If these fluctuations are greater than the actual distance apart of the reference stands, then the resulting ordination will reveal only these fluctuations. It is necessary for any ordination that the sphere of fluctuation for any stand be small in relation to the space occupied by the ordination as a whole. The choice of reference stands should be, therefore, of those stands which are furthest apart and as a consequence, have the greatest sensitivity to over-all compositional change.

AXIS CONSTRUCTION

The ordinate location of points in space by the use of reference stands is illustrated by the ordination of five points whose hypothetical interstand distances are shown in the lower-left of Table 1. The distances, although hypothetical, represent exact spatial distances. They were determined by inverting the estimates of stand similarity which appear in the upper-right of the table so that a high degree of similarity was represented by a low degree of spatial separation. The inversions were accomplished by subtracting each index of similarity from a maximum similarity value of 100.

To locate stands between a pair of selected reference stands, a line connecting the reference stands is drawn to scale on a piece of blank paper, and the position of every other stand is projected onto this line. The projection is accomplished by rotating two arcs representing the distance of the projected stand from each of the reference stands, and then projecting the point of arc intersection perpendicularly onto the axis. Applying the criterion of the greatest degree of spatial separation as determining the choice of reference stands, Table 1 shows stands number 1 and 2 to have a maximum separation of 99.9 units. These stands were selected, therefore, as the x axis reference stands and are placed in Fig. 1 at a distance of 99.9 units. Stand 3 in Table 1 is 70 units from reference stand 1 and the same distance from reference stand 2. Stand 3 is located in Fig. 1, therefore, at the intersection of arcs with radii of 70 units and bases at points 1 and 2. Two such intersections are possible in a two dimensional ordination, and the points of intersection are projected perpendicularly onto the x axis, as shown in Fig. 1, to give an x axis location of 50 units. Stand 4 can also be located along the x axis by the arc intersection and projection technique; it occurs at 62 units along the x axis. Stand 5 is similarly located, after intersection and projection, at 62 x axis units. It can be proven geometrically that a constellation of points in n space can be projected perpendicularly onto the line connecting the two reference points which are furthest distant in the constellation by using the above technique.

TABLE 1. Matrix of hypothetical exact interpoint distances. The upper-right portion of the table shows hypothetical data on point similarity for an exact spatial system. The lower-left portion shows data on similarity which were inverted to show interpoint distance.

Stand No.	1	2	3	4	5
1..........	....	0.1	30	30	30
2..........	99.9	....	30	50	50
3..........	70	70	....	17.8	79.6
4..........	70	50	82.2	....	35.2
5..........	70	50	20.4	64.8	....

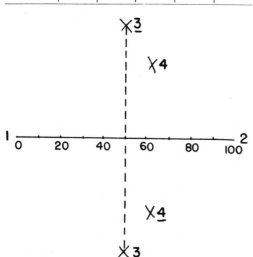

FIG. 1. Demonstration of stand location by the intersection and projection technique. Stand 3 is located at the intersection of arcs with radii of 70 units, and bases at stands 1 and 2. Its projected position on the x axis is midway between stands 1 and 2.

A second axis can be constructed by the same method using a line on the paper erected at a right angle to the x axis. Two new reference stands are selected which are in close proximity on the x axis, but which are nevertheless separated by a great interstand distance. In the matrix in Table 1, stands 3 and 4 fit such criteria showing an x axis separation of 12 units and an interstand distance of 82.2 units. If stand 3 is assigned a location at its upper arc intersection point in Fig. 1, then the interstand distance of 82.2 units of stands 3 and 4 indicates that the proper location of stand 4 is at its lower arc intersection point (intersection of underlined number 4, Fig. 1.). The location of stands 3 and 4 is shown, in relation to the x axis reference stands, in Fig. 2.

Stands 3 and 4 are, therefore, separated by a projected distance of 12 units on the x axis in Fig. 1, but are, nevertheless, separated as shown in Fig. 2 by an interstand distance of 82.2 units. The x axis proximity of stands 3 and 4 and their high spatial separation indicate that they might be used

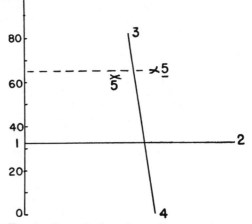

FIG. 2. Demonstration of y axis construction and stand location. Stand 5 is located at one of the two intersections of arcs 20.4 units from reference stand 3 and 64.8 units from reference stand 4. The distance of 50 units from stand 2 to stand 5 indicates the point of intersection to the right is the correct point for projection onto the y axis.

as reference stands for y axis construction. In Fig. 2, a y axis location of stand 5, which is 20.4 units from reference stand 3 and 64.8 units from reference stand 4, is illustrated. The arc intersections of stand 5 with reference stands 3 and 4 shows, after projection, two separate y axis locations, one of which, because the line connecting stands 3 and 4 is not perpendicular to the x axis, is incorrect. To determine the correct position of stand 5, the distance of stand 5 from the reference stands of the x axis should be consulted. The distance of stand 5 from reference stand 2 is 50 units which indicates that the arc intersection to the right (intersection in Fig. 2 with underlined no. 5) which is 50 units from stand 2, is the correct point for projection onto the y axis. Stand 5 is located, therefore, after perpendicular projection at a distance of 65 units on the y axis in Fig. 2.

By following the outlined technique, the information on interpoint distances in Table 1 has been used to indicate the spatial location of the points in a two-dimensional ordinate system. If such distance information necessitated the use of more than 2 dimensions, then further dimensions could be constructed, using the same technique.

APPLICATION OF THE METHOD

X AXIS CONSTRUCTION

The ordination of the upland forests depends upon selecting stands from the complete matrix which can be used as reference points for the location of the other stands. Since the available interstand distances in the complete matrix represent only proximate distances, it is obvious that any resulting stand ordination will only approximate interstand relationships. It is assumed, however, that there is a certain degree of order within the matrix and that the interstand distances are not a series of random numbers. Only a limited number of stand positions for any one stand should, therefore, be possible. As noted before, these positions should fluctuate within a narrow area if an ordination is to present a meaningful approximation of the matrix measurements.

Before the reference stands for the first axis can be selected, coefficient of community values, as found in the complete matrix, must be inverted so that a low coefficient of community value represents a relatively greater spatial separation and a high value represents a close proximity. Although the coefficient has a range from 0 to 100, the error involved in sampling a stand makes it unlikely that any of a series of replicate samples from the same stand will show the maximum value. Two stands were sampled, each 7 times, using the field methods described earlier, and coefficients of community were calculated among these samplings. The mean coefficient of community within each of the 7 replications was 82 which indicated a mean error of index reproducibility of around 20 index units. A value of 80 was, therefore, considered to represent the maximum coefficient value, that is, the value for two indentical stands; the highest value actually found in the complete matrix for 59 stands was 79.

To convert coefficients of community to interpoint distance values, an inversion was accomplished by subtracting each coefficient value from the fixed maximum of 80. Stands which had a coefficient of 0 were separated, therefore, by a maximum distance of 80 units. All subsequent mention of stand interpoint distances refers to the inverted coefficients of community.

Using the criterion of the greatest spatial separation for the choice of reference stands, an examination of the inverted values showed 3 pairs of stands, numbers 35 and 136, 00 and 137, and 33 and 138, to be separated by the maximum distance of 80 units. Since these 3 stand pairs each showed coefficients of community of zero indicating no similarity, their relationship to the other stands in the study was examined to determine whether they were completely unrelated to the other stands. It is suggested that a stand pair with a 0 coefficient of community be used as a reference pair only if each member of the pair shows a value greater than 0 with all stands which are not members of reference pairs. By establishing this criterion, the choice of a stand pair member which shows no relationship to other stands, and which, therefore, contributes nothing to a knowledge of their relative spatial location, will be avoided. This criterion can be met in relation to stand pairs 35-136, 00-137, and 33-138 each member of which shows a relation greater than 0 to every other stand in the study.

The stand pairs selected above appear to represent 2 sets of related stands, numbers 136, 137 and 138 and numbers 00, 33, and 35. The index values between each of the members of each set were highly

significantly correlated, as tested with r. Each of the 3 pairs of reference stands were used, therefore, to ordinate stands along the x axis on the supposition that the use of several sets of stand data might include more information from the matrix than if a separate set were used, and might reduce the fluctuations from non-exact distance measurements.

Since each of the 3 pairs of reference stands were separated by the maximum distance of 80 units, a line of 80 units was drawn connecting each of the reference pairs. The ordination was accomplished by the technique outlined above of arc rotation and of the projection of the point of arc intersection onto the x axis. Stand 89, for example, showed a coefficient of community of 39 and 17 with stands 00 and 137. The inverse of these values, representing spatial separation, is 41 and 63, and stand 89 was located, therefore, at the intersections of arcs with radii of 41 and 63, respectively. These intersections were projected perpendicularly onto the x axis at a position of 25.5 units from reference stand 00 and of 54.5 units from reference stand 137.

The final x axis position of a stand was determined as the median position for the stand in relation to the three pairs of reference stands. There was a close similarity in the x axis positions of the stand in relation to the three pairs of reference stands. Stand 89, for example, had x axis positions of 54.5, 48.5, and 46.5, and was assigned an x axis location of 48.5. The final median values presenting x axis location are shown in column 1 of Table 2.

TABLE 2. Stand locations in three dimensions

Stand No.	ORDINATION AXIS			Stand No.	ORDINATION AXIS		
	X	Y	Z		X	Y	Z
00	79.0	35.5	37.0	93	53.0	54.0	52.0
01	55.0	58.5	41.0	95	4.0	45.75	40.5
03	21.0	55.0	32.0	96	37.5	51.75	33.0
04	32.5	61.25	43.0	100	65.0	30.25	48.0
05	63.0	38.0	50.5	101	62.0	21.0	52.0
06	46.5	33.5	41.5	102	49.5	40.75	45.5
09	45.0	38.75	44.0	103	61.0	33.5	48.0
15	67.5	32.75	58.0	104	68.0	41.0	66.0
16	14.0	39.0	19.0	105	28.0	41.0	48.0
17	21.0	56.5	40.0	106	16.5	44.5	14.0
18	20.5	40.75	28.0	107	47.0	55.5	19.5
19	66.0	27.75	57.5	108	43.0	57.5	28.0
20	63.0	39.5	55.0	109	60.0	60.25	21.0
21	17.0	38.25	13.5	110	18.5	57.75	48.5
23	71.0	53.5	50.5	111	47.0	61.75	33.0
24	56.5	59.0	54.0	112	37.0	42.0	50.0
25	68.5	25.0	59.0	114	31.5	70.75	41.0
26	46.5	61.75	30.5	117	27.0	61.0	45.5
31	76.0	35.25	44.5	118	13.5	66.5	53.0
33	74.0	46.25	39.0	119	18.0	72.5	39.0
35	77.0	38.5	32.0	120	19.5	62.0	28.5
41	11.0	39.0	39.0	121	10.5	64.25	41.0
73	14.5	53.75	31.5	127	55.0	45.75	55.5
85	16.5	62.5	47.5	128	61.0	43.75	32.5
86	71.0	43.5	30.0	136	8.5	47.0	30.0
87	42.0	50.5	47.0	137	7.5	36.75	52.0
88	73.5	37.0	57.0	138	7.5	39.0	34.0
89	48.5	49.0	60.5	151	30.5	47.25	43.0
91	70.0	54.25	39.5	185	47.0	17.5	40.0
92	62.5	25.0	61.5	...	....	.....	....

The choice of reference stands for the second axis is based upon criteria which are, in part, similar to those used in the choice of the first dimension reference stands: stands separated by the greatest interpoint distance and by the least projected x axis distance can be expected, if chosen as reference stands, to give the greatest spatial separation to the other stands. Stand pairs which most closely fit the above criteria are likely to be central in axis location since by the mechanics of the arc intersection and projection technique, the more nearly stands are found toward the center of any axis, the greater is the probability that they will have spatial separation in the new dimension of a relatively great distance. Conversely, stands located towards one of the ends of the axis are less likely to be distantly related, since by sharing a relatively high relationship with the reference stands towards which they are found, they are, therefore, more likely to be related to each other.

A test is suggested for the selection of y axis reference stands in which the value of stand separation on previous axes is subtracted from the index value of interpoint distance, with the highest value considered to be the most suitable. Such a test weights a low separation on previous axes as of equal importance with a high degree of interpoint distance. The importance of choosing reference stands in close proximity on previous axes is illustrated in subsequent z axis construction in which the condition of non-exact interpoint distances makes it impossible to correct for non-perpendicular axes. This test was applied to the 59 stands in the study and stand pair 111 and 185, which are separated by a projected x axis distance of 0 units (Table 2) and by an interpoint distance value of 47 (as inverted from a coefficient of community of 33), gave a maximum value of 47. Another stand pair, 26 and 185, also gave a high value (46.5) by the above test, with separations of 0.5 and 47 respectively. Stands 26 and 111 were found to be highly significantly correlated in their relationship to the other stands, and the use of several reference sets again appeared feasible. Stands 26 and 185, and 111 and 185 were, therefore, selected as y axis reference stands. Since these two sets of stands were separated by only 0.5 and 0.0 units respectively on the x axis, projections after arc rotation were made directly onto the line connecting each reference pair, and this line was considered the y axis. Final y axis location for each stand was determined by taking the mean position for each stand on the two constructed axes. These axis positions are shown in column 2 of Table 2.

Z AXIS CONSTRUCTION

With the completion of the y axis, a search was made for stands which had relatively similar x and y axis positions, but which were, nevertheless, separated by relatively great interstand distances. The same test of maximum axis separation was made as in the choice of the y axis reference stands, and a pair of stands, numbers 89 and 107 were found to give the highest value. Stands 89 and 107 were separated by an interstand distance of 41 units and by projected distances of 0.5 on the x axis and of 8.5 on the y axis for a test value of 32.5. The line connecting reference stands 89 and 107, although not perpendicular to the y axis and, therefore, not exactly parallel to the z axis was, nevertheless, used as a base line onto which to project the arc intersections. This was necessary since the sampling error involved in the area of uncertainty surrounding each stand made it impossible to apply formulae which would correct the effect of a non-perpendicular axis. This error also prohibited the consultation of distances to the x or y axis reference stands, since some stands to be projected onto the z axis were found to be equidistant from the x and y reference stands. No choice could, therefore, be made (as is illustrated in the location of stand 5 in Fig. 2) between upper and lower intersection points. Arc intersections were, therefore, projected directly onto the line connecting reference stands 89 and 107. The projections onto this line were considered z axis stand locations, and are shown in column 3 of Table 2.

RESULTS

Using the values in Table 2, each of the 59 stands studied was located on a two dimensional graph by the intersection of its values on any two of the three axes. In Fig. 3, for example, each point on the graph

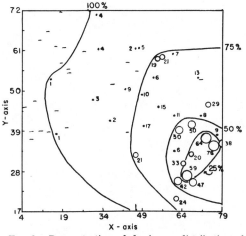

FIG. 3. Demonstration of dominance distribution of *Tilia americana* within the x-y ordination. Each circle or dash represents a stand location. Actual dominance figures in basal area per 100 sq. in. per acre at breast height are given beside each stand location. Values in the upper 25% are represented by the largest circle, values in the 50 to 26% quartile by medium sized circles, values in the 75-51% quartile by small circles, and values in the 100 to 76% quartile by dots. Contour lines are drawn around the 4 quartile lower limits in such a manner as to include all examples of the indicated size class whether or not lesser size class values are present.

represents a stand with its locus determined by its values in Table 2 for the x and y axes. Similar plottings were made for the stand locations on the x and z axes and the y and z axes. These 3 graphs can be thought of as 3 views (front, top, and side) of a three-dimensional cube, within which the stations are located at the intersections of lines projected from each axis. The actual construction of three-dimensional models is very time consuming (Fig. 7).

Once the stands are located in a two- or three-dimensional configuration, it becomes easy to study the behavior of individual species within the stands. In Fig. 3, for example, the actual basal area per acre for *Tilia americana* is plotted on the x-y

[*Editor's Note:* Figures 4 and 5 have been omitted owing to limitations of space.]

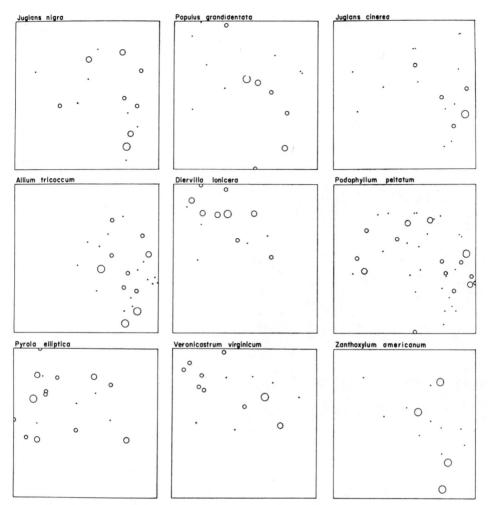

FIG. 6. Frequency behavior (understory species) and dominance behavior (trees) of 9 species not used in ordination construction. Size of circles corresponds to the quartile size class distribution illustrated in FIG. 3. Dominance per acre at the 50% level in sq. in. is as follows: *Juglans cinerea* 2,100; *Juglans nigra* 800; *Populus grandidentata* 1,200. Frequency at the 50% level is as follows: *Allium tricoccum* 10; *Diervilla lonicera* 10; *Podophyllum peltatum* 30; *Pyrola elliptica* 15; *Veronicastrum virginicum* 10; *Zanthoxylum americanum* 20.

graph. The values at each point are the measured basal areas for *Tilia* as taken from the field data for each stand. Those stands in which *Tilia* was absent are indicated by a dash. The basal areas have been put into size classes, as indicated by the circles of different size. In this and all similar figures in the paper, the largest circle includes the top 25% of all of the values; the next size, the third 25%; the smallest circle, the second 25%; and the solid points, the first or lowest 25% of the values. At a glance, therefore, it is apparent that *Tilia* reaches its highest importance in a very small portion of the possible area and that the stands with lesser domi-

nance of this species are spread out from it in a pattern of decreasing occurrence. The contour lines on Fig. 3 have been drawn in such a way as to include all examples of the indicated size class, regardless of whether lesser size classes are also present. They, therefore, indicate the area within which the species may reach the indicated level of domination.

Similar graphs were made for all axis combinations for all species used on the test sheet and for a large number of species not used in the ordination. Fig. 4 illustrates the 3 views for the 8 most important tree species, while Fig. 5 shows x-y views only for the 12 herbs used on the score sheet. Nine species not used in constructing the ordination are shown in Fig. 6 in x-y views.

A three-dimensional representation of the behavior of *Quercus borealis* is given in Fig. 7. The 3 sizes of spheres indicate the top 3 quartiles of dominance per acre. No differentiation is made between the lowest quartile and the stands not containing the species; both are indicated by holes which appear as dots in the figure. A comparison of Fig. 7 with the appropriate views of *Q. borealis* in Fig. 4 will show how these separate presentations may be used to gain a visual image of the ordinations in three-dimensional space.

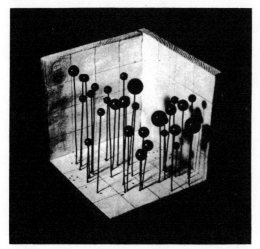

Fig. 7. Three-dimensional model of the dominance behavior of *Quercus borealis* within the ordination. The 3 sizes of spheres indicate the top 3 quartiles of dominance per acre. Stands of the lowest quartile and without the species are represented with holes which appear as dots in the figure. The x axis is on base of model at front from left to right; y axis on base from front to rear; z axis in vertical plane from below to above.

Certain measures of the environment, including soil analyses made upon a pooled sample from 3 random collections of A_1 layer in each stand are given in Table 6 for the stands used in the study. The soil nutrient analyses were made by the State Soils Laboratory, Madison, Wisconsin. Water retaining capacity was determined by the method outlined by Partch (1949).

DISCUSSION

THE MECHANICAL VALIDITY OF THE ORDINATION

The biologic and environmental results of the ordination in Table 2 should be assessed by criteria which are consistent with the assumptions upon which the ordination technique is based. The first of these assumptions is that the degree of compositional similarity between stands can be used as comparative distances to indicate spatial locations for these stands.

The complexity of stand relationship and of the forces which influence community structure is of such magnitude that a matrix of comparative distance cannot, as previously noted, be oriented in a single exact configuration. The complexity of stand relationship is not chaotic, however, and it is assumed that each stand fluctuates within a fairly limited area in its compositional (and spatial) relationships, although this area is enlarged by the sampling error of the techniques used in field survey. The ordination is further based on the assumption that by establishing a set of exact criteria for stand selection, the number of interdependent causal complexes acting within the community can be limited. This limitation increases the probability that reference stands can be selected which are oriented along the lines of major changes in community structure, but it does not necessarily prevent a certain loss of matrix information by the selection of reference stands which reflect, in part, independent causal happenings which are unrelated to the major complexes. It is evident, therefore, that the validity of the ordination should be tested on its ability to approximate (but not exactly reproduce) the estimates of stand similarity, and perhaps to show stand alignments which correlate with the available estimates of physical environmental features.

TABLE 3. Interstand distances of the reference stands from matrix and ordination. The first column lists all pair combinations of reference stands; second column shows coefficients of community inverted to represent interpoint distance; third column shows distances between the stands in the ordination.

Stand Pairs	Matrix Distance	Ordination Distance	Stand Pairs	Matrix Distance	Ordination Distance
00- 35	16	06	89-107	41	41
00- 33	25	12	89-111	35	30
00- 89	41	40	89-185	46	38
00-107	37	42	89- 26	38	33
00-111	48	42	89-136	64	50
00-185	49	37	89-138	64	49
00- 26	58	42	89-137	63	44
00-136	79	72	107-111	25	15
00-138	78	72	107-185	46	43
00-137	80	73	107- 26	25	13
35- 33	21	11	107-136	57	41
35- 89	55	42	107-138	58	45
35-107	48	37	107-137	70	54
35-111	58	38	111-185	47	45
35-185	55	37	111- 26	21	03
35- 26	55	38	111-136	53	41
35-136	80	69	111-138	65	46
35-138	79	69	111-137	70	50
35-137	79	72	185- 26	48	45
33- 89	52	33	185-136	64	49
33-107	51	35	185-138	61	45
33-111	56	32	185-137	69	46
33-185	63	39	26-136	64	41
33- 26	61	33	26-138	59	45
33-136	77	66	26-137	72	51
33-138	80	67	136-138	42	09
33-137	78	68	136-137	34	24
			138-137	30	18

To assess the approximation of ordination distances to coefficient distances, 58 stand pairs were selected at random and their interpoint distances in the ordination compared with their coefficient of community values in the matrix. Interstand distance between two points (with locations x_1, y_1, and z_1, and x_2, y_2, and z_2) was determined by the formula

$$\sqrt{(x_1 - x_2)^2 + (y_1 - y_2)^2 + (z_1 - z_2)^2}.$$ The comparison of the 58 stand pairs showed a correlation value of $-.35$ which is significant at the 1% level. The correlation is negative since ordination distance between stands, which shows low values for a high similarity in composition, was compared with coefficient of community values which show high values for a high similarity. The highly significant correlation demonstrates the tendency for the ordination to approximate the stand relationships in the matrix. The check of ordination distance compared to coefficient distance was also applied to the 11 reference stands. Coefficient of community values were first inverted by the method previously discussed to represent degree of spatial separation and are presented in the first column of Table 3. The second column of the table shows interpoint distance in the ordination as calculated by the above formula. The correlation coefficient of these two distances for 55 stand pairs is $+.73$ which is significant at the 0.1% level.

The test of the ordination as a whole is that it approximates the interstand distance relationships in the matrix of coefficients of community. For any individual axis, however, additional assurance is necessary that it has contributed new and meaningful separations of the stands. A correlation test was, therefore, applied among the stand locations of the 3 axes. Stand locations along the x and y axis and along the y and z axis were found to be uncorrelated. Stand locations on the x and z axis were, however, correlated at the 5% level, though not at the 1% level. It was, therefore, possible that the z axis repeated, in part, information previously revealed in the x axis. It was decided that this repetition was not sufficiently great to justify discarding the z axis for the following reasons: (1) As will be demonstrated, the z axis showed meaningful separations of species midpoint locations which were not available on previous axes. (2) The x-z species distribution patterns in Figs. 4 & 9 showed little tendency toward a linear arrangement which would result if the axes were perfectly correlated. (3) Of 10 environmental measurements which were tested with each axis, 7 were correlated with the x axis, but, of these 7, only 2 were also correlated with the z axis. One of these two correlations was of a ratio which had a different basis on the z axis than on the x axis. There was also an environmental feature which correlated with the z axis, but not with the x axis.

The third test of ordination validity is whether the x, y, and z axes lead to a randomization of stand location. Such a randomization would obscure any differences in species or environmental behavior. It

is apparent that if this were the case, then the species midpoints on each of these axes, as shown in Table 4, would have been in the same location, which they clearly are not. A random unordered stand orientation would probably also result in few or no environmental correlations, but as seen from Table 7, this does not happen. Every environmental feature is correlated with at least one of the axes.

TABLE 4. Location of species midpoints on ordination axes. Midpoints are mean axis locations of dominance values and represent point at which species reaches its optimum importance with respect to size.

Species	X	Y	Z
Quercus macrocarpa	16.3	46.4	38.3
Quercus velutina	17.4	49.3	41.0
Carya ovata	30.1	44.8	29.5
Prunus serotina	31.5	46.5	30.1
Quercus alba	35.5	46.7	32.9
Quercus borealis	40.9	51.8	42.7
Ulmus americana	44.8	33.5	41.9
Populus grandidentata	47.6	45.4	49.4
Juglans nigra	56.6	41.0	43.4
Ostrya virginiana	57.7	34.9	50.4
Fraxinus americana	62.2	31.8	45.5
Juglans cinerea	63.1	40.4	43.7
Carya cordiformis	63.3	43.6	42.7
Tilia americana	64.9	37.5	49.9
Ulmus rubra	67.8	42.9	38.3
Acer saccharum	68.4	40.5	48.8

The determination of species midpoints referred to above was made by finding the mean quantitative behavior (in this case, absolute basal area per acre) in each of 10 equal gradient sections, weighting the mean value by axis position, summing these weighted values, and dividing by the sum of the quantitative behaviors. These midpoint values for dominance per acre are shown in Table 4. They indicate the point at which each species reaches its optimum importance, at least with respect to size.

The differing relationships of species with each other along the 3 gradients can be seen from Table 4. Along the x axis, for example, *Acer saccharum* and *Ulmus rubra* occupy almost identical positions, and both are separated from *Ostrya virginiana* by over 10 units, yet, on the z axis, *Acer* and *Ulmus* are separated by over 10 units, while *Acer* is less than 2 units distant from *Ostrya*. Similarly, *Ulmus americana* and *Quercus borealis* which are less than 4 units distant on the x axis, are separated by over 18 units on the y axis, while *Juglans nigra* and *Ostrya virginiana* which are 1.1 units apart on the x axis, are separated by 6.1 and 7.0 units on the y and z axis, respectively.

The distances between the basal area per acre midpoints of the species is shown, for 3 dimensions, in the upper-right of Table 5. This table can be used as a basis for a spatial ordination of the species which is comparable to the patterns presented in deVries (1953). A drawing of such an ordination with midpoint locations the same as on the 3 axes of Table 4

Table 5. Species midpoints—interpoint distances in three dimensions. The upper-right of the table shows the distances between the dominance midpoints of species in three dimensions. The lower-left indicates whether there was greatest separation of midpoints in the first, second, or third dimension.

	Q.m.	Q.v.	C.o.	P.s.	Q.a.	Q.b.	U.a.	P.g.	J.n.	O v.	F.a.	J.c.	C.c.	T.a.	U.r.	A.s.
Quercus macrocarpa	. .	4.1	16.4	17.3	19.9	25.6	31.4	33.2	41.0	44.7	48.7	47.5	47.3	50.7	51.6	53.5
Quercus velutina	2	. . .	17.7	18.0	20.0	23.7	31.7	31.6	40.2	43.8	48.3	46.6	46.3	49.6	50.8	52.3
Carya ovata	1	1	. . . .	2.3	6.7	18.4	22.3	26.5	30.1	36.0	38.1	36.2	35.7	41.0	38.8	43.1
Prunus serotina	1	1	2	. . . .	4.8	16.6	22.0	25.1	28.9	35.2	37.4	34.9	34.3	39.8	37.4	41.8
Quercus alba	1	1	1	1	. . . .	12.3	18.5	20.5	24.2	30.7	33.1	30.3	29.8	32.5	33.0	37.1
Quercus borealis	1	1	3	3	3	. . . .	18.7	11.4	19.1	25.0	29.3	24.9	23.8	28.8	28.7	30.3
Ulmus americana	1	1	1	1	2	2	. . . .	14.3	14.1	15.5	17.9	19.6	21.1	22.0	25.1	25.5
Populus grandidentata	1	1	3	3	3	3	2	. . . .	11.7	14.6	20.3	17.3	17.2	19.0	23.2	21.4
Juglans nigra	1	1	1	1	1	1	1	1	. . . .	9.3	10.9	6.5	7.2	11.1	12.4	13.0
Ostrya virginiana	1	1	1	1	1	2	2	2	3	. . . .	7.4	10.2	12.9	7.5	17.7	12.2
Fraxinus americana	1	1	1	1	1	1	1	1	2	3	. . . .	8.8	12.2	7.6	14.3	11.2
Juglans cinerea	1	1	1	1	1	1	1	1	1	3	2	. . . .	3.3	6.3	7.6	7.3
Carya cordiformis	1	1	1	1	1	1	1	1	1	2	2	2	. . . .	3	9.6	8.5
Tilia americana	1	1	1	1	1	1	1	1	1	1	2	3	3	. . . .	13.1	4.7
Ulmus rubra	1	1	1	1	1	1	1	1	1	3	2	3	1	3	. . . .	10.8
Acer saccharum	1	1	1	1	1	1	1	1	1	1	2	1	3	1	3	. . . .

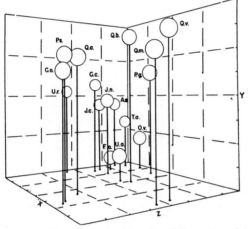

Fig. 8. Three-dimensional drawing of the center point locations of dominance behavior of tree species. Determination of center point locations explained in text. Point on frame closest to reader is lowest value for all 3 axes. The traditional position of the axes was changed to prevent the hiding of some species in the drawing.

is presented in Fig. 8. Both the upper-right of Table 5 and Fig. 8 show community relationships of species by suggesting an estimate of the relative correspondence of their stand locations; the less distant is the degree of midpoint separation, the more likely are the species to occur in the same stands.

The lower left of Table 5 lists the axis for each stand pair on which there is the greatest spatial separation in their dominance midpoints. In spite of the greater importance of the x axis (with a maximum distance between midpoints of 52.1 as compared to 20.0 and 20.9 for the y and z axes), there are 33 of a total of 120 stand pairs which give a greater separa-

tion on the y or z axis than on the x axis. These separations, in many cases, complemented the results of field observations, and gave indication that a biologic interpretation of the meaning of the axes might be possible.

The Biologic Validity of the Ordination

The ordination, therefore, by the use of a technique which is open to modification, gave one (but certainly not the only possible) approximation to the information on stand similarity in the matrix of coefficients of community. This approximation was made in 3 dimensions which were demonstrated to give species locations and environmental correlations which were non-random and non-repetitive. The ultimate test of the value of the ordination is, however, a biologic one, and succeeding discussion will attempt to utilize the ordination in examining the nature of the community and of its factorial relationships.

The views of species distribution in Figs. 4, 5, and 6 can be used in creating a mental image of the species as they appear in three dimensions, an image which will reveal the same effect as that of the 3 dimensional pattern in Fig. 7. From these visualizations and from Fig. 7, it is evident that each species shows all or part of an atmospheric distribution, that is, one in which there is an increasing concentration (number of points) and importance (size of points) of the species as the center of the distribution is reached. Away from the center, the decrease in numbers and sizes of the points is not always uniform in all directions but the species distributions, nevertheless, suggest that an idealized distribution would show a concentration and size of points diminishing outward in all directions from a dense center to an area beyond a sparse periphery where the points no longer occur.

An atmospheric distribution is a form which can be expected from an extension of the frequently expressed concept of ecologic amplitude into more than 2 dimensions. This concept postulates a mini-

TABLE 6. Environmental measurements by stands. Average canopy estimates from a number of stations within each stand. Soil analyses from pooled samples of 3 random collections of the A_1 layer in each stand. Nutrient values in pounds per acre at a soil depth of 7 in.; W.R.C. is water retaining capacity; OM is organic matter.

Stand No.	Percent Canopy	Depth A_1 (inch)	pH	WRC%	Ca (lbs.)	K (lbs.)	P (lbs.)	NH₄ (lbs.)	OM%	1/10 Ca K
00	96	4.25	7.0	53	......	280	50	..	....	...
01	99	4.0	7.3	85	6,000	185	45	20	....	3.2
03	65	4.25	5.8	93	4,000	255	60	30	....	1.6
04	70	2.5	6.0	75	50	160	60	45	....	0.3
05	88	5.25	7.0	87	8,000	230	60	20	....	3.5
06	98	3.0	7.0	96	8,000	250	95	30	....	3.2
09	90	2.5	7.4	77	12,000	200	105	20	7.0	6.0
15	..	3.5	6.0	58	......	180	40	..	6.0	...
16	..	3.0	5.6	36	2,400	100	50	..	5.0	2.4
17	78	1.7	5.7	100	6,000	130	40	40	5.2	4.6
18	..	3.0	7.0	64	......	220	70	..	....	...
19	..	7.0	6.7	85	9,400	140	50	..	....	6.7
20	97	8.0	7.2	54	6,000	230	135	20	....	2.6
21	..	3.0	5.4	48	1,100	150	60	..	....	0.7
23	..	2.5	6.8	107	10,000	130	60	..	9.0	7.7
24	..	4.0	5.4	67	4,000	180	40	..	4.0	2.2
25	97	3.5	6.0	71	600	155	130	40	8.5	0.4
26	88	2.25	6.2	93	5,000	200	75	20	....	2.5
31	87	3.25	7.3	75	10,000	430	200	10	....	2.3
33	97	6.0	7.4	66	10,000	370	80	30	....	2.7
35	88	3.5	6.1	75	5,000	190	70	10	....	2.6
41	82	1.5	4.9	79	3,000	260	50	25	....	1.2
73	..	....	...	...	......	...	...	..	....	...
85	..	....	...	...	......	...	...	..	....	...
86	98	5.0	6.9	57	5,500	185	60	15	....	3.0
87	85	....	6.2	53	5,000	190	70	30	....	2.6
88	98	....	7.7	83	10,000	130	30	10	....	7.7
89	..	6.0	8.0	42	11,000	250	...	..	....	4.4
91	..	1.0	...	107	......	240	...	..	9.0	...
92	..	....	...	38	......	...	...	..	....	...
93	..	1.0	6.0	125	15,000	260	...	..	13.5	5.7
95	70	1.5	7.0	68	3,400	240	50	..	4.0	1.4
96	..	0.5	5.8	79	2,300	160	...	..	6.5	1.4
100	..	3.0	7.0	94	15,000	180	...	..	11.5	8.3
101	..	1.5	7.0	94	7,000	160	...	..	7.5	4.3
102	..	4.0	6.3	72	......	60	...	..	5.5	...
103	95	2.5	7.0	67	6,000	205	10	20	....	2.9
104	94	3.0	6.9	61	6,000	190	60	10	....	3.2
105	72	1.0	7.1	66	4,500	185	60	25	5.5	2.4
106	92	3.0	7.0	68	5,000	190	80	25	5.5	2.6
107	..	2.0	5.5	56	1,000	150	...	..	2.5	0.7
108	..	5.0	5.7	56	6,200	265	75	60	....	2.3
109	80	3.5	7.4	84	8,000	340	80	35	....	2.4
110	85	3.0	6.5	60	8,000	200	60	25	....	4.0
111	..	0.5	5.5	67	1,300	140	...	..	2.0	0.9
112	85	1.25	6.0	85	1,500	180	60	50	....	0.8
114	..	....	6.0	123	4,200	320	...	..	12.5	1.3
117	80	3.5	7.1	134	10,000	270	80	25	....	3.8
118	83	3.0	6.4	113	7,000	205	60	20	15.0	3.4
119	65	1.5	6.9	139	6,000	250	30	20	....	2.4
120	..	....	...	68	......	...	...	..	....	...
121	..	1.0	6.5	74	5,000	170	...	..	6.5	2.9
127	..	4.0	...	107	......	...	...	..	....	...
128	..	1.0	...	162	......	...	...	..	....	...
136	..	5.0	7.1	89	8,000	185	20	15	....	4.3
137	71	1.5	5.5	54	2,000	180	30	15	....	1.1
138	..	1.0	5.6	81	6,000	180	40	30	....	3.3
151	..	1.5	5.3	133	6,000	180	30	35	....	3.3
185	..	0.25	5.8	88	5,000	180	40	15	....	2.8

mum, optimum, and maximum behavior for each species in relation to the dynamics of community structure, and has often been demonstrated in 2 dimensions by contour-shaped ("solid normal") patterns and in one dimension by bell-shaped ("normal") patterns. It can be shown that a compression of an atmospheric distribution into 2 dimensions will give contour-shaped patterns as shown in Fig. 3 and in Wiedemann (1929), Ramensky (1930), Pogrebñjak (1955), and Whittaker (1956). Further compression into one dimension will yield the bell-shaped patterns demonstrated in the original linear treatment of the upland forest (Curtis and McIntosh 1951).

Fig. 6, which shows species patterns similar in form to the two preceding figures, demonstrates the relevance of the ordination to the entire plant population of the upland hardwoods. The 9 species in Fig. 6 include all of the minor tree species for which adequate data were available and an unbiased selection of shrub and herb species. Although none of the species in Fig. 6 contributed to the placement of stands in the ordination, they, nevertheless, show the same atmospheric distributions outlined above. The ability of the ordination to give meaningful patterns to species not used in its construction is considered as both a basic test of the usefulness of the ordination and as a demonstration of the feasibility of gradient construction by the consideration of less than the total species complement.

Figs. 4 through 6 show each species to have an individual pattern, different in size and location, although fairly similar in shape to those of other species. The distribution and the relationship of the patterns within the ordination is clearly one of continuous variation, as was previously demonstrated in the linear continuum of Curtis and McIntosh. The species used in the ordination, as well as those in Fig. 6 which were examined after the ordination was completed, can be described, therefore, as having patterned, non-random distributions within the prescribed geographic, environmental, and physiognomic limits of the study. Each of these distributions moves outward from central areas of high density to peripheral areas of sparse density, and this movement reveals along 1, 2, or 3 dimensions corresponding bell-shaped, contour-shaped, or atmospheric distributions. Each species has a separate area of location, and within this area its distribution is interspersed to varying degrees with other species distributions so that there is a continuous change in stand composition from any part of the ordination to any other part.

The above description supports, to a large degree, conclusions from work completed in a diversity of geographic regions and vegetations, including the studies of Gleason (1926), Vorobyov & Pogrebñjak (1929), Ramensky (1930), Sörenson (1948), Sjörs (1950), Curtis & McIntosh (1951), Whittaker (1951), deVries (1953), Goodall (1954a), Guinochet (1954), Webb (1954), Churchill (1955), Horikawa & Okutomi (1955), Poore (1956), and Hewetson (1956). It is suggested that evidence for the individualist theory of species distribution and for the continuum nature of community structure is now sufficiently compelling to require studies concerned with community structure to examine the relative continuity or discontinuity of their material and to use quantitative methods which will permit this examination. At a minimum, the examination would include a sampling of at least one analytic character in a sufficient number of stands to allow comparisons of quantitative composition. If an apparent grouping of stands into a discrete unit (i.e. association, etc.) is suspected, then this unit should be tested against samples of related vegetations to determine whether there are separate groups of stands with a certain range of variation within each group, or whether this variation is great enough to obscure the boundaries between the groups. If the latter is true, the application of ordination methods is necessary. A reasonable approach to the treatment of phytosociologic material about which little is known might be to check carefully the homogeneity of each stand, and then to apply an ordination technique. If clumps of stands are shown along the resultant gradients, it would then be possible to regard these clumps as castes (associations, etc.) and determine the suitable parameters necessary for the future classification of each caste.

One reason for reexamining the upland hardwoods of Wisconsin was the diversity of interpretations which various readers gave the original paper. Thus, some correspondents questioned if the linear continuum was not a statistically advanced restatement of succession, while others (Horikawa & Okutomi 1955) regarded it as a demonstration of relationships which were independent of succession. The continuous compositional variation which was demonstrated was at times assumed to apply to spatial transition as seen in the field (Churchill 1955), as contrasted to theoretic variation in the structure of communities regardless of their microgeographic relationships. In clarification of the above interpretations, it should be evident from the present study that the dimensions of the ordination are purely compositional and cannot necessarily be directly related to factors or to complexes of factors. Successional change is only one among many causal forces which have shaped the species distribution patterns of Figs. 4 through 6. The dimensions represent an approximation of the changes in compositional structure which are present within the community and are not spatial transitions as they exist in the field, although, as noted by Gleason (1926) and as is evident to many field workers, there are often natural areas where the salient features of a continuum can be observed.

The Nature of the Gradients

It is likely that the degree to which there are separations in more than one dimension was in part dependent upon the qualifications used to differentiate the community initially. Had sufficient knowledge been available before the study was begun, it might have been possible to eliminate those stands from the study which had relatively poor subsoil drainage and one of the additional dimensions might, thereby, have been eliminated. On the other hand, had there been an expansion of the data to include poorly drained forests and those subject to inundation, then a more complete picture of the southern forest as a whole could have been given. The view of Ashby (1948), that the most important aspect of community study is the original delineation of the study area, is very applicable to the present study.

As previously noted, the 3 dimensions of the ordination represent compositional gradients which are not likely to be related to any single causal agency. The probability that every factor is a constant influence on the structure of a stand, and the interrelated nature of biotic and physical factors, prohibits the identification of single causal mechanisms. In spite of these limitations, however, certain over-all patterns are evident in the 3 axes, patterns which relate to broad bio-physical complexes and to history. It is noteworthy that some of these patterns of species and of environmental features had not been suspected before the application of multidimensional technique.

Thus, in general, the x axis duplicates the original linear continuum of Curtis & McIntosh and shows a complex of conditions which include gradients from higher to lower light intensity and evaporation, gradients from lower to higher soil moisture and relative humidity, and gradients from more to less widely fluctuating soil and air temperature. The order of species along the x axis is basically determined by an over-all linear direction in the many paths of community development within the upland forest. These paths follow a network of successional patterns which are mainly related neither to primary nor secondary succession but to recovery from past disturbance. This disturbance, in the form of fire, reduced the forest in many places to a savanna or barrens condition in which there were scattered oaks and/or oak brush and roots (Cottam 1949; Bray 1955). During this reduction, the more terminal species which were also the more fire susceptible, were replaced by less terminal and by initial species which were the least fire susceptible. The coincidence that *Quercus macrocarpa* is both the most initial and fire resistant species, and that *Acer saccharum* is the most terminal and fire susceptible

species, suggests that the ultimate explanation for the composition of a forest stand in upland Wisconsin is largely an historic one. The longer a stand has been free of fire (and other disturbance forces) and the more favorable the habitat in which it occurs, the more likely it can develop to a maple-basswood forest. The x axis shows, therefore, mainly the relationship of the community to major past disturbance factors and to its own developmental recovery from these factors.

To examine correlations with physical measurements of the environment, a check was made, using r, the correlation coefficient, of all available environmental data with each of the three dimensions. The results are shown in Table 7. Since both environmental data and stand locations showed approximately normal distributions, the significance of correlation was checked with the t test by examining the hypothesis that correlation $(\rho) = 0$. The values in Table 7 are the probability that $\rho = 0$, and that there is, therefore, no correlation. Thus, a probability of $< .01$ is a basis for rejecting the hypothesis and is considered a highly significant correlation. A value of $< .05$ is considered a significant correlation, while values of $> .05$ are considered to represent no significant correlation. By the above interpretation, for example, percent canopy is positively correlated with the x axis at a highly significant level, while it is not correlated with either the y or z axis.

The values in Table 7 show a highly significant correlation between percent canopy, depth of A_1, organic matter, pH, Ca, P, $\dfrac{.1 Ca}{K}$ and the x axis. Water retaining capacity, K and NH_4 are not, however, correlated with the x axis. An examination of the values in Table 6 indicated the majority of the features related to the x axis, were positively correlated with each other. It is highly probable that these correlations represent the measured aspects of an increasingly mesic environment which accompany the successional recovery of the forest from past disturbance. While only a general interpretation can be made of this recovery at present, it is clear that the dynamics of factorial interrelationships along the x axis offer a broad area for future research.

The y axis seems to be correlated, in part, with the influence of surface and sub-surface drainage, and, consequently, with a soil moisture complex, with internal soil air space and aeration probably also

TABLE 7. Environmental correlations with three ordination axes. A_1 is depth of A_1 in inches; W.R.C. is water retaining capacity; O.M. is organic matter; Values in body of table are the probability that there is no correlation.

	Percent Canopy	A_1	pH	W.R.C.	Ca	K	P	NH_4	O.M.	$\dfrac{.1\,Ca}{K}$
x.........	+ <.001	+ <.01	+ <.01	>.05	+ <.001	>.05	+ <.001	>.05	+ <.01	+ <.001
y.........	>.05	>.05	>.05	+ <.001	>.05	>.05	>.05	+ <.01	>.05	>.05
z.........	>.05	>.05	>.05	>.05	>.05	>.05	− <.05	>.05	+ <.05	+ <.01

influential. The 3 tree species with midpoints towards the lower end of the y axis in Table 4 are *Fraxinus americana*, *Ulmus americana*, and *Ostrya virginiana* and are, of the species occurring in the upland forests, the most tolerant of poor drainage and aeration. Species which are found toward the upper end of the y axis are predominantly oaks, such as *Quercus borealis*, *Q. alba*, *Q. macrocarpa*, and *Q. velutina*, which are intolerant of inundation and poor aeration. Species with midpoints towards the center of the axis, such as *Acer saccharum*, *Juglans nigra*, *Ulmus rubra*, and *Carya cordiformis* are mesic species with an intermediate tolerance of poor drainage and aeration.

Two environmental features, W.R.C. and NH_4 show a highly significant positive correlation with the y axis. Water retaining capacity is represented with its highest values occurring towards the upper end of the axis assumed above to have the better internal drainage. The anomaly of high W.R.C. associated with good drainage is perhaps explained by the differences in past history of the stands along the y axis. It was found that W.R.C. has a significant negative correlation with depth of the A_1 layer. This correlation is apparently related to the circumstance that stands with high water-retaining capacities have a sharply demarcated boundary between the A_1 and A_2 layers, while the opposite stands tend to have a diffuse boundary, with the organic matter gradually decreasing in amount. These differences represent trends towards mor humus and typical podzols on the one hand and mull humus and gray-brown podzolic or brown forest soils on the other. They probably reflect past history to the extent that stands at the upper end of the y axis (mor humus, high W.R.C.) have been occupied by mixed conifer-hardwood forests in more recent postglacial times than stands at the other end, which may have developed on savannas or grasslands with a deep layer of incorporated humus. Substantiating evidence for this was seen in the distribution patterns of *Diervilla lonicera*, *Goodyera pubescens*, *Maianthemum canadense*, *Pteridium aquilinum* and *Pyrola elliptica* which closely matched the distribution of the higher values of water retaining capacity. All of these species currently reach their optimum in the conifer forests of northern Wisconsin and are only incidental members of the southern forest under discussion. The mean W.R.C. in stands in which 3 or more of the above species occurred in the quadrat samples is 103 as compared to a mean of 78 for the remaining stands.

The strong positive correlation of NH_4 with the y axis is not related to pH, since these two features show no correlation with each other. There is, however, a likelihood that the NH_4 correlation is, in part, controlled by the soil moisture and aeration complex discussed above in relation to tree distribution. The assumed poorly aerated soils towards the lower part of the y axis have, perhaps, a relatively higher proportion of denitrifying bacteria. With an increase in internal drainage and in soil oxygen along the y axis, there is a proportional increase in nitrifying bacteria and, therefore, an indirect increase in NH_4. The y axis, therefore, appears to represent a complex of internal soil drainage and aeration factors which is coincident with a cliseral historic factor reflecting post-glacial vegetational changes.

The z axis apparently represents, in part, the influence of recent disturbance, with species which benefit by disturbance being separated from the species with which they are proximate on other axes. Both *Carya ovata* and *Prunus serotina* at the lower end of the z axis are gap phase species (Watt 1947, Bray 1956b) which take advantage of oak wilt gaps and of grazing in initial forests. *Populus grandidentata* is another gap phase species and is found toward the upper end of the z axis. Its ability to enter areas opened by fire has long been noted (Chamberlin 1877) and observed in the field. *Ulmus rubra*, and to a lesser extent *Carya cordiformis*, apparently do well in some disturbed intermediate and early terminal forests and, as a result, are pulled away from their proximity to *Acer saccharum* to which they are closely adjacent on the previous axes. *Quercus alba* and *Quercus borealis* are separated on the z axis to a greater extent than on the x or y axes, a separation which might be related to the ability of red oak to increase in importance in maple-basswood forests which have been burned, forests from which white oak had been eliminated earlier by successional developments.

The values in Table 7 show 3 environmental features, K, O.M., and $\dfrac{.1Ca}{K}$ to be correlated with the z axis. Potassium shows a significant negative correlation which might be related to amount and content of loess, in that K is characteristically high in loessial soils in the Middle West and stands towards the upper end of the z axis might have an increasingly shallow or absent loessial layer. The highly significant positive correlation between the ratio $\dfrac{.1Ca}{K}$ and the z axis is illustrated by values which more than double from one end of the axis to the other. This correlation is the result of a Ca content which is not correlated along the axis coupled with the decreasing K content noted above. There is, therefore, a consequent increase in the ratio. Along the x axis there is a similar $\dfrac{.1Ca}{K}$ correlation, but it is produced by an uncorrelated K content which is accompanied by an increasing Ca content. No connection is directly apparent between this ratio and the growth of plants although it may reflect amount of available K. Thus, availability of K is considerably decreased with increased Ca, if K is tending toward low levels of fertility (Lutz & Chandler 1946). Such low fertility levels are possible in the decreasing K concentration along the z axis which is accompanied by a uniformly high Ca level. A similar non-availability

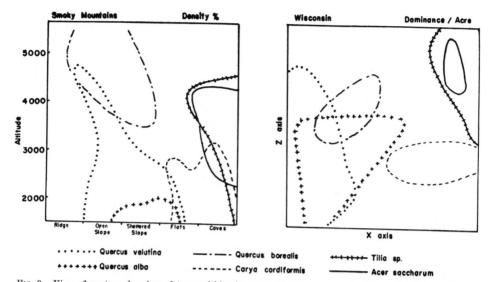

FIG. 9. View of contour drawing of trees within the x-z ordination compared with nomograms of the same or similar species in the Smoky Mountains (Whittaker 1956). Contour lines in the Wisconsin ordinations are made at the 25% (upper) size class level. Dominance values at this level in sq. in. are as follows: *Acer saccharum* 12,000; *Carya cordiformis* 400; *Quercus alba* 8,500; *Quercus borealis* 11,400; *Quercus velutina* 6,-000; *Tilia americana* 5,700.

of K correlated with the x axis is not as likely, since the non-correlated K concentration along this axis indicates there is no area where low K values are consistently found with high Ca values.

The reasons for the significant positive z axis correlation with organic matter are not readily apparent. The overwhelming probability that all the measured environmental factors plus numerous non-measured factors are interacting with each other and with biotic forces, and that plants are responding to the interactions rather than to single factors indicates that attempts to pinpoint the axes too closely are doomed to failure.

The difficulty of assigning definite factorial meanings to the ordination axes is further emphasized by a consideration of Fig. 9. This represents a comparison of the two-dimensional behavior patterns of a series of tree species which occurred in the Wisconsin study and also in Whittaker's 1956 analysis of the Great Smoky Mountains vegetation. The species of Tilia are slightly different in the two regions. The patterns on the Great Smoky graph are taken from Whittaker's nomograms. They portray either the 10% contour of species density or the highest value attained by the species if it was less than 10% and thus represent favorable or optimum conditions for growth. The patterns on the Wisconsin graph are the contours of the top quartile of dominances per acre. The absolute levels in square inches per acre for this 75% level are indicated in the legend of the figure. This may seem an unfair comparison since only selected contours are used, but a study of the full set of curves in Whit-

taker's original graphs in relation to the full diagrams for the Wisconsin data only strengthens the comparison.

With the exception of *Quercus alba*, the patterns in the two graphs are remarkably similar, especially if the axes of the Wisconsin graph are rotated slightly to the right. The exact meaning of this coincidence is not clear. A relationship between the x axis and Whittaker's moisture gradient seems reasonable enough, but there is no immediately evident connection between the factors usually associated with elevation and any possible interpretation of the z axis. All of the stands in the Wisconsin study were within a few hundred feet of each other in elevation and did not differ significantly in average temperature as measured by nearby weather stations.

The similarity in species distribution patterns between Wisconsin and Tennessee suggests that the biotic forces within a community have a high degree of homeostatic stability in their interactions with other causal agencies. Such stability is a feature of the dynamics of open systems in which, although a wide variety of factorial configurations may exist, there are apparently only a limited number of system structures which are possible. In biotic communities a limitation in structure means each species is confined to a definite community position in relation to other species with the unlikehood that all combinations of species can occur. It is the limited number of stand compositions which permits the application of quantitative methods to community classification.

Further Uses of the Ordination

Although the main purpose of the study was to investigate the phytosociology of community structure, the ordination results can be used as a basis for further research and for investigations in related fields. The value of the original upland hardwoods continuum in animal ecology (Bond 1955) and in plant ecology (Tresner, Backus & Curtis 1954; Gilbert & Curtis 1953; Hale 1955) has been demonstrated. The construction of additional dimensions should facilitate any further work which is done by providing a more complex pattern of community structure. Thus, studies of life histories of individual species can be given an exact community background against which to correlate those changes in the morphology, vitality, manner and means of dissemination, and perhaps taxonomy and phenology of the species which are determined by differences in community composition. Other important phytosociologic features which can be directly related to differences in community structure are (1) the distribution patterns of species, both as to variation in population means and in manner of dispersion, as demonstrated in Whitford (1949) and (2) the variation in interspecific association relationships.

The nature of the causal role of factors shaping a community is an ultimate goal in ecology. The determination of community structure, as was suggested earlier, is perhaps a key to these causal relationships. Goodall (1954b) notes, "It is the high correlation between different environmental factors that often suggests a deceptively simple relationship between plant distribution and the environment. There is much to be said for the view that the complexes of environmental factors determining plant distribution can be indicated and measured better indirectly, through the plants themselves, than by direct physical measurements; this is, of course, the idea behind the use of "phytometers" in agricultural meteorology, and the attention devoted to phenology." By separating unlike stands along several different compositional gradients, possible extremes in causal reactions are determined and the location of crucial junctures of change in community structure is possible. Elaborate and expensive tests can then be made at these junctures which would have been impossible to apply to the community as a whole. A further interpretation of the southern upland hardwoods would then, perhaps, be able to demonstrate the interactive nature of a series of factors as they change the structure of the community and, in turn, are changed in their own expression.

SUMMARY

The relationships of 59 stands of upland hardwood forest in southern Wisconsin were studied by means of a new ordination technique. Quantitative measurements of 26 different species were used as the basis of the ordination. These 26 species included the 12 most important trees and 14 herbs and shrubs. The quantitative characters used were absolute density per acre and absolute basal area per acre for the trees and simple frequency for the herbs and shrubs. These measurements were arranged on a "score" sheet for each stand after having been put on a comparable and relative basis. The degree of phytosociologic similarity of any stand to each of the other stands was assessed by Gleason's coefficient of community. Such coefficients were calculated for each stand with each other stand (1711 values).

It was considered that the degree of similarity of two stands as shown by their coefficient could be translated into a spatial pattern in which the inverse of the coefficient was equated with linear distance. Using groups of stands which were least similar as end points of an axis, the relative positions of all other stands were located along this axis by a geometric technique of arc projection, with the radii of the arcs given by the inverted coefficients of similarity. When this was done, it was seen that some stands were equidistant from the two end groups but nevertheless were themselves unlike. Using the most distantly related of these, a second axis was erected by the same method as the first. The position of all stands on this new axis was then determined. Further inspection revealed that a few stands were about equally spaced between the ends of both the new axis and the first axis but were still dissimilar. These were used to produce a third axis. The three axes were at right angles to each other, so the positions of any given stand on each axis could be projected to give a three-dimensional locus for that stand. When all stands were thus located in three-dimensional space, a highly significant correlation was found between their actual, measured spacings and the original coefficient of community values between them.

Given the framework of the spatial distribution of the stands, the behavior of individual species was readily examined by plotting a measured phytosociologic value of a species at the loci of the stands of its occurrence. It was found that all species (including an unbiased selection of those not used in the original 26 on which the ordination was based), formed atmospheric distributions with high or optimum values in a restricted portion of the array, surrounded by decreasing values in all directions. No two species showed the same location, but each was interspersed to varying degree with other species, in a continuously changing pattern.

The three axes of the ordination are compositional gradients, and their structure is interpreted as not the result of a causal determination by the physical environment but of an interaction between organisms and environment. A preliminary attempt was made to describe patterns of this interaction although an exact description remains for future study. Thus, the major (x) axis appeared to represent, in part, patterns of developmental recovery from major past disturbances. This recovery was correlated with features of an increasingly shaded and mesic environment including canopy cover and the following soil features: depth of A_1 layer, organic matter, pH,

Ca, P, and $\dfrac{1\ Ca.}{K}$. The second axis was related, in part, with surface and subsurface drainage and with soil aeration and it showed correlation with soil water retaining capacity and NH_4. The third axis was apparently related to recent disturbance and the influence of gap phase replacement, and it showed correlation with the soil feautres of $K, \dfrac{.1\ Ca,}{K}$ and organic matter. The possibility that the axes may have different interaction patterns than suggested above is evident in the similarity between the two-dimensional distributions of a number of tree species in the Wisconsin stands and in Whittaker's Smoky Mountain stands which were arranged along topographic and altitudinal gradients.

LITERATURE CITED

Agrell, I. 1945. The Collemboles in nests of warm-blooded animals, with a method for sociological analysis. Kungl. Fysiograf. Sallisk, Handl. **56**: 1-19.

Ashby, E. 1948. Statistical ecology. II—A reassessment. Bot. Rev. **14**: 222-234.

Bond, R. R. 1955. Ecological distribution of breeding birds in the upland forests of southern Wisconsin. Ph.D. Thesis. Univ. of Wisc.

Bray, J. R. 1955. The savanna vegetation of Wisconsin and an application of the concepts order and complexity to the field of ecology. Ph.D. Thesis. Univ. of Wisc.

———. 1956a. A study of mutual occurrence of plant species. Ecology **37**: 21-28.

———. 1956b. Gap phase replacement in a maple-basswood forest. Ecology **37**: 598-600.

Brown, R. T. & J. T. Curtis. 1952. The upland conifer-hardwood forests of northern Wisconsin. Ecol. Monog. **22**: 217-234.

Cain, S. A. 1944. Foundations of plant geography. New York.

Chamberlin, T. C. 1877. Native vegetation of eastern Wisconsin. *In* Chamberlin's Geology of Wisconsin. **2**: 176-187.

Churchill, E. D. 1955. Phytosociological and environmental characteristics of some plant communities in the Umiat region of Alaska. Ecology **36**: 606-627.

Clements, F. E. & G. W. Goldsmith. 1924. The phytometer method in ecology; the plant and community as instruments. Carnegie Inst. Wash. Publ. **356**.

Clifford, H. T. & F. E. Binet. 1954. A quantitative study of a presumed hybrid swarm between *Eucalyptus elaeophora* and *E. goniocalyx*. Austral. Jour. Bot. **2**: 325-337.

Cole, L. C. 1949. The measurement of interspecific association. Ecology **30**: 411-424.

Cottam, G. 1949. The phytosociology of an oak woods in southwestern Wisconsin. Ecology **30**: 271-287.

Cottam, G. & J. T. Curtis. 1949. A method for making rapid surveys of woodlands by means of pairs of randomly selected trees. Ecology **30**: 101-104.

Cottam, G., J. T. Curtis, & B. W. Hale. 1953. Some sampling characteristics of a population of randomly dispersed individuals. Ecology **34**: 741-757.

Curtis, J. T. & R. P. McIntosh. 1951. An upland forest continuum in the prairie-forest border region of Wisconsin. Ecology **32**: 476-496.

Daubenmire, R. 1954. Vegetation classification. Veröffentl. des Geobot. Inst. Rübel. **29**: 29-34.

deVries, D. M. 1953. Objective combinations of species. Acta Bot. Neerlandica **1**: 497-499.

———. 1954. Constellation of frequent herbage plants, based on their correlation in occurrence. Vegetatio **5 & 6**: 105-111.

Dice, L. R. 1948. Relationship between frequency index and population density. Ecology **29**: 389-391.

Fisher, R. A. 1936. The use of multiple measurements in taxonomic problems. Ann. Eugen., Lond. **7**: 179-188.

Forbes, S. A. 1907a. An ornithological cross-section of Illinois in autumn. Ill. State Lab. Nat. Hist. Bull. **7**: 305-335.

———. 1907b. On the local distribution of certain Illinois fishes: An essay in statistical ecology. Ill. State Lab. Nat. Hist. Bull. **7**: 273-303.

———. 1925. Method of determining and measuring the associative relations of species. Science **61**: 524.

Gilbert, M. L. 1952. The phytosociology of the understory vegetation of the upland forests of Wisconsin. Ph.D. Thesis. Univ. of Wisc.

Gilbert, M. L. & J. T. Curtis. 1953. Relation of the understory to the upland forest in the prairie-forest border region of Wisconsin. Wisc. Acad. Sci. Arts Lett. Trans. **42**: 183-195.

Gleason, H. A. 1910. The vegetation of the inland sand deposits of Illinois. Ill. State Lab. Nat. Hist. Bull. **9**: 23-174.

———. 1920. Some applications of the quadrat method. Torrey Bot. Club Bull. **47**: 21-33.

———. 1926. The individualistic concept of the plant association. Torrey Bot. Club Bull. **53**: 7-26.

———. 1952. The new Britton and Brown illustrated flora of the northeastern United States and adjacent Canada. Lancaster, Penna.: Lancaster Press.

Goodall, D. W. 1953a. Objective methods for the classification of vegetation. I. The use of positive interspecific correlation. Austral. Jour. Bot. **1**: 39-63.

———. 1953b. Factor analysis in plant sociology. Paper read to Third Internat. Biometric Conf. in Bellagio. Sept. 4, 1953.

———. 1954a. Vegetational classification and vegetational continua. Angew. Pflanzensoziologie, Wien. Festschrift Aichinger **1**: 168-182.

———. 1954b. Objective methods for the classification of vegetation. III. An essay in the use of factor analysis. Austral. Jour. Bot. **2**: 304-324.

Gregg, J. R. 1954. The language of taxonomy. New York.

Guinochet, M. 1954 Sur les fondements statistiques de la phytosociologie et quelques unes de leurs conséquences. Veröffentl. des Geobot. Inst. Rübel. **29**: 41-67.

———. 1955. Logique et dynamique du peuplement végétal. Paris: Masson et Cie.

Hale, M. E., Jr. 1955. Phytosociology of corticolous cryptogams in the upland forest of southern Wisconsin. Ecology **36**: 45-63.

Hansen, H. M. 1930. Studies on the vegetation of Iceland. In The Botany of Iceland (ed. Rosenvinge & Warming). **3**(10):1-186. Copenhagen.

Hanson, H. C. 1955. Characteristics of the *Stipa comata-Bouteloua gracilis-Bouteloua curtipendula* association in northern Colorado. Ecology **36**: 269-280.

Hewetson, C. E. 1956. A discussion on the "climax" concept in relation to the tropical rain and deciduous forest. Empire Forestry Rev. **35**: 274-291.

Horikawa, Y. & K. Okutomi. 1955. The continuum of the vegetation on the slopes of Mt. Shiroyama, Iwakuni City, Prov. Suwo. The Seibutsugakkaishi **6**: 8-17. (Japanese, English summary.)

Hosokawa, T. 1955-1956. An introduction of 2 x 2 table methods into the studies of the structure of plant communities (on the structure of the beech forests, Mt. Hiko of S. W. Japan). Jap. Jour. Ecol. **5**: 58-62; 93-100; 150-153. (Japanese, English summary.)

Hughes, R. E. 1954. The application of multivariate analysis to (a) problems of classification in ecology; (b) the study of the interrelationships of the plant community and environment. Congr. Int'l. de Bot. Rap. et Comm. **8**: (sect. 7/8): 16-18.

Isaacson, S. L. 1954. Problems in classifying populations. In Statistics and Mathematics in Biology. (O. Kempthorne, et al., eds.). Ames, Iowa.

Kato, M., M. Toriumi & T. Matsuda. 1955. Mosquito larvae at Mt. Kago-bo near Wakuya, Miyagi Prefecture, with special reference to the habitat segregation. Ecol. Rev. **14**: 35-39. (Japanese, English summary.)

Kucera, C. L. & R. E. McDermott. 1955. Sugar maple-basswood studies in the forest-prairie transition of central Missouri. Amer. Midland Nat. **54**: 495-503.

Kulczyński, S. 1927. Zespoły roślin w Pieninach (Die Pflanzenassoziation der Pieninen). Internat. Acad. Polon. Sci., Lettr. Bull., Classe Sci. Math. et Nat., ser. B. Sci. Nat. Suppl. **2**: 1927: 57-203.

Lippmaa, T. 1939. The unistratal concept of plant communities. Amer. Midland Nat. **21**: 111-145.

Lorenz, J. R. 1858. Allgemeine Resultate aus der pflanzengeographischen und genetischen Untersuchung der Moore in präalpinen Hügellande Salzburg's. Flora **41**: 209 et seq.

Lutz, H. J. & R. F. Chandler. 1946. Forest soils. New York.

McIntosh, R. P. 1957. The York Woods, a case history of forest succession in southern Wisconsin. Ecology **38**: 29-37.

Mitsudera, M. 1924. Studies on phytometer II-Phytometer method in communities. Jap. Jour. Ecol. **4**: 113-120. (Japanese, English summary.)

Motyka, J., B. Dobrzański & S. Zawadzki. 1950. Wstępne badania nad łąkami południowowschodniej Lubelszczyzny. (Preliminary studies on meadows in the southeast of the province Lublin.) Univ. Mariae Curie-Skłodowska Ann. Sect. E. **5** (13): 367-447.

Nash, C. B. 1950. Associations between fish species in tributaries and shore waters of western Lake Erie. Ecology **31**: 561-566.

Oosting, H. J. 1956. The study of plant communities. 2d ed. San Francisco: Freeman.

Parmalee, G. W. 1953. The oak upland community in southern Michigan. Ecol. Soc. Amer. Bull. **34**: 84. (Abst.).

Partch, M. L. 1949. Habitat studies of soil moisture in relation to plants and plant communities. Ph.D. Thesis. Univ. of Wisc.

Pirie, N. W. 1937. The meaninglessness of the terms 'life' and 'living'. In Perspectives in Biochemistry, (eds. J. Needham & D. E. Green). Cambridge.

Pogrebnjak, P. S. 1930. Über die Methodik der Standortsuntersuchungen in Verbindung mit den Waldtypen. Internatl. Congr. Forestry Exp. Stations, Stockholm Proc. **1929**: 455-471.

————. 1955. Osnovi lesnoj tipologie (Foundations of forest typology). Publ. Acad. of Sci. of the Ukrainian S. S. R., Kiev.

Poore, M. E. D. 1956. The use of phytosociological methods in ecological investigations. IV. General discussion of phytosociological problems. Jour. Ecol. **44**: 28-50.

Raabe, E. W. 1952. Über den "Affinitätswert" in der Pflanzensoziologie. Vegetatio **4**: 53-68.

Ramensky, L. G. 1930. Zur Methodik der vergleichenden Bearbeitung und Ordnung von Pflanzenlisten und anderen Objekten, die durch mehrere, verschiedenartig wirkende Faktoren bestimmt werden. Beitr. z. Biol. der Pflanz. **18**: 269-304.

————. 1952. On some principal aspects of present day geobotany. Bot. Zhur. **37**: 181-201. (Russian.)

Randall, W. E. 1951. Interrelations of autecological characteristics of woodland herbs. Ecol. Soc. Amer. Bull. **32**: 57. (Abst.)

Shreve, F. 1942. The desert vegetation of North America. Bot. Rev. **8**: 195-246.

Sjörs, H. 1950. Regional studies in North Swedish mire vegetation. Botaniska Notiser. **1950** (2): 173-222.

Sörenson, Th. 1948. A method for establishing groups of equal magnitude in plant sociology based on similarity of species content. Acta K. Danske Vidensk. Selsk. Biol. Skr. J. **5**: 1-34.

Spalding, V. M. 1909. Problems of local distribution in arid regions. Amer. Nat. **43**: 472-486.

Stewart, G. & W. Keller. 1936. A correlative method for ecology as exemplified by studies of native desert vegetation. Ecology **17**: 500-514.

Thurstone, L. L. 1947. Multiple-factor analysis. Chicago: The Univ. of Chicago Press.

Torgerson, W. S. 1952. Multidimensional scaling: I. Theory and method. Psychometrika **17**(4): 401-419.

Tresner, H. D., M. P. Backus & J. T. Curtis. 1954. Soil microfungi in relation to the hardwood forest continuum in southern Wisconsin. Mycologia **46**: 314-333.

Tucker, L. G. 1956. Personal communication.

Tuomikoski, R. 1942. Untersuchunger über die Untervegetation der Bruckmoore in Ostfinnland. I. Zur Methodik der pflanzensoziologischen Systematik. Soc. Zool.-Bot. Fenn. Vanamo. Ann. Bot. **17**: 1-203.

Vorobyov, D. V. & P. S. Pogrebnjak. 1929. Lisovyi tipologitchnyi visnatchnik ukrainskogo polissya. (A key to Ukrainian forest types.) Trudi z lisovoi dosvidoni spravy na Ukraini **11**.

Watt, A. S. 1947. Pattern and process in the plant community. Jour. Ecol. **35**: 1-22.

Webb, D. A. 1954. Is the classification of plant communities either possible or desirable? Bot. Tidsskr. **51:** 362-370.

Whitford, P. B. 1949. Distribution of woodland plants in relation to succession and clonal growth. Ecology **30:** 199-208.

Whittaker, R. H. 1951. A criticism of the plant association and climatic climax concepts. Northw. Sci. **25:** 17-31.

————. 1952. A study of summer foliage insect communities in the Great Smoky Mountains. Ecol. Monog. **23:** 1-44.

————. 1954. Plant populations and the basis of plant indication. Angewandte Pflanzensoziologie, Wien. Festschrift Aichinger **1:** 183-206.

————. 1956. Vegetation of the Great Smoky Mountains. Ecol. Monog. **26:** 1-80.

Wiedemann, E. 1929. Die ertragskundliche und waldbauliche Brauchbarkeit der Waldtypen nach Cajander im sachsischer Erzgebirge. Allgemeine Forst und Jagd-Zeitung. **105:** 247-254.

Williams, C. B. 1954. The statistical outlook in relation to ecology. Jour. Ecol. **42:** 1-13.

Womble, W. H. 1951. Differential systematics. Science **114:** 315-322.

17

Copyright © 1972 by the British Ecological Society
Reprinted from *J. Ecol.* **60**:305–324 (1972)

THE APPLICATION OF QUANTITATIVE METHODS TO VEGETATION SURVEY

III. A RE-EXAMINATION OF RAIN FOREST DATA FROM BRUNEI

By M. P. AUSTIN,*‡ P. S. ASHTON† AND P. GREIG-SMITH*

**School of Plant Biology, University College of North Wales, Bangor, and
†Department of Botany, University of Aberdeen*

INTRODUCTION

Since the original work on the rain forests of Brunei was published (Ashton 1964), understanding of the numerical analysis of vegetation data has increased considerably (e.g. Lambert & Dale 1964; Williams, Lambert & Lance 1966; Lambert & Williams 1966; Orloci 1966, 1967; Austin & Orloci 1966; Webb *et al.* 1967a, b; Greig-Smith, Austin & Whitmore 1967; Austin & Greig-Smith 1968). The Brunei work is one of the most detailed studies of tropical forest, and relatively complete soil data have now been collected from the plots. In view of this, and the increased knowledge of technique, it was considered worthwhile to re-examine the data using recently developed methods.

DATA

The vegetation data are from two areas, Andulau Forest Reserve and Kuala Belalong in Brunei (see Ashton (1964) for further details of location, etc.). Within each area fifty 1-acre (0·4 ha) plots were surveyed and the girths of all trees exceeding 12 in. (30 cm) girth were recorded for 0·1 acre (0·04 ha) divisions of the plots. Species density and presence can therefore be determined for plots of any multiple of 0·1 acre. Soil samples were collected from 8–12 in. (20–30 cm) and 20–24 in. (51–61 cm) depths at nine randomized sites within each plot. Groups of three adjacent samples were bulked before analysis and the average of the three resulting values used. Soil data available for comparison are: litter and humus depth and colour, depth of permanent water table, presence or absence of mottling and its depth; for the two depths sampled: soil colour (Munsell code), porosity measured as log rate of percolation through a standardized sample, mechanical fractions, pH, total phosphorus, nitrogen, and 'reserve' calcium, magnesium, potassium and Group III (iron and aluminium) elements. Total phosphorus was determined by the method of Fogg & Wilkinson (1958) after digestion with perchloric acid; total nitrogen was determined by the Kjeldahl method. 'Reserve' quantities were determined as described by Bailey (1967); the analysis represents that fraction extracted intermediate between a strong acid treatment and a sodium carbonate fusion and in these particular soils closely approaches total values. These methods were chosen partially because of their ease and low cost, but also because exchangeable values in these soils are so low and variable as to be meaningless at low sampling intensities. In view of the relatively

326   ‡ Present address: C.S.I.R.O. Division of Land Research, P.O. Box 109, Canberra City, A.C.T. 2601, Australia.

rapid rate of release of elements bound in feldspar and elsewhere in these mineral soils, it was considered that total figures would be biologically more meaningful in relation to stable primary woody vegetation. 'Reserve' estimates have been found to be closely correlated with production in herbaceous crops in Sarawak (I. M. Scott, personal communication).

METHODS

The methods used for analysis of vegetation were association-analysis (Williams & Lambert 1959, 1960), principal component analysis (P.C.A.) of the weighted similarity coefficient matrix (Orloci 1966), and optimal agglomeration using single cycle fusions (Orloci 1967; Austin & Greig-Smith 1968): for the soils data P.C.A. of the correlation matrix (R-analysis) and calculation of component scores (Austin 1968) were used. The vegetation data used were density per 1-acre (0·4-ha) plot for those species which occurred in at least three plots, considering the two areas separately, and whose identification and delimitation were accurate. This gave 136 out of a total of 420 species in the Belalong plots (the ten least frequent of these were eliminated from the analysis because of limitations of computer storage) and 121 out of 472 species in the Andulau plots.

RESULTS

Comparison with original analysis

The results of the association-analysis, using $\Sigma \frac{\chi^2}{N}$ as the association parameter and a $\chi^2_{max} = 6·64$ stopping rule, on the two separate areas are shown in Fig. 1. The two sets of data were also ordinated using density standardized in standard deviate units. The distribution of the association-analysis groups on the ordination is shown in Fig. 2; Table 1 summarizes the topography and composition of the association-analysis groups. The results confirm the general conclusions of Ashton (1964), though certain aspects require modification.

In the Belalong results (Fig. 2a) ridges above 500 m are clearly distinct from plots at lower altitudes, and are separated into shale lithosols and clay latosols on the third axis; the latter separated on the first axis of the original Bray and Curtis ordination of Ashton (1964). It appears that the original ordination emphasized the soil difference at the expense of another, unspecified, factor more directly related to altitude. There was no indication from the original ordination of a single distinct Temburong river bank stand; no environmental explanation for this isolation was found but reference to the vegetation data indicates that, by using standard deviates, the single high density value of the rare (in relation to these plots) rheophytic tree *Dipterocarpus oblongifolius* Bl. may have had a disproportionate influence (cf. Austin & Greig-Smith 1968). The Andulau results suggest that the distinction between hillside and ridge is even less definite than originally suggested, and other differences may be more important. Certain association-analysis groups have stands occurring on both hillsides and alluvium though the ordination makes a clear distinction between the stands. When a three-dimensional model of the ordination is constructed, two gradients are clearly seen, one in the alluvial stands and the other in the hill stands, and are recognizable in Fig. 2(b).

Austin & Greig-Smith (1968) have commented on these results with regard to their relative heterogeneity. Further analysis of the Belalong data has not been attempted as the number of stands in each group is limited and ordination within them unlikely to be

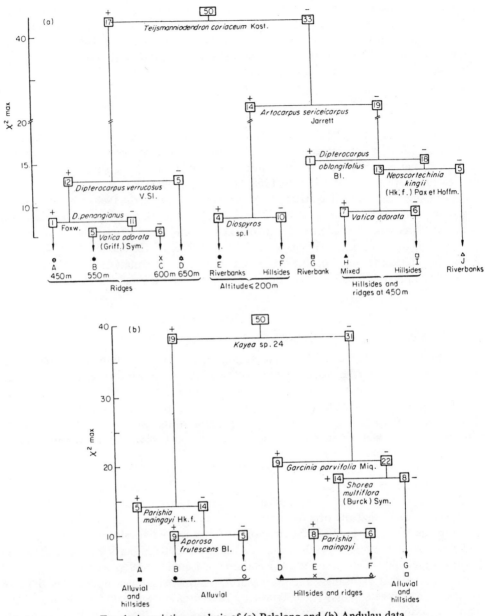

Fig. 1. Association-analysis of (a) Belalong and (b) Andulau data.

informative. The analysis has been continued with the Andulau data which are less heterogeneous topographically. Standard deviates may not be the most satisfactory standardization for tropical forest data (Austin & Greig-Smith 1968). The Andulau data were therefore re-analysed using presence data (Fig. 3). As expected for hetero-geneous data (Austin & Greig-Smith 1968) the results show little difference from those obtained using standardized quantitative data. Compared with Ashton's (1964) ordina-tion the positions of the main groups of the alluvial stands are transposed in relation to hill stands; stands 1–9 (B, Fig. 1b) appear less similar to the hill stands than stands 11–16

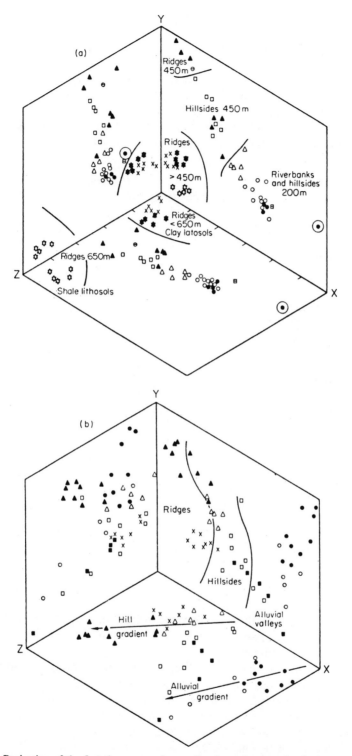

Fig. 2. Projection of the first three axes of an ordination using tree density data standardized in standard deviate units. The distribution of the association-analysis groups is shown by the symbols given in Fig. 1. (a) Belalong. Note the gradient in stands below 500 m and the district grouping of the ridge stands above this altitude on the first two axes. The Temburong stand with high density of *Dipterocarpus oblongifolius* Bl. is circled (see text). (b) Andulau. With the exception of one hillside stand associated with the ridges, the projection on the first and second axes positions the stands along a topographic gradient.

(A,C, Fig. 1b) in the present ordination. The x axis of the original ordination was related to soil sand content. The alluvial plots show this on the z axis of the present ordination. The y axis of the original ordination is loosely related to the x axis of the present one; as a result the x/z diagonal of the present ordination is closely comparable with the x/y diagonal of the original. The present x axis distinguishes stands on alluvium completely from those on residual soils; environmental data now available confirm a grouping by soil nutrient status (see below) along this axis but do not support the original

Table 1. *Relationship of vegetation groups and topography showing number of sites in each group*

	A	B	C	D	E	F	G			
ANDULAU										
Alluvial valley bottoms sites 1–20	3	9	5	–	–	–	3			
Hillsides sites 21–35	2	–	–	4	2	2	5			
Ridges sites 36–50	–	–	–	5	6	4	–			

Association-analysis groups

	A	B	C	D	E	F	G	H	I	J
BELALONG										
River banks (Temburong) sites 1–5	–	–	–	–	1	–	1	–	–	3
River banks (Belalong) sites 6–10	–	–	–	–	3	–	–	–		2
Hillsides at 200 m sites 11–20	–	–	–	–	–	10	–	–	–	–
Hillsides at 450 m sites 21–30	–	–	–	–	–	–	–	4	6	–
Ridges at 450 m sites 31–34	1	–	–	–	–	–	–	3	–	–
Ridges at 550 m sites 35–40	–	5	1	–	–	–	–	–	–	–
Ridges at 600 m sites 41–45	–	–	5	–	–	–	–	–	–	–
Ridges at 650 m sites 46–50	–	–	–	5	–	–	–	–	–	–

interpretation of an inter-relationship between this and drainage. The marked variation in the hill stands which is apparent on the y axis of the present ordination was not revealed by the Bray and Curtis ordination.

Association-analysis is unlikely to be effective when the numbers of stands are small; with $\chi_{max} = 6.64$ as the stopping rule, heterogeneity within groups of less than eleven stands will not be detected. As there are a number of such groups in these data, it was considered worthwhile to check the classification by means of a polythetic agglomerative technique. The presence data were therefore analysed using the optimal agglomeration technique of Orloci (1967) (Fig. 4). Optimal agglomeration produced substantially the same groupings, apart from the not unexpected redistribution of some stands of the negatively defined association-analysis group G amongst other groups. On the ordination (Fig. 3) stand 10 is placed some distance away from other members of the groups to which it is assigned in the two classifications; examination of field notes shows that this stand is heterogeneous, containing both alluvial and hill vegetation and soils.

The correlation between vegetation and soils can be examined in more detail by means of a P.C.A. of the soil properties (Austin 1968), though limitations in this procedure are discussed later in this paper. Stand 15 was omitted from this analysis as the soil samples showed a peat soil (see below). The analysis is complicated by the logical dependency of the mechanical fractions, percentage silt, clay and sand. Inclusion of all three as

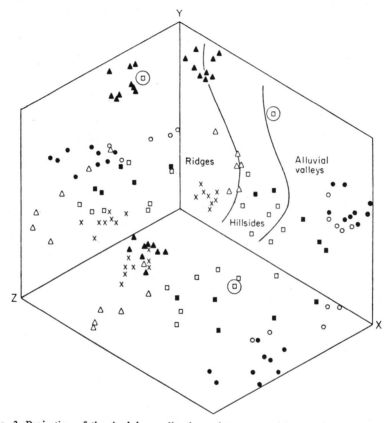

Fig. 3. Projection of the Andulau ordination using presence/absence data. Two hillside stands now occur in the ridge grouping. Symbols as in Fig. 2. Stand 10 is circled (see text).

variables will automatically weight the analysis in favour of variation in mechanical properties. This can be avoided by substituting two orthogonal variables for the three measures. A P.C.A. of the correlation matrix between these three measures for each depth was therefore carried out. The component scores for the two major components were calculated and used in the general soil P.C.A. The two components can conveniently be interpreted as (1) degree of sandiness, and (2), the relative proportions of silt and clay regardless of the percentage sand.

A P.C.A. of a correlation matrix of all soil variables, including the mechanical fraction components, was then carried out and the component scores calculated. The first component (Table 2) has high loadings on the sand component and mineral nutrients in general, and can be considered a measure of alluvial richness. When the scores are plotted

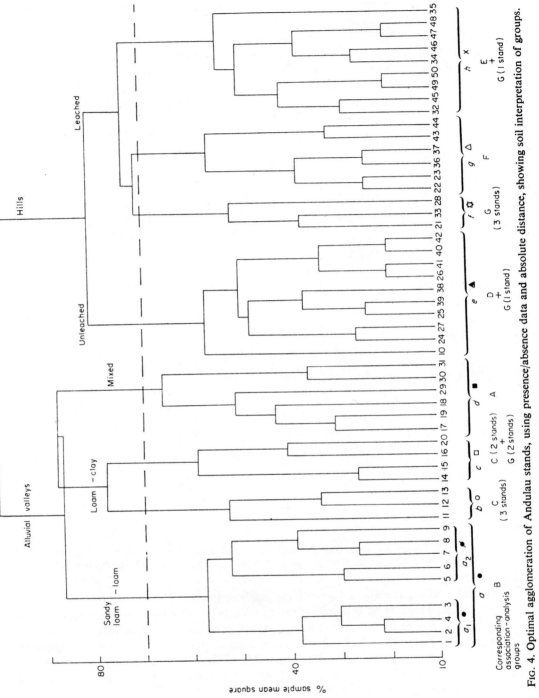

Fig. 4. Optimal agglomeration of Andulau stands, using presence/absence data and absolute distance, showing soil interpretation of groups.

Table 2. *Loadings of variables* (v) *on first component of soil properties (all except stand 15)*

	8–12 in. (20–30 cm)	20–24 in. (51–61 cm)
pH	−0·268	−0·678
Phosphorus (ppm)	0·891	0·673
Calcium (ppm)	0·499	–
Magnesium (ppm)	0·848	0·666
Potassium (ppm)	0·882	0·649
% nitrogen	0·870	0·376
% Group III elements	0·802	0·252
First mechanical fraction component	0·914	−0·739
Second mechanical fraction component	–	–
Munsell colour: 1st parameter (hue)	0·151	0·811
Munsell colour: 2nd parameter (value)	−0·182	0·531
Munsell colour: 3rd parameter (chroma)	−0·637	0·902
Log penetration	0·530	0·835
Humus depth	0·302	
Humus colour	−0·284	

$$-, < |0·1|. \ \Sigma v^2 = \lambda.$$

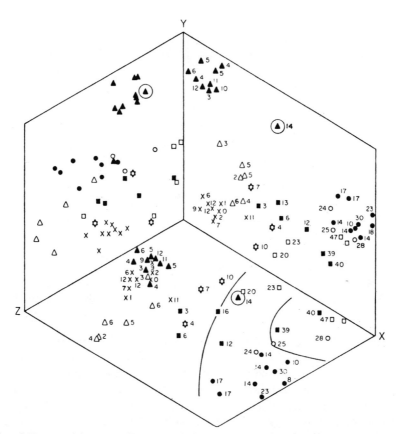

FIG. 5. Presence/absence ordination as in Fig. 3 showing optimal agglomeration groups (symbols as in Fig. 4). Scores for the first principal component (score +13) of the environmental (soil) ordination plotted on the vegetational ordination. Stand 10 is circled (see text).

on the vegetation ordination (Fig. 5), the distinction between the vegetation of alluvial and hill soils is seen to be correlated with the soil variation. The relationships of the other components are unlikely to be informative in view of the heterogeneity indicated by the first component; detailed consideration requires separate study of within-group variation.

Analysis of alluvial stands

Ordination is inefficient when applied to heterogeneous data (Greig-Smith *et al.* 1967). The data have therefore been split into two on the basis of the obvious environmental

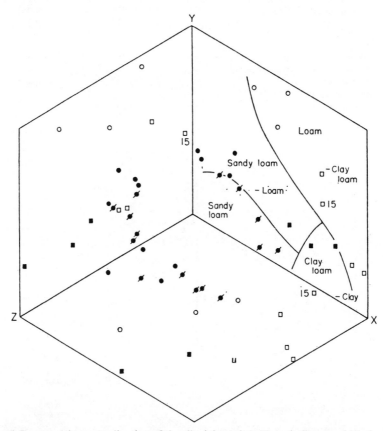

FIG. 6. Presence/absence ordination of the alluvial stands 1–20, excluding stand 10, showing the optimal agglomeration groups. Symbols as in Fig. 4. Note position of stand 15 (see text).

heterogeneity between alluvial and hill stands. Ordination of the twenty alluvial stands confirmed the distinctness of stand 10. The ordination was therefore repeated after exclusion of this stand and the results are shown in Fig. 6. The optimal agglomeration groups are distinct on the ordination. Stand 15, however, is known to have a distinct soil type, peat over clay (at least for the points sampled). This soil difference is not reflected in the vegetation. If it is assumed that stand 15 is heterogeneous with respect to soils and contains the only patch of the peat/clay soil in the stands sampled, then, on the frequency criterion (>3) for species to be included in the ordination, no characteristic species for such a soil would have been included. Stand 15 would then appear similar to stands having the other soils present in stand 15. There is no obvious solution to the recognition and inclusion of rare indicator species in numerical studies (cf. Austin & Greig-Smith 1968).

Examination of the results of applying numerical techniques to the soil properties of the alluvial stands (excluding stand 15) revealed that correlation with the vegetation types could be obtained using only the soil mechanical fractions at the 8–12 in. (20–30 cm) depth. Soil variation at the 20–24 in. (51–61 cm) depth showed very little relationship to the vegetational variation. This. is not surprising in view of the relatively impeded drainage and shallow rooting on most of these soils, and the heterogeneity with depths of the alluvial deposits, already discussed by Ashton (1964). The same procedure was adopted for the mechanical fractions as was used previously. Two orthogonal components were extracted and the values (scores) for these calculated for the eighteen stands. The

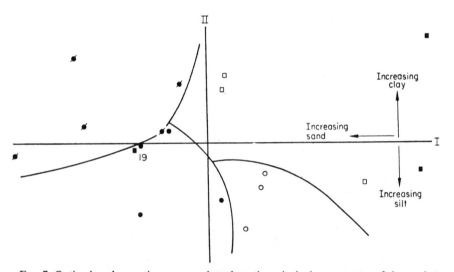

FIG. 7. Optimal agglomeration groups plotted on the principal components of the mechanical fractions (% silt, sand and clay) at 8–12 in. (20–30 cm) of the alluvial stands (stands 10 and 15 excluded). Note the anomalous position of stand 19 (see text).

same interpretation as previously can be applied to the components. The optimal agglomeration groups are plotted on these two axes of soil variation in Fig. 7. The vegetation groups, with one exception, show a very clear zonation with increasing sand and the varying proportions of silt and clay, though the correlation with position on the vegetation ordination is not exact. This ordination gives no support to the original interpretation (Ashton 1964) of variation in the alluvial stands as being due to drainage and water-table effects but this was based on inadequate environmental data. The variation found in this ordination can be detected on the y axis of the original ordination, which suggests that the difference may have arisen from an attempt to interpret minor variations on the inefficient original ordination of heterogeneous data (cf. Greig-Smith *et al.* 1967).

Ashton (1964) contended that his ordination had unequivocably shown that floristic variation within tropical forest was more related to soil gradients than to chance and opportunity of establishment. This conclusion has been criticized in that no comparison was made of stands on a uniform topography. The present ordination of the alluvial stands takes account of this criticism and substantiates the original claim, at least in relation to these data.

To examine this problem further, each 1-acre (0·4-ha) alluvial plot was subdivided into five 2 × 1-chain (0·08-ha) subplots. An association-analysis and an ordination of qualitative data, excluding those species occurring in less than 5% of the subplots, were carried out for the 100 subplots. Both indicate the distinctness of all five of the subplots of plot 10. These subplots were therefore excluded and the analyses repeated. The association-analysis gave six groups at a level of $\chi^2_{max} = 10·5$, and these are shown on the ordination

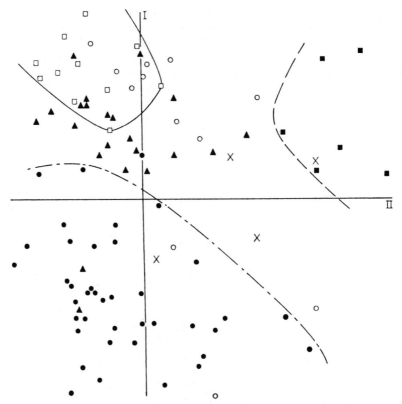

FIG. 8. Ordination of the alluvial subplots (two axes only) using presence/absence data and showing the groups obtained from an association-analysis of the same data. Groups defined as follows: (1) ×, + *Hydnocarpus kunstleri* (King)Warb. + *Baccaurea pyriformis* Gage; (2) ●, + *H. kunstleri* − *B. pyriformis*; (3) ■, − *H. kunstleri* + *B. pyriformis*; (4) ○, − *H. kunstleri* − *B. pyriformis* + *Aporosa frutescens* Br.; (5) ▲, − *H. kunstleri* − *B. pyriformis* − *A. frutescens* + *Kayea* sp. 24; (6) □, − all four species.

in Fig. 8. The first axis separates the sandy soil group (2) from the rest, while the second axis principally separates group 3. Group 3 is composed of subplots of plots 11 and 19. The third axis, not shown here, separates groups 4 and 5 from the rest, with 4 above, and 5 below, the plane defined by axes one and two. Group 4 contains subplots from plots 5–6 and 17,18; the interpretation of this group is not certain. With this exception, the results reflect those shown in Fig. 6. They emphasize the anomalous behaviour of plot 19, which appears from Fig. 8 to be at the opposite extreme of a gradient from the clay plots 14–16 and 20, with plots 17 and 18 intermediate. Plot 19 was sited near the head of a small valley, where the soil surface was liable to frequent flooding by silty fast-flowing water, leading to rapid deposition of mineral sediments and removal of humus. Plots 14–16 were among those with a more or less permanently high water-table and this axis, and the third axis in Fig. 6, may possibly be related to frequency and duration of flooding though there is no close correlation between them and the water-table data available; a predictable relationship seems to exist, however, with depth of humus. The arrangement of these particular plots (plot 20 excepted) closely repeats their position on the *x* axis of the original ordination (1964). It therefore seems that the present ordinations distinguish on separate axes two environmental gradients which were identifiable but confused and their importance reversed on the first two axes of the original ordination, presumably as a result of including both hill and alluvial stands.

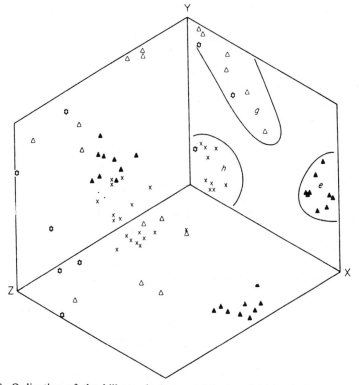

Fig. 9. Ordination of the hill stands (presence/absence data) excluding stands 29–31 showing the optimal agglomeration groups. Symbols as in Fig. 4.

Analysis of hill stands

Ordination of the vegetation data of the hill stands revealed heterogeneity; stands 29–31 were extremely distinct on the third axis. They occur in the optimal agglomeration group *d* with stands 17–19, which were anomalous in the alluvial ordination; there is, however, no evidence to suggest that these stands are heterogeneous in themselves in the same way as stand 10. It has not proved possible to interpret this group, though they are possibly associated with a different lithology (see below). When re-ordinated without these stands, four major groups of stands equivalent to the major groups of hill stands seen in the optimal agglomeration classification (Fig. 4) are apparent (Fig. 9). Each group comprises both hillsides and ridges with no obvious correlations with the soil data available.

The application of an environmental ordination to the soil variables poses problems. The soil properties may be heterogeneous but as they are continuous variables, as opposed to the qualitative vegetation variables, it is unlikely that the results will correspond closely to the vegetation even apart from the problem of nonlinear correlation. An attempt to examine the possible use of environmental ordination in a potentially heterogeneous situation was made. The problems of assessment are considerable, resulting from the subjective choice of variables for inclusion, the lack of any necessary correlation between the size of environmental component and its importance for determining vegetation variation, and the probability that any relationship between the vegetation and soils will be a complex nonlinear joint function of the soil components (Austin 1968). These qualifications mean that without other independent evidence the interpretation of results must remain as tentative hypotheses.

Three P.C.A.'s of the soil variables were carried out.

The first analysis included all variables at both depths with the exceptions of litter and humus depth and colour, which were possibly vegetation-dependent. Examination of the results suggested that components I, II and VI provided the projection showing the greatest visual correlation with the stand arrangement of the vegetation ordination. The communalities of the variables on these components were determined, the five variables with the lowest communalities (8–12 in. (20–30 cm) depth: pH, second Munsell parameter; 20–24 in. (51–61 cm) depth: calcium, second mechanical fraction component and log penetration) were eliminated from the analysis and the P.C.A. repeated. When a later component is selected for plotting, to the exclusion of some earlier ones, in this way, and some variables are then eliminated, two difficulties arise: (1) the loadings and component scores of the later component will have been influenced by the eliminated components; and (2) soil properties apparently related to the vegetation may be eliminated because they are less important in describing the soil variation.

Component scores are often expressed such that the scores are proportional to the amount of the total variation in the correlation matrix accounted for by the component. The size of the component may not, however, reflect the relationship it may have with external variables as indicated above; calculation of scores of an environmental ordination on this basis may then distort a relationship. It would seem more appropriate to standardize all environmental components such that the sum of squares of the scores equals one to ensure that all are judged on the same basis, as in Fig. 10.

Components I–III (Fig. 10a, b) of the reduced matrix distinguish the optimal agglomeration groups very clearly with the exception of certain stands in group *h* (see later). The first component can again be interpreted as a general soil-fertility component and separates the vegetation groups to some extent (Fig. 10b). A correlation between the first axis of the vegetation ordination and the axis at 45° to the axes two and three of the soil ordination is apparent (Fig. 10a). If the loadings are rotated through 45°, it is possible to indicate which soil variables are related to this axis of variation, i.e. component II′ (Table 3). The rotated component has high loadings on potassium content, first and third Munsell colour parameters and log penetration (8–12 in. (20–30 cm)) and pH (20–24 in. (51–61 cm)). Negative scores indicate stands with high potassium and low rate of penetration, and positive scores stands with low potassium, relatively intense yellow-red colour and high calcium, magnesium, iron and aluminium (Table 3).

These variables suggest a leaching gradient with negative scores indicating a loss of nutrients in the upper levels of the soil independent of the general fertility level, though the negative loadings for potassium would not appear to support this interpretation.

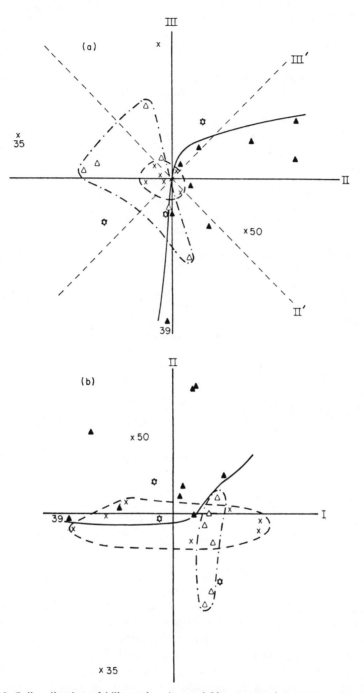

FIG. 10. Soil ordination of hill stands using variables measured at 8–12 in. (20–30 cm) and 20–24 in. (51–61 cm). (a) Stand scores on the second and third components of the soil P.C.A. See text for discussion of rotated axes II′ and III′. (b) Stand scores on the first and second components. Optimal agglomeration groups: *h*, ×; *e*, ▲; *g*, △; *b*, ✿.

This can be explained if the potassium occurs in the form of a stable mineral, while calcium and magnesium occur as less stable minerals. There is some evidence from similar Bornean soils to suggest that mica may be the dominant clay mineral; however, the clay content of the soils studied is low and the mineralogy of the silt and sand fraction

Table 3. *Loadings of variables* (v) *on rotated axis II' of soil properties of hill stands*

	8–12 in. (20–30 cm)	20–24 in. (51–61 cm)
pH	*	0·366
Phosphorus (ppm)	—	0·113
Calcium (ppm)	0·224	*
Magnesium (ppm)	0·164	—
Potassium (ppm)	−0·272	−0·265
% nitrogen	0·130	—
% Group III elements	0·137	—
First mechanical fraction component	−0·137	0·119
Second mechanical fraction component	−0·135	*
Munsell colour: 1st parameter (hue)	−0·275	−0·224
Munsell colour: 2nd parameter (value)	*	−0·275
Munsell colour: 3rd parameter (chroma)	0·268	0·439
Log penetration	−0·251	*

—, < |0·1| ; *, not included. $\Sigma v^2 = 1·0$.

Table 4. *Mean values* (±95% *confidence limits*) *of measures of leaching in hill plot groups* e, g *and* h

	Group		
	e	g	h
Colour intensity (third Munsell parameter) at 8–12 in. (20–30 cm)	6·14 (±1·00)	5·38 (±0·88)	5·67 (±0·48)
Ratio 8–12 in./20–24 in. (51–61 cm)	0·77 (±0·03)	0·68 (+0·05, −0·04)	0·76 (±0·04)
% Group III elements at 8–12 in.	3·72 (±0·48)	2·69 (±0·77)	3·46 (±0·60)
Ratio 8–12 in./20–24 in.	0·76 (+0·08, −0·07)	0·54 (+0·10, −0·08)	0·65 (±0·02)
% clay fraction at 8–12 in.	16·75 (±2·66)	13·57 (±2·12)	14·68 (±3·13)
Ratio 8–12 in./20–24 in.	0·80 (±0·04)	0·69 (+0·06, −0·05)	0·67 (±0·05)

in unknown. A particularly stable type of mica is muscovite (Brewer 1964), which contains relatively high potassium and very little calcium or magnesium (Deer, Howie & Zussman 1962). Leaching would then result in a loss of calcium and magnesium and a relative increase in potassium. If it could be established that muscovite was a common mineral in these soils then the leaching gradient could be accepted as a reasonable hypothesis. Table 4 shows the differences in some of the variables which are associated with the 'leaching gradient'. The degree of differentiation between colour intensity, Group III elements (Al, Fe) and the clay fraction at the upper and lower levels in the profiles provides some supporting evidence of a difference in leaching intensity, the ratios being highest in the 'unleached' group e plots (Table 4).

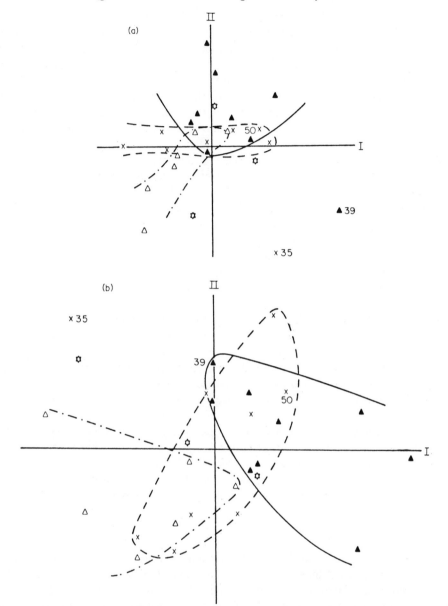

FIG. 11. Soil ordinations using data from the 8–12 in. (20–30 cm) level only. Symbols as in Fig. 10. (a) All soil variables. (b) Five variables having the highest communalities.

On the basis of this tentative hypothesis, the vegetational variation can be interpreted as follows: group *e* (positive scores on component II′) is characteristic of relatively unleached sites on richer soils (negative scores on component I, Fig. 10b); group *h* occurs over the whole range of fertility levels but only on intermediately leached soils (Fig. 10a), while group *g* occurs on poor, leached soils. A correlation between vegetation and soils has been detected and a tentative interpretation made but it is apparent that the inter-relationship is complex and not a simple monotonic function.

The second analysis examined whether using only the soil variables at the 8–12 in. (20–30 cm) level gave better correlation with the vegetation, on the assumption that the deeper soil was relatively unimportant in affecting vegetation variation.

The results of using all the 8–12 in. (20–30 cm) level variables only are given in Fig. 11(a). The anomalous position of stand 50 in the previous analysis is no longer apparent, suggesting that the distinctness was due to an anomalous value for a variable at the 20–24 in. (51–61 cm) level ; this is not identifiable with certainty, though values for the first Munsell parameter were lower than in other plots in the group, and those for the second Munsell parameter exceptionally low. Examination of the values for the variables for stands 35 and 39 indicates that the soil mechanical fraction values and total nitrogen are unusual in stand 35, while stand 39 is similarly anomalous in its soil nitrogen content, accounting for their extreme positions.

The third ordination (Fig. 11b) assumed that only the most important (high communalities) variables at the 8–12 in. (20–30 cm) level were related to the vegetational variation, other variables contributing only 'noise'. It shows a greater difference than the previous one from the ordination using both depths. The five variables having the highest communalities (with the exception of magnesium) on the 8–12 in. (20–30 cm) level P.C.A. were taken as the variables for this P.C.A.; these were first mechanical fraction component, percentage group III elements, first and third Munsell parameters and log rate of penetration. The results indicate a leaching gradient very clearly, but no longer reflect the position of the stands in the vegetational ordination although the vegetation groups are distinct.

The problems posed by these three ordinations are complex. Each, with exceptions, distinguishes between the vegetation groups, though none shows a close correlation with the pattern of similarities between groups which is seen in the vegetational ordination. The soil ordinations are not independent of each other and are therefore difficult to compare. If the measurements used in the third ordination only had been available, the results shown in Fig. 11(b) (III and subsequent axes show no relationship to the vegetational variation) might well have been accepted without further study. The results obtained from environmental ordinations depend on the number and choice of variables included and it is unlikely that definitive hypotheses are possible from this type of approach. In this case, potassium, soil colour and penetration rate are related to vegetational variation, but are insufficient to explain the differences, while general soil fertility plays 'some role'. Degree of leaching provides a possible hypothesis but further study of the environment is needed.

DISCUSSION

The results indicate that, contrary to some claims, variation in the species composition of trees of girth greater than 12 in. (30 cm) in tropical lowland forest on sites of similar topography is associated with variation in soils. By the use of these numerical techniques, definite hypotheses regarding six of the eight groups derived from the optimal agglomeration classification (Fig. 4) can be put forward. The vegetation on alluvial sites (groups a–c) may be determined by soil texture, which in this case is related to soil 'richness' (nutrient status) and possibly to periodicity of flooding. The vegetation on the hills (groups e, g, h) is associated with a leaching gradient within relatively uniform textured soils. Group d appears unrelated to the alluvial gradient, while f appears unrelated to the hill gradient; group d was not included in the hill environmental ordinations though it contained as many hill as alluvial plots. With the exception of 29 and 31, all plots within these two groups were sited on a separate northern boundary ridge or alluvium adjacent to it; mechanical analysis supported field observations that the substratum there was different and the soil richer in clay than in the other hill plots, though there was a high degree of heterogeneity. The conclusions from the more recently developed techniques do not directly contradict those of Ashton (1964) but provide much more effective hypothesis-generating descriptions.

These results and those of the previous paper in this series (Austin & Greig-Smith 1968) are different from those of Poore (1968), who found no perceptible variation of vegetation in relation to hill topography in Malaya. It is important to try and resolve these differences, which appear to arise from the interaction of two problems, the ecological assumptions regarding possible environmental factors and the limitations of numerical methods.

Topography is often assumed to be a relatively minor environmental factor in tropical rain forest. Poore comments on the occurrence of swampy areas and describes the topography as '. . . or gently sloping hills intersected by a network of steep-sided, clearly defined drainage channels . . . height from valley bottom to hill top is usually about 20 m and never more than 30 m . . . slopes rarely reach a gradient of 30° . . .'.

However, Austin & Greig-Smith (1968) demonstrated that topographic variation of similar magnitude of relief (115 ft (35 m)), which Nicholson (1965) suggests is associated with a soil difference, was related to vegetational variation. The results obtained on variation within the alluvial and hill stands of the Andulau data provide considerable evidence of a soil catena with associated changes in vegetation. Poore assumes that the topography and soils were sufficiently uniform although he mentions a catenary sequence.*

The effect of including varying topographic sites in one plot on the numerical analyses may well have obscured environmental correlations nonetheless. The recognition of several environmental groupings of species in Poore's inverse analysis, bearing in mind the limitations of the methods (see below), suggests that his conclusions may need to be reconsidered.

Austin & Greig-Smith (1968) showed that a minimum number of species was required to provide sufficient information for an effective ordination. They also demonstrated, from an ordination based on randomly chosen species, that an over-defined data matrix with a random element may give an uninterpretable ordination dominated by the random element. There is no reason to suppose that similar effects do not occur with classification techniques. Poore's analysis of nineteen species in 651 plots (area B) is likely to have contained insufficient information per plot for meaningful results to be obtained (though the mean density of individuals is not given, the mean density of individuals per plot for *all* species in area A was only 4·6). χ^2 is known to have a skewed distribution when low expectancies occur in the 2×2 table and this will influence the results of association-analysis when there are large numbers of rare species (Goldsmith 1967; W. T. Williams, personal communication) and give rise to 'chaining'. Inverse analysis is particularly sensitive to abundance effects (Williams & Lambert 1961). Both these phenomena can be seen in Poore's analysis of the plots of area A; all nineteen commonest species and thirty-three of the forty-four constant species occur in the first six groups. Austin & Greig-Smith (unpublished) found similar effects in a normal association-analysis of the Sepilok data (Austin & Greig-Smith 1968) using all species. An analysis using only the most abundant species gave meaningful results correlating with those obtained by ordination. However, the environmental variation between Sepilok plots may well be greater than

---

* Soils are generally very deep and uniform on the mature physiography of Malaya in comparison with those of northern Borneo. Species with distinctly different edaphic ranges in Borneo can occur together in the same forest type in Malaya, where the number of species per unit area is consistently higher. It seems possible therefore that the generally more fertile mature Malayan soils do indeed exert less rigorous differentiating pressures on the vegetation.

that between Poore's plots, as it is in Brunei. Austin (personal observation) found differences in soil texture which correlated with the first axis of the ordination, and were unrelated to the topographic variation which was associated with positions along the second axis. Clearly the limitations of the techniques must be borne in mind when reaching ecological conclusions, particularly when negative results are being interpreted.

The major distinction between these studies, however, appears to lie in the lower girth class limits of Ashton (1964) and Austin & Greig-Smith (1968) (12 in. (30 cm)), and Poore (1968) (3 ft (91 cm)). As a result, for any given plot size and considering only the most abundant species, the data of Ashton and Austin & Greig-Smith will have contained more vegetational information. Increasing the plot size to provide adequate information for larger girth sizes increases the environmental heterogeneity and thus leads to no gain in information. It may be that the large 'minimal area' commonly assumed for tropical rain forest is the area which provides an adequate sample of the environmental heterogeneity to be found in any given habitat. The results presented in this paper indicate that, given sufficient information, plots 2×1 chain (40×20 m) can be used for ecological studies.

Examination of the soil results suggests many interesting possibilities but more knowledge of the variation in tropical rain forest soils is needed before any extensive study of their influence on vegetation can be considered. The indication of soil catenas in the three areas in Brunei, Sabah (Nicholson 1965) and Malaya (Poore 1968) which have been intensively studied should be followed up. In summary it appears that further careful work is required before it is unequivocably demonstrated that the *majority* of species are influenced by soil factors in tropical rain forest.

ACKNOWLEDGMENTS

We are indebted to John Bailey, formerly Soils Chemist, Agriculture Department, Kuching, Sarawak, and his staff for analysis of soils and to the staff of the Computing Laboratory, University College of North Wales, especially B. Sanderson, for assistance with the analysis of the data.

The analysis of the data was part of a study of the applicability of quantitative methods to tropical forest, supported by a grant from the Science Research Council.

SUMMARY

Vegetation and soil data from two rain forest sites in Brunei, Borneo, which have been previously subjected to ordination (Ashton 1964), have been re-examined by more recently developed methods.

The vegetation data from each site have been separately subjected to association-analysis, principal components analysis of the weighted similarity coefficient matrix using density and, for one site, presence, and, for one site, optimal agglomeration analysis using single cycle fusions. The major resultant plot groupings from the more homogeneous of the two sites were then individually subjected to principal components analysis; one group was subdivided into units of 0·2 ac(0·08 ha) and re-analysed using the same method.

The soil data from another of the more homogeneous plot groups, which had proved more difficult to interpret directly by the vegetation ordination, was subjected to three principal component analyses of the correlation matrix (R-analysis), using successively reduced numbers of soil variables, and compared by superimposing the optimal agglomeration groupings.

The results overall confirmed Ashton's earlier conclusion that variation within this rain forest at the scales analysed was more related to soil gradients than to chance and opportunity of establishment, but they provided interpretations that differed in detail from those of the earlier study, which was based on inadequate environmental information. The more efficient techniques used here have allowed meaningful interpretations of vegetation/soil inter-relationships in plots as small as 0·08 ha and have provided more effective hypothesis-generating descriptions. The results of the present analyses have differed more or less from one another, emphasizing the complexity of the underlying relationships, which need not be linear, and emphasizing in particular the difficulties of ordinating environmental parameters. Further careful work is required before it can be unequivocally demonstrated that the majority of species are influenced by soil factors in tropical forest; particular emphasis should now be given to collections of detailed soil data.

REFERENCES

Ashton, P. S. (1964). Ecological studies in the mixed Dipterocarp forests of Brunei State. *Oxf. For. Mem.* 25.

Austin, M. P. (1968). An ordination study of a chalk grassland community. *J. Ecol.* 56, 739–57.

Austin, M. P. & Greig-Smith, P. (1968). The application of quantitative methods to vegetation survey. II. Some methodological problems of data from rain forest. *J. Ecol.* 56, 827–44.

Austin, M. P. & Orloci, L. (1966). Geometric models in ecology. II. An evaluation of some ordination techniques. *J. Ecol.* 54, 217–27.

Bailey, J. M. (1967). Chemical changes in a Sarawak soil after fertilization and crop growth. *Pl. Soil*, 27, 33–52.

Brewer, R. (1964). *Fabric and Mineral Analysis of Soils.* Wylie, New York.

Deer, W. A., Howie, R. A. & Zussman, J. (1962). *Rock-forming Minerals. III. Sheet Silicates.* Longman, London.

Fogg, D. N. & Wilkinson, N. T. (1958). The colorimetric determination of phosphorus. *Analyst, Lond.* 83, 406–14.

Goldsmith, F. B. (1967). *Some aspects of the vegetation of sea cliffs.* Ph.D. thesis, University of Wales.

Greig-Smith, P., Austin, M. P. & Whitmore, T. C. (1967). Quantitative methods in vegetation survey. I. Association-analysis and principal component ordination of rain forest. *J. Ecol.* 55, 483–503.

Lambert, J. M. & Dale, M. B. (1964). The use of statistics in phytosociology. *Adv. ecol. Res.* 2, 59–99.

Lambert, J. M. & Williams, W. T. (1966). Multivariate methods in plant ecology. VI. Comparison of information-analysis and association-analysis. *J. Ecol.* 54, 635–64.

Nicholson, D. I. (1965). A study of virgin forest near Sandakan, North Borneo. *Symposium on Ecological Research in Humid Tropics Vegetation*, pp. 67–87. Government of Sarawak and UNESCO Science Cooperation Office for Southeast Asia.

Orloci, L. (1966). Geometric models in ecology. I. The theory and application of some ordination methods. *J. Ecol.* 54, 193–215.

Orloci, L. (1967). An agglomerative method for classification of plant communities. *J. Ecol.* 55, 193–206.

Poore, M. E. D. (1968). Studies in Malaysian rain forest. I. The forest on Triassic sediments in Jengka Forest Reserve. *J. Ecol.* 56, 143–96.

Webb, L. J., Tracey, J. G., Williams, W. T. & Lance, G. N. (1967a). Studies in the numerical analysis of complex rain forest communities. I. A comparison of methods applicable to site/species data. *J. Ecol.* 55, 171–91.

Webb, L. J., Tracey, J. G., Williams, W. T. & Lance, G. N. (1967b). Studies in the numerical analysis of complex rain forest communities. II. The problem of species sampling. *J. Ecol.* 55, 525–38.

Williams, W. T. & Lambert, J. M. (1959). Multivariate methods in plant ecology. I. Association-analysis in plant communities. *J. Ecol.* 47, 83–101.

Williams, W. T. & Lambert, J. M. (1960). Multivariate methods in plant ecology. II. The use of an electronic digital computer for association-analysis. *J. Ecol.* 48, 689–710.

Williams, W. T. & Lambert, J. M. (1961). Multivariate methods in plant ecology. III. Inverse association-analysis. *J. Ecol.* 49, 717–29.

Williams, W. T., Lambert, J. M. & Lance, G. N. (1966). Multivariate methods in plant ecology. V. Similarity analyses and information-analysis. *J. Ecol.* 54, 427–45.

Part IV

RECENT PERSPECTIVES

Editor's Comments
on Papers 18 and 19

18   VAN DER MAAREL
The Braun-Blanquet Approach in Perspective

19   WHITTAKER
The Population Structure of Vegetation

Phytosociology has multiple functions in modern ecology. It continues its early tradition of describing vegetation in diverse parts of the world. In many areas, particularly the tropics, description has an urgency due to the fact that much of the area may be destroyed before adequate phytosociological analysis is possible. It is somewhat ironical that the wellsprings of phytosociology in the work of Humboldt, Warming (see Goodland 1975), and Schimper were heavily influenced by tropical vegetation but that the major traditions of phytosociology developed in temperate areas much modified by human activity and simpler than the richest of tropical vegetation. Even the phytosociology of North America was constrained by extensive human influence on its vegetation prior to the development of modern phytosociology. H. A. Gleason lamented, in 1910, that the vegetation of Illinois was largely replaced by cornfields, and F. E. Clements recognized much of the western grasslands of the United States as a *"disclimax"* caused by centuries of cattle grazing.

Phytosociologists have always found it necessary to distinguish between "natural" vegetation, which is the vegetation cover in the absence of human influence, and other vegetation that is under human influence. Some assumed that the vegetation would return to the climax or stable natural communities if human influence was eliminated. Daubenmire, for example, envisioned the climax as the potential vegetation on a site, in the absence of disturbance, which he reconstructed from scattered remnants of primeval vegetation. European phytosociologists recognized that

most of the areas they have studied have been disturbed by human actions for millenia and that there is little likelihood of re-
constructing vegetation in its prehuman conditions. Tüxen (1956)
proposed the concept of "potential natural vegetation" as the vegetation that would become established in an area in the absence
of man under present climatic and site conditions but without an
assumption that it was a return to climax conditions existing prior
to human disturbance.

Description, classification, and mapping of vegetation is a
continuing and important function of phytosociology. Adequate
understanding of vegetation or the more encompassing ecosystem of which it is a part, it is hoped, will serve as a basis for management and planning of the earth's landscapes, for effective estimates of the productive capacity of particular regions and sites,
and for prediction of the consequences of given human impacts.
Many planning and legal aspects of particular types of land use (e.g.,
wetlands) require more effective methods of delimiting, classifying, and mapping vegetation and relating it in specific ways to the
physical environment and to animal populations.

Basic to the wish and need to use phytosociological insights
in management and planning is the necessity to study the intrinsic
relationships of vegetational communities and their distribution
over the landscape. This requires fundamental research on properties of vegetation such as diversity and stability of communities,
and study of the life history and population phenomena that are
the bases of successional and spatial changes in the distributions
of plant species and the aggregations which they form.

Van der Maarel (Paper 18) reviewed the development and
current state of the Braun-Blanquet school. There is little doubt
that the concepts and methods developed by Braun-Blanquet,
and generations of his associates and students, have been and are
a major influence on phytosociology; it seems that some of the
other traditions and schools of Continental Europe, outside of Russia, have converged on the Braun-Blanquet school (Trass and
Malmer 1973).

One of the major continuing offshoots of the Braun-Blanquet
tradition is the effort to establish a code of phytosociological nomenclature to establish a standard system of names predicated on
the Braun-Blanquet association as the basic rank and following
a rule of priority (Barkman et al. 1976). Similar codes have been
proposed before but have not been generally accepted and applied. A part of the current effort anticipates the ultimate step in
syntaxonomy of plant communities, a Prodromus or collected

treatment of European plant communities according to the Braun-Blanquet system.

Perhaps more significant is the widespread use of numerical methods of ordination and classification (Goodall 1973) and, as noted in Part III, their use in sorting the tables of relevés that are the traditional basis of the Braun-Blanquet classification. Van der Maarel noted that early pioneer stages of vegetation are most effectively classified while later, more stable stages (climax) are most effectively ordinated. This suggests that the pioneer stages are more discrete while later stages tend to continuous variation. This relationship has been the burden of much discussion in recent years.

Whittaker (Paper 19) discussed some of the developments in American plant ecology, notably gradient analysis and dominance-diversity studies of communities. He reviewed developments in quantitative analysis of plant communities seen in Papers 12–18 and related these to the concept of niche recently popularized among animal ecologists, who are currently much involved in problems of community analysis. Whittaker suggested that the earlier dichotomy seen between methods of classification and ordination or gradient analysis should be replaced by a view of these as complementary and potentially partners that may often be used together. There is certainly much indication of convergence in method and, it may be hoped, in concept between traditions of diverse origin. Although Whittaker stressed complementarity and convergence, the discussion following his paper suggests that there are still some barriers to complete understanding. R. Tüxen commented that the audience understands Whittaker's clear English, but added, "it is very, very difficult for us, fieldmen, to follow you in your theoretical ideas." Other questions and comments suggested that some of the perennial problems of phytosociology are still with us, for example, the question of sample size and homogeneity of the vegetation and that the continuity of vegetation commonly reported is an artifact of these. This seems unlikely, as many different sampling and ordination approaches have produced similar results. Another concern addressed by Whittaker is the common observation of obvious discontinuity of vegetation in the field. This is, of course, familiar and expected where environments change or where the vegetation is markedly modified by humans. Vegetation in the field may demonstrate varying degrees of continuity or discontinuity.

Phytosociologists may agree that vegetation is substantially continuous but prefer to classify the most distinctive and recognizable aggregations of species and habitats as associations, as for example Tüxen (Paper 11). Other positions are seen in Poore (1955), who recognized repeating groups in a less formalized way as *"noda,"* in Williams and Lambert (1961), who used numerical clustering techniques in *"nodal analysis"* to classify groups of samples based on association of species populations, and in Whittaker (Paper 19), who suggested that groups of species whose modes fall close together on a gradient might be recognized informally as *"commodia."* Divergences of method and treatment of vegetational data still persist, although it may be hoped that the polarity of earlier eras has been reduced (Dansereau 1968).

The traditional concerns and methods of phytosociology converge with some of the recent interests of animal ecologists in community and the species niche which are the subject of *Niche: Theory and Application,* Benchmark Papers in Ecology, Volume 3, edited by Whittaker and Levin. Some of the best examples of this convergence are the papers by Terborgh (Paper 24) and James (Paper 25) in that volume. Terborgh analyzed the distribution of bird species along an altitudinal gradient as Whittaker had done many years before in his study of insect and tree distribution in the Great Smoky Mountains. James used a variety of methods, including principal components, which produced a three-dimensional map of bird species according to habitat much as Austin et al. (Paper 17 in this volume) show tree distributions in three dimensions.

Much has been written about the animal niche in multidimensional space. Most of the work developing from this fruitful concept is a form of direct gradient analysis showing the response curves of animal populations along selected gradients of food size or environmental factors such as vegetation or temperature. Niche studies, like much phytosociological work, show that species respond individualistically on gradients and that they usually form a series of curves having their population centers, or optimal responses, at different positions on the gradient. Species are interpreted as having evolved toward differential distribution along gradients and, as a consequence, form a complex continuum of overlapping species response curves scattered along the gradient (Cody 1975). One of the major factors involved in the evolution of patterns of niche dispersion is believed to be competition.

The niche, by definition, is the position the species occupies within the community which, like the phytosociologists community, is homogeneous and occupies a homogeneous habitat *(biotope).* A commonly unrecognized problem which the student of the animal niche shares with the tradition phytosociologist is, therefore, "What is the community?" ("Was ist eine Gesellschaft.")

Modern phytosociology now has an embarrassment of techniques for data collection and analysis. Not the least of its problems is to assess the relative merits of these and to clarify the ecological meaning of the results, be they clusters or orders, and to relate them to the environment in such ways that the causal relations between vegetation and the physiological and population responses of its component species may be elucidated.

REFERENCES

Barkman, J., J. Moravec, and S. Rauschert. 1976. Code of Phytosociological Nomenclature. *Vegetatio* **32**:131–185.

Cody, M. L. 1975. Towards a Theory of Continental Species Diversities. In *Ecology and Evolution of Communities,* eds. M. L. Cody and J. M. Diamond, pp. 214–257. Cambridge, Mass: Belknap Press of Harvard University Press.

Dansereau, P. (ed.). 1968. The Continuum Concept of Vegetation: Responses. *Bot. Rev.* **34**:315–332.

Goodall, D. W. 1973. Numerical Classification. In *Ordination and Classification of Vegetation,* ed. R. H. Whittaker, pp. 577–615. The Hague, The Netherlands: W. Junk Publ.

Goodland, R. J. 1975. The Tropical Origin of Ecology: Eugen Warming's Jubilee. *Oikos* **25**:240–245.

Poore, M. E. D. 1955. The Use of Phytosociological Methods in Ecological Investigations. IV. General Discussion of Phytosociological Problems. *J. Ecol.* **44**:28–50.

Trass, H., and N. Malmer. 1973. Northern European Approaches to Classification. In *Ordination and Classification of Vegetation,* ed. R. H. Whittaker, pp. 531–574. The Hague, The Netherlands: W. Junk Publ.

Tüxen, R. 1956. Die heutige potentielle naturliche Grundlage der Landwirtschaft. *Angew. Pflanzensoziol. (Stolzenau, Weser)* **13**:5–42.

Williams, W. T., and J. M. Lambert. 1961. Nodal Analysis of Associated Populations. *Nature* **191**(4784):202.

18

Copyright © 1975 by Dr. W. Junk N. V.-Publishers

Reprinted from *Vegetatio* **30**(3):213–219 (1975)

THE BRAUN-BLANQUET APPROACH IN PERSPECTIVE

E. VAN DER MAAREL

Division of Geobotany, University of Nijmegen, Toernooiveld, Nijmegen, The Netherlands.

Keywords:

Braun-Blanquet, Classification, History of phytosociology, Numerical phytosociology, Ordination, Phytosociology

Introduction

The 30th volume of Vegetatio has to some extent been devoted to J. Braun-Blanquet and this contribution will conclude our scientific celebration of Braun-Blanquet's 90th birthday. In addition to Lebrun's French and Pignatti's German hommage I would like to convey my tribute in the English language.

Although I became familiar with the Braun-Blanquet approach over 20 years ago – mainly through V. Westhoff, whose continuing stimulus I greatly acknowledge – it was only during the preparation of our contribution for the Handbook of Vegetation Science (Westhoff & van der Maarel 1973) that I became fully aware of the historical significance of Braun-Blanquet's life and work.

From this historical study it became all the more clear that the oeuvre of Braun-Blanquet has a crucial position in the history of phytosociology. This position might even be compared with that of J. S. Bach's oeuvre in the history of music. Bach's music is generally recognised as a completion of the Baroque period in its synthesis of forms and styles of predecessors like Palestrina, J. P. Sweelinck, H. Schütz, J. Pachelbel and A. Scarlatti and contemporaries like F. Couperin and A. Vivaldi. At the same time Bach considerably influenced various great romanticists including Mozart and Beethoven[*] and still later composers.

Braun-Blanquet's phytosociology is a synthesis, a completion of ideas and methods from the beginning of the 19th century including those of A. von Humboldt, O. Heer,

H. Lecoq, A. Kerner von Marilaun and R. Hult, of the 'fathers of the Zürich-Montpellier school', C. Schröter, H. Brockmann-Jerosch, E. Rübel, Ch. Flahault and J. Pavillard, and of many contemporaries including W. Lüdi, E. Furrer, H. Jenny, R. Gradmann, A. Tansley, C. Raunkiaer, R. Nordhagen, A. K. Cajander and P. Jaccard.

As such Braun's work is an end-point which was already fully achieved around 1920. – It seems rather unrealistic to qualify this work in terms of epigonism as has been done from time to time and recently by Braun's old rival H. Gams (1972).

At the same time Braun's phytosociology has inspired lots of younger scientists. This inspiration has been provided particularly in Braun's 'personal' institute, the S.I.G.M.A. the International Station for Mediterranean and Alpine Geobotany at Montpellier. Many thousands of vegetation studies, to which R. Tüxen's series Excerpta Botanica Sectio Sociologica and Bibliographica Syntaxonomica are excellent guides, many hundreds of essays and textbooks, most of which are mentioned in van der Maarel, Tüxen & Westhoff (1970) and Westhoff & van der Maarel (1973), and last but not least thirty volumes of Vegetatio provide ample evidence of the far-reaching impact of Braun-Blanquet's intellectual system and also of its flexibility and its aptness for further development. Of all pupils and collaborators one stands out emphatically: R. Tüxen.

Braun has also influenced many vegetation scientists from other Schools and approaches. Particularly the amalgation with numerical plant ecology and gradient analysis seems to be fruitful. With this development we have to connect the name of R. H. Whittaker above others. As we pointed out in the Handbook paper the floristic-sociological approach of phytosociology has been so largely shaped by Braun that neither of the current names for the approach,

---

[*] For this information I relied on various sources, including a Dutch History of Music by Höweler (1951). To avoid any misunderstanding: Bach may be considered the outstanding composer of this time, while at the same time the genius of Händel, D. Scarlatti and other contemporaries is recognised!

including French-Swiss and Zürich-Montpellier school, is appropriate. – We also consider the term school as inadequate because of the evident openness of the approach. – I may repeat here our suggestion to speak of the Braun-Blanquet approach.

Essence of the Braun-Blanquet approach

From Braun's early papers, especially the 1921 one on 'principles of a systematics of plant communities on a floristic basis', we may trace three main ideas in the approach (Westhoff & van der Maarel 1973):
– Plant communities are conceived as types of vegetation, recognised by their floristic composition. The full species compositions of communities better express their relationships to one another and to their environments than any other characteristic.
– Amongst the species that make up the floristic composition of a community, some are more sensitive expressions of a given relationship than others. For practical classification (and indication of environment) the approach seeks to use those species whose ecological relationships make them most effective indicators. These diagnostic species comprise character-species, differential-species, and constant companions and together they form the characteristic species combination.
– Diagnostic species are used to organise communities into a hierarchical classification of which the association is the basic unit. The hierarchy is not merely necessary but invaluable for the understanding and communication of community relationships that it makes possible.

This typological approach implies the clear distinction between abstract units, particularly associations, and concrete representative stands in the field. Although the name association individual has been used for the latter, Braun-Blanquet has never adopted an organismal concept of the plant community, as was sometimes believed. Braun (e.g. 1951, 1964) even rejected this view because in the plant community 'the organismal character, the centralised organisation and the specialisation of functions are failing'. His views come very close to what may be called the integrational concept (cf Rowe 1961, Goodall 1963, Langford & Buell 1969). As we concluded in our Handbook paper we may well escape confusion when we speak of a concrete phytocoenose and an abstrat phytocoenon (cf Westhoff 1951, 1970).

Another implication of the typological approach is the recognition of phytocoenoses in the field which need a description (relevé) before being assigned to a phytocoenon. Again there is a misunderstanding, which I explained at some length in an earlier paper (van der Maarel 1966). It is sometimes thought that the relevés from the Braun-Blanquet approach are merely single plot analyses, where the single plot is chosen more or less haphazard within a stand. Such a procedure is then rejected in favour of a multiple plot analysis. In fact, however the selection of the stand takes place so as to make it opportune to analyse the entire stand, i.e. only a limited area is considered sufficiently uniform as to be considered a phytocoenose. Thus a phytocoenose description is neither a single plot nor a multiple plot, but rather an 'overall plot' analysis. Hence the critical step in the procedure is the selection of stands.

A third misunderstanding concerns the supposed believe of the Braun-Blanquet disciples in the general discontinuity in the field, i.e. the occurrence of well defined stands, easy to recognise as distinct from the surrounding stands and moreover easy to assign to one particular community type. This believe is further supposed to lead to the idea of the Braun-Blanquet system as a natural classification. This believe is by far not universal. Most Braun-Blanquet followers are aware of the existence of continuities between stands as well as between types. (cf. a comment by Braun-Blanquet and Tüxen on the Vegetatio paper by Goodall 1963, which reads in translation: 'We are convinced that the plant cover of the earth in all of its dimensions can be divided into phytosociological groupings of higher and lower rank; their delineation may be either sharp or less sharp and gliding'. In fact there is much agreement on the approach advocated by Whittaker (1970) which stresses the population structure of vegetation and defines the phytocoenose as a 'system of interacting, niche-differentiated and partially competative species'. It is generally recognised in the Braun-Blanquet approach that with the words of Whittaker (1956) 'Because of environmental interruptions and some relative discontinuities inherent in vegetation itself, the pattern may also be considered a complex mixture of continuity and relative discontinuity'.

A typological concept does not imply a general recognition of discontinuity. This was clearly expressed by Tüxen (1955) who interpreted phytosociological classification through the concept of types as ideal concepts, recognised in an empirical way from 'correlation concentrates', i.e. groups of correlated characters. That which is evident and characteristic of a type is always its nucleus, not its periphery; types are not pigeonholes but foci in a field of variation. The expression 'ideal concepts' reminds us to the indication 'idealistic view of the plant community' given by Poore (1964) to characterise the typological Braun-Blanquet approach.

This idea was elaborated in various later papers, notably by Von Glahn (1968), who distinguished three aspects in the type-concept: the vegetation type as identity, as maximal correlative concentration and as systematic category. His reasoning comes rather close to the theory of classification developed by Whittaker (1962).

As to the naturalness of the Braun-Blanquet classification system, one might consider the system natural in the sense of dealing with a large number of generalisation purposes (cf. Poore 1962). However, Braun himself has always kept distance, (e.g. 1951 and 1959, p. 147: 'the question of the naturalness of such a division (i.e. the floristic-sociological system) seems superfluous. In the sense of Kant a system is a whole, ordered according to certain principles. Whether nature as such forms a system we cannot decide. This decision is of no significance for science'.

An essential element in the practical approach is the relevé procedure itself. Apart from possible considerations on the selection of stands (cf. Westhoff & van der Maarel 1973) there has been comparatively little discussion on the analysis.

The relatively careful description of structure, the relatively rough but very effective combined estimation of cover-abundance and the estimation of the sociability of all participating species, the systematical description of the superficial features of the site, are the main elements of the analysis, which have hardly changed since they were proposed (Braun-Blanquet 1921) and which still form a very powerful tool in phytosociology. It is still my opinion (e.g. van der Maarel 1966) that refinements of the crucial cover-abundance scale as have been frequently suggested are rather useless, mainly because the relative inaccuracy is rapidly increasing with further refinements of the scale whilst naturally occurring temporal changes in the performance of species must warn us for an overemphasis of quantitative differences. In fact as long as we know very little about the relationship between species-performance and environmental variation and fluctuation we should rather over-emphasise presence against absence of species.

A final remark on the essence of the approach is concerned with the profound awareness of the temporal changes in community structure and composition, which can be found already in the earliest papers by Braun. It has appeared from many studies on succession that the Braun-Blanquet approach allows a careful observation and interpretation of successional developments. One major idea resulting from the dynamical view of the plant community is that of the so called sociological progression, suggested by Braun-Blanquet in his crucial 1921 paper. As we pointed out in the Handbook paper this idea of an arrangement of vegetation types according to the level of structural and organisational development as reflected in stratification, growth form complexity, etc. is very similar to the recent ideas about the strategy of ecosystem development (cf. Odum 1969), Despite the criticisms of this idea, sociological progression may still be considered as a major idea in the interpretation of community structure and a welcome aid to the classification hierarchy.

Main periods in the Braun-Blanquet approach

In order to appreciate the historical significance of Braun's work as well as possible future lines of development in the approach we may specify periods in the development of the Braun-Blanquet approach. These periods will now be discussed briefly.

1. *Precursors of phytosociology* (ca. 1830–1890).
Many ideas of Braun-Blanquet's phytosociology go back to O. Heer's (1835) vegetation monograph of the Sernftal in Switzerland (1845). Heer had a clear notion of the plant community as a type, its interrelations with the environment and the diagnostic value of plant species with a fidelity to one particular type of environment. Heer distinguished 30 'localities' (vegetation-site complexes that resemble our present-day ecosystems!). Each locality type is characterised by certain environmental characteristics, which may be found back in the type's name and characteristic species. It often can be recognised easily as a modern association. E.g. Heer's locality of the 'Schneetälchen' (small depressions prolongedly filled with snow) corresponds with the *Salicetum herbaceae*. Heer also paid attention to the performance of species, particularly their sociability, for which he used a ten-point scale.

A following important study was by Lecoq (1844) who developed a quantitative measure which was essentially a combined cover-abundance estimation. The same author (1854) defined a plant association, 'association végétale', very much along the lines Braun elaborated later, including the use of faithful species. Thus with Lecoq and Heer phytosociology clearly has a French-Swiss origin! Of the many authors who published on particular vegetations Kerner von Marilaun (1863) deserves special mention. Contributors from N. Europe to be mentioned here are Von Post (1862) and Hult (1881). Von Post was one of the first to stress that vegetation itself is the object of study and hence the basis of a classification. Most earlier authors relied partly or wholly on the environment for their classification. Von Post also used a hierarchy of vegetation types. Hult considerably refined the analysis through careful distinction of vegetation layers and a five-point scale for cover. (See Trass & Malmer 1973 for further information).

2. Early phytosociology (ca. 1890–1910)

This is the period of the founders of the Zürich-Montpellier school, already mentioned above. Important contributions from this period for the development of the Braun-Blanquet approach included – the use of constant species derived from such tables, esp. by Brockmann-Jerosch (1907) and – the consequent use of tables in which similar stand descriptions were collected (e.g. Stebler & Schröter 1893); the use of an extensive classification hierarchy as well as a clear notion of synecology as a branch of vegetation science. (e.g. Schröter & Kirchner 1902), Pavillard (cf. 1935). From N. Europe the contribution of Cajander (e.g. 1903) concerning the different degrees of stability in communities was of some relevance. Also Raunkiaer's (cf. 1934) work on life forms and the law of frequency was of influence. Developments Braun did not take over from this period were the emphasis on physiognomy and structure in classification and the specification of the fidelity concept with constancy as a prerequisite.

This period is logically concluded by the adaption during the 1910 International Botanical Congress at Brussels of the definition by Flahault & Schröter (1910) of the association as 'a plant community of definite floristic composition, presenting a uniform physiognomy and growing in uniform habitat conditions. The association is the fundamental unit of synecology.'

It is remarkable that Braun-Blanquet, though a pupil from this Zürich-Montpellier school did not adopt as much ideas from this period as he did from the earlier one! It seems that the most important stimulus for the development of the approach in this period came from the German Gradmann (1909) and this concerned the use of character-species.

3. Establishment of the Braun-Blanquet approach (ca 1910–1930)

Between 1913 and 1921 Braun introduced his approach in a number of papers of which only the concluding and most important 1921 one, is referred to. In this essay we find a plea for a floristic instead of an ecological classification system, the introduction of fidelity and fidelity degrees, the analytical scales for cover-abundance and sociability, considerations on the dynamogenitical importance of species, on periodicity and vitality, on outline of an association diagnosis including a biological (life form) spectrum the outline of a hierarchy and finally the sociological progression.

In a subsequent paper (1925) the hierarchy was further developed, and the differential species introduced (on the suggestion by Koch 1925).

This period was concluded by the first edition in 1928 of Braun's magnum opus Pflanzensoziologie, which appeared in an English edition Plant Sociology four years later. The often quoted – but also often neglected – surprise of this book is the thorough treatment of synecology. Besides being a mile-stone in the development of the approach this book is with his later editions included, therefore a continuous demonstration of the ecological potention of the approach – as well as its founder! –

4. Consolidation of the approach (ca 1930–1950)

This period is characterised by the personal instruction of many scholars in the S.I.G.M.A. and the spread of the approach over most of Europe. The atmosphere of this period is very vividly described by Braun himself (1968). It was also a period of scientific activity with found it major expression in the series S.I.G.M.A. communications, by Braun and many of his pupils. By 1950 the number of communications had passed the 100. This period may finally be characterised by the gradual fusion of the Uppsala approach under its conciliatory leader G. E. Du Rietz (e.g. 1921, 1936) and the junction of its fundamental unit, the sociation, with the Braun-Blanquet classification hierarchy. (cf. Westhoff & van der Maarel 1973, Trass & Malmer 1973).

5. Full organisation of phytosociology (ca 1950–1970)

Although the S.I.G.M.A. continued to act as a centre of research and tuition a second centre developed around R. Tüxen, which has gradually become the headquarters of the organisation of phytosociology. From the early thirties until 1963 Tüxen was director of the Centre (later State Institute) for Vegetation Mapping at Stolzenau, and ever since leader of his personal 'Workshop for theoretical and applied phytosociology'. Since the early fifties he also acts as secretary-general of the International

Society for Vegetation Science (with the non-committal English name International Society for Plant Geography and Ecology, which, however will soon be adapted). Tüxen's major achievements (vividly described by Braun-Blanquet 1969) include:
– The performance of a systematic description of all plant communities in a large region, namely his (1937) Plant Communities of NW Germany. This monograph was followed by a lot of similar works, of which Oberdorfer's (1957) is an outstanding and fully documented example. By means of such monographs it became possible to achieve a synecological and synchorological synthesis of very large areas, as is beautifully exemplified by Ellenberg's (1963) Vegetation of Central Europe and the Alps.

The consistent application of phytosociology to agriculture, forestry, hydrology and physical planning, particularly through large-scale vegetation maps. This branch of phytosociology has no doubt largely contributed to the acceptance of the Braun-Blanquet approach as a scientific enterprise. The thorough mapping of vegetation involved a refinement of the scale of observation, as a result of which many new community types were distinguished in a much more detailed hierarchy and with more emphasis on the structural uniformity of types. A further deepening of the approach was provided by the development of the concept of potential natural vegetation in relation to the patterns of replacement communities existing in cultural landscapes (Tüxen 1956). The Handbook paper treats these aspects in more detail.
– The promotion of the approach in other continents through tuition of scientists from all over the world and particularly through the series of Symposia of the International Society, held at Stolzenau and, from 1964 onwards, in Rinteln. Of these Symposia especially the 1964 one on phytosociological systematics, the 1966 one on community morphology and the 1970 one on basic problems and methods in phytosociology are of relevance for the universalisation of the Braun-Blanquet approach (Tüxen 1968, 1970, van der Maarel & Tüxen 1972).

Besides the Braun-Blanquet approach appeared to be more and more applicable to other than vascular plant communities. Barkman's (1958) monograph on epiphytic cryptogam communities may be mentioned as an outstanding example. Last but not least we mention Vegetatio as a characteristic of this period. Founded 1948 and led by Braun-Blanquet and Tüxen for many years it served as an official organ of the International Society and after some years of uncertainty and change, it continues to do so.

Present lines of development

Since about 1970 some new tendencies can be recognised. First of all there is the ultimate step in the organisation of syntaxonomy initiated by Tüxen. Syntaxonomical literature is collected in the series Bibliographica Syntaxonomica. The results of a collective treatment of European plant communities is realised in the project Prodromus of the European Plant Communities. (A first outcome is the syntaxonomy and synecology of the Spartinetea communities (Beeftink & Géhu 1973). The syntaxonomical achievements are accompanied by an active Committee on Nomenclature with J. Moravec as secretary which is to produce a nomenclatural code.

A second development is the use of multivariate methods. Both numerical classification and ordination have proved to be succesful tools in phytosociological synthesis (cf. van Groenewoud 1965, Ivimey-Cook 1966, van der Maarel 1969, 1974, Moore et al. 1970, Whittaker 1972, Pignatti 1975). A special Working Group for Data-Processing is elaborating these techniques, particularly for the treatment of large data-sets. With reference to our paper and various others in Part V of the Handbook of Vegetation Science, edited by R. H. Whittaker as well as the recent book by Orlóci (1975) we may conclude that the basis elements in the description and interpretation of plant communities according to the Braun-Blanquet approach are very well supported by numerical techniques. According to an idea expressed earlier (van der Maarel 1966) there may be a relation between the optimum effectiveness of classification and ordination and the general maturity of the community type. Along the line of sociological progression, i.e. from pioneer to mature stages there seems to be a decrease in classification effectiveness and an increase in ordination effectiveness.

Some perspectives

It is not easy to predict further developments of phytosociology. Pignatti (this issue) sketches our present position as being on the junction of separating roads, along which the above mentioned two present developments run. He foresees a danger in either development: the prodromus activities may lead to an undesired formalism and a deviation from the ecological basis of phytosociology; the numerical activities may lead to an undesired level of abstraction.

In my opinion the danger in these developments is real, but could be diminished when we would be able to let these roads run parallel by promoting close connections between the two activities.

Even with a close connection both developments may suffer from a certain sterility, unless a synthesis could be achieved with current ecological theory. In this respect I would like to repeat the name of Whittaker who in various recent contributions (1970, 1972) has indicated the perspective of integrating the recent theory of the population structure of vegetation with the essentially typological Braun-Blanquet approach.

I would like to add a joined perspective of integrating ecological theory on the space-time relationships between vegetation and environment as were discussed during the First International Congress of Ecology under the common heading of diversity and stability. With van Leeuwen (1966) I would specify the need for further research along two parallel lines: accurate description of environmental variation as well as of environmental dynamics and, in connection with both lines, the study of the response of plant populations and through them of the vegetation, our ultimate object of study.

Only with a sound ecological theory and with a close connection between general ecology and phytosociology, our science will really proceed on the basis of its achievements indicated above, which we so largely owe to the founder of modern phytosociology J. Braun-Blanquet.

References

Barkman, J. J. 1958. Phytosociology and Ecology of Cryptogamic Epiphytes. Van Gorcum, Assen. 628 pp.

Beeftink, W. G. & J.-M. Géhu. 1973. Spartinetea maritimae. Prodrome des Groupements Végétaux de Europe. Fasc. 1. Cramer, Lehre. 43 pp.

Braun-Blanquet, J. 1921. Prinzipien einer Systematik der Pflanzengesellschaften auf floristischer Grundlage. Jahrb. St. Gallen Naturw. Ges. 57: 305–351.

Braun-Blanquet, J. 1925. Zur Wertung der Gesellschaftstreue in der Pflanzensoziologie. Vierteljahrsschr. Naturf. Ges. Zürich 70: 122–149.

Braun-Blanquet, J. 1928. Pflanzensoziologie. Grundzüge der Vegetationskunde. Biologische Studienbücher 7. 1. Ed. Berlin. x + 330 pp.

Braun-Blanquet, J. 1932. Plant Sociology. (Transl. by G. D. Fuller and H. S. Conard). New York, xviii + 439 pp. Reprint 1966.

Braun-Blanquet, J. 1951. Pflanzensoziologie. Grundzüge der Vegetationskunde, 2nd. ed. Springer, Wien. 631 pp.

Braun-Blanquet, J. 1959. Grundfragen und Aufgaben der Pflanzensoziologie. Vistas in Botany 1: 145–171.

Braun-Blanquet, J. 1964. Pflanzensoziologie, Grundzüge der Vegetationskunde, 3rd ed. Springer, Wien-New York. 865 pp.

Braun-Blanquet, J. 1968. L'école phytosociologique Zuricho-Montpelliéraine et la S.I.G.M.A. Vegetatio 16: 1–78.

Braun-Blanquet, J. 1969. Reinhold Tüxen, Meister-Pflanzensoziologe. Vegetatio 17: 1–25.

Brockmann-Jerosch, H. 1907. Die Pflanzengesellschaften der Schweizeralpen. I. Die Flora des Puschlav und ihre Pflanzengesellschaften. Engelmann, Leipzig. 438 pp.

Cajander, A. K. 1903. Beiträge zur Kenntniss der Vegetation der Alluvionen des nördlichen Eurasiens. I. Die Alluvionen des unteren Lena-Thales. Acta Soc. Sci. Fenn. 32 (1): 1–182.

Du Rietz, G. E. 1921. Zur methodologischen Grundlage der modernen Pflanzensoziologie. Akad. Abhandl. Uppsala. 272 pp.

Du Rietz, G. E. 1936. Classification and nomenclature of vegetation units 1930–1935. Svensk. Bot. Tidskr. 30: 580–589.

Duvigneaud, P. 1946. La variabilité des associations végétales. Bull. Soc. Roy. Bot. Belg. 78: 107–134.

Ellenberg, H. 1963. Vegetation Mitteleuropas mit den Alpen. In, 'Einführung in die Phytologie' IV-2, ed, H. Walter, Stuttgart, 943 pp.

Flahault, Ch. & C. Schröter. 1910. Rapport sur la nomenclature phytogéographique. (Phytogeographische Nomenklatur). Actes III Congr. int. bot. Bruxelles 1: 131–164.

Gams, H. 1972. Die floren- and vegetationsgeschichtliche Erforschung der Alpen. Ber. Dt. Bot. Ges. 85: 7–10.

Glahn, H. von. 1968. Der Begriff des Vegetationstyps im Rahmen eines allgemeinen naturwissenschaftlichen Typenbegriffes. In, 'Pflanzensoziologische Systematik', ed. R. Tüxen, Ber. Symp. Int. Ver. Vegetationskunde, Stolzenau 1964: 1–13. Junk, The Hague.

Goodall, D. W. 1963. The continuum and the individualistic association. Vegetatio 11: 297–316.

Gradmann, R. 1909. Über Begriffsbildung in der Lehre von der Pflanzenformationen. Bot. Jb. 43, Beibl. 99: 91–103. Reprinted 1940.

Groenewoud, H. van. 1965. Ordination and classification of Swiss and Canandian forests by various biometric and other methods. Ber. Geobot. Inst. Rübel Zürich 35: 28–102.

Heer, O. 1835. Die Vegetationsverhältnisse des südöstlichen Teiles des Kantons Glarus; ein Versuch die pflanzengeographischen Erscheinungen der Alpen aus klimatischen und Bodenverhältnissen abzuleiten. Mitt. Theor. Erdkunde Zürich.

Höweler, C. 1951. Inleiding tot de Muziekgeschiedenis. 5e druk. H. J. Paris, Amsterdam. xii + 476 pp.

Hult, R. 1981. Försök till analytik behandling af växtformationern. Meddn. Soc. Fauh. Flor. Fenn. 8. Helsinki: 155 pp.

Ivimey-Cook, R. B. & M. C. F. Proctor. 1966. The application of association-analysis to phytosociology. J. Ecol. 54: 179–192.

Kerner von Marilaun, A. 1863. Das Pflanzenleben der Donauländer. Innsbruck, 348 pp.

Koch, W. 1925. Die vegetationseinheiten der Linthebene unter Berücksichtigung der Verhältnisse in der N.O. Schweitz. Jb. St. Gall. Naturw. Ges. 61 (2): 1–146.

Langford, A. N. & M. F. Buell. 1969. Integration, identity and stability in the plant association. Adv. Ecol. Res. 6: 83–135.

Lecoq, H. 1844. Traité des plantes fourragères, ou flore des prairies naturelles et artificielles de la France. Paris, 620 pp.

Lecoq, H. 1954. Etude sur la géographie botanique de l'Europe. Vol. 1, Paris.

Leeuwen, C. G. van. 1966. A relation theoretical approach to pattern and process in vegetation. Wentia 15: 25–46.

Maarel, E. van der. 1966. Dutch studies on coastal sand and dune vegetation, especially in the Delta region. Wentia 15: 47–82.

Maarel, E. van der. 1969. On the use of ordination models in phytosociology. Vegetatio 19: 21–46.

Maarel, E. van der. 1974. The Working Group for Data-Processing of the International Society for Plant Geography and Ecology in 1972–1973. Vegetatio 29: 63–67.

Maarel, E. van der, R. Tüxen & V. Westhoff. 1950. Bibliographie pflanzensoziologischer Lehrbücher. Excerpta Bot. Sect. B. 11: 86–160.

Maarel, E. van der & R. Tüxen (eds). 1972. Grundfragen und Methoden in der Pflanzensoziologie (Basic problems and methods in phytosociology). Ber. Symp. int. Ver. Vegetationskunde, Rinteln 1970: xix + 523 pp. Junk, Den Haag.

Moore, J. J., O. Fitzsimons, E. Lambe & J. White. 1970. A comparison and evaluation of some phytosociological techniques. Vegetatio 20: 1–20.

Oberdorfer, E. 1957. Süddeutsche Pflanzengesellschaften. Jena, XXXVIII + 564 pp.

Odum, E. P. 1969. The strategy of ecosystem development. Science 164: 262–270.

Orlóci, L. 1975. Multivariate analysis in vegetation research. Junk, The Hague. VIII + 276 pp.

Pavillard, J. 1935. The present status of the plant-association. Bot. Rev. 1: 210–232.

Pignatti, E. & S. Pignatti. 1975. Syntaxonomy of the Sesleria varia-grasslands of the calcareous Alps. Vegetatio 30: 5–14.

Poore, M. E. D. 1962. The method of successive approximation in descriptive ecology. Adv. Ecol. Res. 1: 35–68.

Poore, M. E. D. 1964. Integration in the plant community. J. Ecol. 52 (supp.) 213–226.

Post, H. von. 1862. Försök till en systematisk uppställning af vextställena i mellersta Sverige. Bonnier, Stockholm. 42 pp.

Raunkiaer, C. 1934. The life forms of plants and statistical plant geography. Oxford. xvi + 632 pp.

Schröter, C. & O. Kirchner. 1902. Die Vegetation des Bodensees. Schr. Ver. Gesch. Bodensees Lindau 9 (2): 1–86.

Stebler, F. G. & C. Schröter. 1893. Beiträge zur Kenntnis der Matten und Weiden der Schweiz. 10. Versuch einer Übersicht über die Wiesentypen der Schweiz. Landw. Jahrb. Schweiz 6: 95–212, Bern.

Trass, H. & N. Malmer. 1973. North European approaches to classification. In Handbook of Vegetation Science (ed. R. Tüxen. Part V Ordination and classification of Communities (Ed. R. H. Whittaker), p. 529–574. Junk, The Hague.

Tüxen, R. 1937. Die Pflanzengesellschaften Nordwestdeutschlands. Mitt. Flor.-Soz. Arbeitsgem. Niedersachsen 3: 1–170.

Tüxen, R. 1955. Das System der nordwestdeutschen Pflanzengesellschaften. Mitt. Flor.-Soz. Arbeitsgem., Stolzenau, N.F. 5: 155–176.

Tüxen, R. 1956. Die heutige potentielle natürliche Vegetation als Gegenstand der Vegetationskartierung. Angew. PflSoziol., Stolzenau 13: 5–42.

Tüxen, R. 1968. Pflanzensoziologische Systematik. Ber. Symp. Int. Ver. Vegetationskunde, Stolzenau 1964. xii + 348 pp. Junk, The Hague.

Tüxen, R. (ed). 1970. Gesellschaftsmorphologie. Ber. Symp. Int. Ver. Vegetationskunde, Rinteln 1966. Junk, The Hague. xvi + 360 pp.

Westhoff, V. 1951. An analysis of some concepts and terms in vegetation study or phytocoenology. Synthese 8: 104–206.

Westhoff, V. 1970. Vegetation study as a branch of biological science. Misc. Papers Landbouwhogeschool Wageningen 5: 11–30.

Westhoff, V. & E. van der Maarel. 1973. The Braun-Blanquet approach. In Handbook of Vegetation Science (Ed. R. Tüxen) Part V. Ordination and Classification of Communities (ed. R. H. Whittaker), p. 617–726. Junk, The Hague.

Whittaker, R. H. 1956. Vegetation of the Great Smoky Mountains. Ecol. Monogr. 26: 1–80.

Whittaker, R. H. 1962. Classification of natural communities. Bot. Rev. 28: 1–239.

Whittaker, R. H. 1967. Gradient analysis of vegetation. Biol. Rev. London 42: 207–264.

Whittaker, R. H. 1970. The population structure of vegetation. In R. Tüxen (ed.). Gesellschaftsmorphologie. Ber. Symp. Int. Ver. Vegetationskunde, Rinteln 1966: 39–59, Junk, The Hague.

Whittaker, R. H. 1972. Convergens of ordination and classification. In 'Basic problems and methods in phytosociology', ed. E. van der Maarel & R. Tüxen, Ber. Symp. int. Ver. Vegetationskunde Rinteln 1970: 39–57. Junk, The Hague.

19

**Reprinted from *Gesellschaftsmorphologie (Strukturforschung)*, Ber. Symp. Int. Ver.
Vegetationskunde, Rinteln 1966, R. Tüxen, ed., Dr. W. Junk N. V.-Publishers, 1970,
pp. 39–59**

THE POPULATION STRUCTURE OF VEGETATION*

by

R. H. WHITTAKER

Cornell University Ithaca, New York U.S.A.

INTRODUCTION

It is a pleasure to be able to discuss some recent developments in American plant ecology before European phytosociologists. I have only one regret – that the long, partially separate developments of English-language ecology and Continental phytosociology have led to the depth of difference that we observe, with scientists on the two continents often approaching the same phenomena through different techniques, concepts, and perspectives. I think, though, that the development of ecology as a science may have been enriched by these differences (WHITTAKER 1962), and that mutual understanding is now more important than agreement.

Toward this end I should like to discuss current work in two areas of plant ecology – gradient analysis and dominance-diversity studies – as they relate to one another, to problems of classifying plant communities, and to some concepts of phytosociology. My concern is with the population structure of plant communities – the manner in which plant populations are organized or related to one another within particular communities and along environmental gradients – and what this population structure implies for the interpretation and classification of communities.

DIRECT GRADIENT ANALYSIS

Gradient analysis seeks to understand vegetation by studying relationships among gradients or variables on three levels – environmental factors, species populations, and community characteristics (WHITTAKER 1951, 1956, 1967). When these variables are studied along a major environmental gradient which is accepted as given, as a basis for arranging and interpreting the data, the approach is direct gradient analysis. For example, elevation in mountains may be used as a basis for arranging samples from plant communities in sequence in transect tables that represent the way plant populations and community charac-

---

* Research carried out at Brookhaven National Laboratory under the auspices of the U.S. Atomic Energy Commission.

teristics change along the elevation gradient. Table I is such a transect; Fig. I shows curves of species populations and community characteristics along another elevation gradient.

Table I.

Mean tree stratum coverage per cents for elevation belts, Santa Catalina Mountains, Arizona (WHITTAKER and NIERING 1965)

Transect steps Elevation in meters	Elev. Weights	I over 3000	II 2700- 3000	III 2400- 2700	IV 2100- 2400	V 1800- 2100	VI 1500- 1800	VII 1200- 1500	VIII 900- 1200	IX 750- 900	X below 750
Number of samples		10	22	50	50	50	50	50	50	50	15
Picea engelmanni	1	56	13								
Abies lasiocarpa	1	26	7	6							
Populus tremuloides	2	5	1	2	0.1						
Salix scouleriana	2	0.1	0.2	0.1	0.1						
Robinia neomexicana	3	0.1	0.3	2	0.7	0.1					
Pseudotsuga menziesii	3	3	35	28	20	7	0.1				
Pinus strobiformis	3	9	22	12	6	4	0.1				
Abies concolor	3		15	13	7	0.8					
Acer glabrum	3		1	1	0.1	0.1					
Quercus gambelli	4		0.1	0.1	0.5	0.1					
Pinus ponderosa	4		.7	21	38	10	2				
Acer grandidentatum	4			1	4						
Arbutus arizonica	4			0.2	0.4	3					
Quercus rugosa	5			2	2	3	1				
Quercus hypoleucoides	5			5	13	19	1				
Quercus arizonica	5			0.1	1	3	4	2			
Alnus oblongifolia	5				2	3	6				
Juniperus deppeana	6				0.01	2	3	1	0.01		
Pinus chihuahuana	5					1	0.1				
Pinus cembroides	6					6	6	0.1			
Juglans major	6					0.1	0.4	0.2	0.2		
Quercus emoryi	6					0.8	1	1	0.2		
Cupressus arizonica	6						7	3			
Fraxinus velutina	7						0.5	0.5	2		
Platanus wrightii	7						1	1	3		
Quercus oblongifolia	7						0.2	2	0.2		
Vauquelinia californica	7						0.04	0.3	0.2		
Fouquieria splendens	8						0.1	0.6	1.2	0.2	0.6
Prosopis juliflora	8						.04	0.3	3	1.5	0.1
Cercidium floridum	8							0.1	0.3	.01	
Acacia greggii	8							.04	2	0.4	0.1
Carnegiea gigantea	8							0.1	1	0.2	0.4
Cercidium microphyllum	9							0.1	3	4	5
Acacia constricta	9								0.1	0.6	0.3
Olneya tesota	10										0.4
Total tree coverage		99.2	101.6	93.5	94.9	63.0	33.6	12.3	16.4	6.9	6.9
Weighted averages		1.30	2.64	3.24	3.83	4.55	5.56	6.35	7.69	8.67	8.88
Percentage similarity											
to sample 1		100	33.1	21.0	9.6	8.8	0.3	0	0	0	0
to sample 6		0.3	3.3	6.8	11.2	39.8	100	40.0	9.2	0.7	0.7
to sample 10		0	^	0	0	0	0.7	9.8	36.9	71.0	100

Results of interest from such work: (1) The curves formed when densities (or some other measurement of importance) of a species population are followed along an undisturbed and uninterrupted environmental gradient are generally bell-shaped, apparently binomial, in form. (2) The centers or modes and the limits of these curves along the gradient differ. The species do not form groups of associates with closely similar distributions; in general as would be predicted by the „principle of species individuality" of RAMENSKY (1925) and GLEASON (1926) no two species have the same distribution. (3) Because of this, and because species populations overlap broadly and taper gradually from maximum density to rarity and absence, plant communities in general intergrade continuously along uninterrupted environmental gradients (as also stated as the principle of community continuity by RAMENSKY and GLEASON).

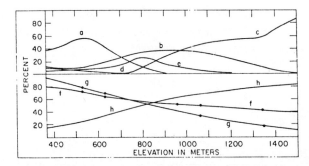

Fig. 1. Curves for species populations and community characteristics along the elevation gradient on south-facing slopes bearing pine forests in the Great Smoky Mountains, Tennessee. Above, major tree species populations in percentages of stems over 1 cm dbh: *a, Pinus virginiana; b, P. rigida; c, P. pungens; d, Quercus marilandica; e, Q. coccinea.* Below, trends in community characteristics: *f,* species diversity of vascular plants in sample quadrats (in per cent of a maximum of 44); *g,* tree-stratum above-ground net annual production (in per cent of a maximum of 1200 g/m²/yr); *h,* shrub-stratum coverage per cents. (WHITTAKER 1956, 1965, 1966, 1967).

Three associations might well be recognized among the communities of Fig. 1 – one at low elevations with *Pinus virginiana, Quercus marilandica,* and others as characteristic species, one at middle elevations with *Pinus rigida* and *Quercus coccinea,* and one at high elevations with *P. pungens* and other species. The three associations form an ecological series in the sense of Finnish and Russian phytosociologists in relation to the elevation gradient. The point here is not merely that there are mixed or transitional communities between the associations of this series. One may, rather, regard the associations as parts we choose to distinguish within a single community-continuum (WHITTAKER 1951, 1956, 1967; CURTIS and MCINTOSH 1951; BROWN and CURTIS 1952). (4) Continuous change along the gradient appears not only in species distributions, but also in trends or gradients of community characteristics (Fig. 1).

These observations, and those we may draw from a transect including a larger number of species (Fig. 2), relate to a number of other concepts.

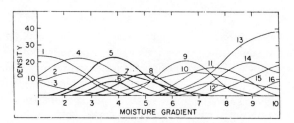

Fig. 2. Distributions of plant species populations along a topographic moisture gradient. (The curves are generalized; for actual data see WHITTAKER 1951, 1956, 1960; WHITTAKER and NIERING 1965, 1968). Species 5, 6, 7, and 8 form a commodal group which might be used as an ecological group, or as a character-species group for an association that extends from step 3 to step 6 along the moisture gradient. Within this association species 1 and 2, versus species 9 and 10, might serve as differential-species for subassociations of moister and drier environments.

1) The community continuum is a coenocline. (The concept and terms were suggested independently by WHITTAKER (1960, 1967) and VAN DER MAAREL (1960), MAAREL and LEERTOUWER (1967, syncline); coenocline has been preferred because the other term is used in geology.)

2) A major environmental gradient – such as elevation, the topographic moisture gradient, etc. – is by no means a single-factor gradient. Along such gradients many characteristics of environment change together; these are gradients of environmental complexes and may be termed complex-gradients (WHITTAKER 1956, 1967). A complex-gradient and its corresponding coenocline together form a gradient of ecosystems, in CLEMENTS' (1936) term, an ecocline.

3) Within the coenocline we may recognize one aspect of the continuous change, that of species composition, as a compositional gradient (BRAY and CURTIS 1957). We can measure relative distance of samples apart along the gradient, comparing the samples by coefficient of community, percentage similarity, or related measurements (Table I). Such measurements express ecological distance (WHITTAKER 1952, 1967), relative separation of samples along environmental (or successional or disturbance) gradients, as indicated by relative similarity of community composition.

4) If the distributions of species are individualistic, what do we mean when we say species are associated? Primarily that their distributional centers – the modes or peaks of the bell-shaped curves – are close together along an environmental gradient, and hence the species tend to occur together in the same communities. Species with their modes near one another may be termed commodal (WHITTAKER 1956, Fig. 2). Character-species groupings, and ecological groups in the sense of ELLENBERG (1948, 1950, 1952), are commodal groupings (which may correspond to one another) used for different purposes of classification of communities and indication of environment.

5) Positions along an ecocline may be defined by various means, including commodal groups, groups of differential-species with their limits close together (Fig. 2), and measurements of ecological distance in a compositional gradient. Weighted averages of the representation of ecological groups in samples are also a most effective approach to recognition of positions along the coenocline (ELLENBERG 1948; WHITTAKER 1951, 1954; CURTIS and McINTOSH 1951; ROWE 1956; GOFF and COTTAM 1967). Table I illustrates the expression of changing position along a compositional gradient in both weighted averages and percentage similarity values. Because the coenocline and complex-gradient are parallel (and functionally related), we may use these measurements to indicate position along the environmental complex-gradient. Such is the basis of ELLENBERG's (1948, 1950, 1952) indicator applications, and the basis of arranging samples into transects in some work on gradient analysis (WHITTAKER 1951, 1956, 1960; CURTIS and McINTOSH 1951; WHITTAKER and NIERING 1965; BROWN and CURTIS 1952; CURTIS 1955; WARING and MAJOR 1964).

We may carry work in direct gradient analysis on to study relations of species and communities to two environmental gradients – elevation and topographic moisture in mountains, say. These gradients become the axes of a chart, and vegetation samples are plotted on the chart in relation to them. When the samples have been classified, and boundaries between the community-types are drawn on the chart, the result is a mosaic diagram (WHITTAKER 1951, 1956, 1960; WHITTAKER and NIERING 1965, 1968) such as the background chart of Figs. 3 and 4. Similar charts relating European plant communities to one another and major gradients have been used by ELLENBERG (1963) and others. These charts permit us to relate to one another environmental complex-gradients (the axes), species populations (Fig. 3), community trends (Fig. 4), and community-types.

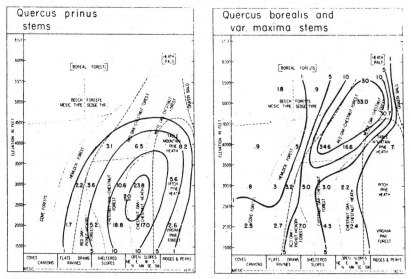

Fig. 3.   Population charts for two tree species in the Great Smoky Mountains, plotted on a „mosaic diagram" of the vegetation pattern (WHITTAKER 1956). Data are percentages of tree stems over 1 cm dbh. The population contours for *Quercus prinus*, left, outline a typical binominal solid, those for *Q. borealis* on the right a more complex pattern with two ecotypes, var. *borealis* at high elevations and var. *maxima* at low.

Densities or other importance values for species populations may be plotted on such charts, and the values bounded with population contour lines (Fig. 3). The figure for *Quercus prinus* is typical of such figures, which may be visualized as population hills, binomial solids. A transect through one of these in any direction produces one of the bell-shaped curves shown in Figs. 1 and 2. Some of these population solids are more complex in form, consisting of partially or wholly separate subpopulations or ecotypes (for example *Q. borealis* in Fig. 3). A vegetation pattern may be conceived in terms of many of these population solids, their centers scattered in the environmental „space" that the figure represents, each with its boundaries differently outlined, broadly overlapping with one another. These many species form a complex and continuous population

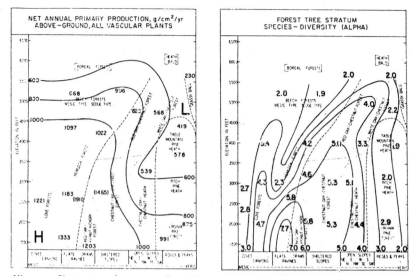

Fig. 4. Patterns of community characteristics in the Great Smoky Mountains, plotted on a ,,mosaic diagram" of the vegetation pattern. Left, net primary production, decreasing from low-elevation moist environments toward high elevations and dry environments (WHITTAKER 1966); right, species diversities (*alpha* value of FISHER et al. 1943), decreasing from low-elevation submesic environments toward high elevations, dry, and very moist environments (WHITTAKER 1956).

pattern corresponding to the pattern of environmental gradients. This concept, of communities as forming complex population patterns in relation to environmental patterns, I think a most fundamental development from gradient analysis.

INDIRECT GRADIENT ANALYSIS

In direct gradient analysis, major environmental gradients are accepted as given, as axes on the basis of which we arrange samples into transects or patterns and study the relations of populations and communities to environments. Suppose, however, that we approach the problem without prior assumptions about major environmental gradients. Vegetation samples are taken and compared with one another in such ways as to cause the major directions of community variation to emerge from the data. This procedure, in which the approach to environmental gradients is indirect, may be termed indirect gradient analysis (WHITTAKER 1967).

Some work of the Wisconsin school and others (BRAY and CURTIS 1957; MAYCOCK and CURTIS 1960; BEALS and COTTAM 1960; LOUCKS 1962; AYYAD and DIX 1964; AUSTIN and ORLOCI 1966; ORLOCI 1966) illustrates the approach. A set of samples are taken from the vegetation of a landscape. These samples are compared with one another in all possible combinations by percentage similarity or some comparable measurement. A triangular table, a matrix of sample similarity values results; Table II illustrates such, for a simple case with only ten samples (each of which, in this case, is actually an average of a number of relevés

365

Table 2.

Coefficients of community for ten forest associations (see Fig. 5) in Poland
(FRYDMAN and WHITTAKER 1968).

Association	1	2	3	4	5	6	7	8	9	10
1	-	38.6	13.3	10.3	7.1	6.3	4.9	5.1	6.0	0
2			19.4	20.0	22.2	10.1	6.9	12.2	7.6	2.1
3				28.4	8.5	4.8	15.8	27.8	26.9	7.8
4					13.4	19.5	22.6	42.3	24.6	9.2
5						32.4	17.2	12.4	7.5	3.2
6							21.5	23.8	16.5	3.7
7								46.6	39.8	13.4
8									39.0	15.4
9										32.9
Mean	9.2	13.9	15.3	19.0	12.4	13.9	18.9	22.5	20.1	8.8

representing an association). For each sample we obtain also the sum of
its similarity values with all other samples. The sample which has the
lowest similarity sum is „most extreme" in our set, most distant from
other samples along some direction of community variation; in Table 2
this is sample 10. This sample, and another which is lowest in similarity
to it (sample 1) may be used as the two end-points of our first axis of
vegetational variation. The similarity values of other samples to these
two permit us to arrange all other samples in sequence, from those like
the first end-point sample through those with almost equal similarity
values to both end-points to those most like the second end-point. The
manner in which similarity values change along a gradient is illustrated
in Table 1, especially by parts of the transect (steps 1 to 6, and steps 6
to 10) since zero similarities result from comparing more distant samples
along this long gradient. We have, at this point, arranged or ordinated
our samples into something like a transect, but on the basis of difference
and similarity in their species composition, not of a known environmental
gradient.

Among the samples which lie near the middle of this first axis, we may
choose two which are least like one another and use these as end-points
for a second axis (samples 3 and 6 in Fig. 5). All the samples may again
be arranged in relation to this second axis; and in some cases we may seek
a third pair of end-point samples and a third arrangement. Fig. 5 represents the result of arranging the ten samples of Table 1 in relation to two
axes. This arrangement or ordination (GOODALL 1954b) represents the
samples as located by relative positions in a range, or field, of vegetational
variation as expressed in our two axes. The axes are directions of change
in composition of vegetation samples; they are consequently compositional gradients. Samples are arranged by ecological distances along
these axes. We have stated that a compositional gradient is one aspect
of a coenocline or community-gradient, and that the coenocline parallels
a complex-gradient of environment, as part of an ecocline. We would

expect then, that Fig. 5 would represent an arrangement of our samples in relation to a pattern of major environmental gradients. It is a different approach to the same end as Figs. 3 and 4 – recognizing a pattern of plant communities in relation to a pattern of environments.

We may study the environmental pattern by plotting data for environmental measurements at the points for samples in Fig. 5. When we

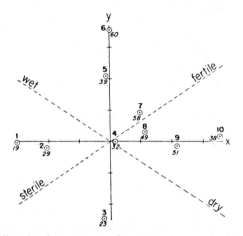

Fig. 5.   An ordination in two axes of ten samples representing forest associations in Poland (data of FRYDMAN and WHITTAKER 1968). Samples are ordinated by coefficients of community (Table 2); unit distances of ten per cent change in coefficient of community are marked on the axes. Numbers for associations are above the points; numbers below the points are species diversities (mean numbers of plant species in relevés. Associations represented: 1, Sphagnetum medii; 2, Pineto–Vaccinietum uliginosi; 3, Pineto–Vaccinietum myrtilli; 4, Abietum polonicum; 5, Cariceto elongatae–Alnetum; 6, Circaeo–Alnetum; 7, Querceto–Carpinetum medioeuropaeum; 8, Fagetum carpaticum; 9, Querceto–Potentilletum albae; 10, Coryleto–Peucedanetum cervariae.

do this, we find that soil moisture and soil fertility change across the pattern in the directions indicated. Our axes do in fact represent compositional gradients corresponding to environmental complex-gradients, but the axes are oblique in relation to the major soils gradients as we usually think of these. We can also plot population data for species at the points, and obtain population distributions suggestive of those in Fig. 3. We can plot community characteristics, such as the numbers of species in samples entered in Fig. 5. Richness in species increases toward the more fertile soils, though it is lower in the dry site of sample 10. Fig. 5 thus, like Figs. 3 and 4, represents a pattern of communities and ecosystems, a pattern in which we may relate to one another gradients and patterns of environmental factors, species populations, and community characteristics.

We can also think of Fig. 5 as representing a hyperspace – an abstract space defined by our compositional gradients as abstract axes (GOODALL 1963; WHITTAKER 1967). In terms of this hyperspace we can now rephrase some of the questions which have been approached by direct gradient analysis: (1) Are species clustered or dispersed in the hyperspace? Do we observe groups of species in which, because associated species are closely similar in distribution, form clusters with fewer species between these clusters? Or are the species scattered through the hyperspace, as the principle of species individuality might suggest? (2) Are samples clustered or dispersed in the hyperspace? We may plot a much larger number of samples into the hyperspace than were used for Fig. 5. (The samples must be chosen in a way which avoids subjective preference for samples typical of associations and against samples which are intermediate or transitional). Do these samples then fall into natural clusters representing associations, with relatively few transitional samples lying between these clusters? Or are the samples scattered in the hyperspace as we might expect from the principle of community continuity?

Results from research in indirect gradient analysis in general support the scattering of both species and samples in the hyperspace (WHIT-TAKER 1967; McINTOSH 1967). An effective study of direct interest to Europeans was carried out by DAGNELIE (1960, 1962) using factor analysis. Factor analysis and principal component analysis are mathematically more advanced ways of deriving (from a table of distributional similarities or correlations of species, rather than of sample similarities) „extracted factors" of community variation, corresponding to the axes of the Wisconsin ordination (GOODALL 1954a; AUSTIN and ORLOCI 1966;

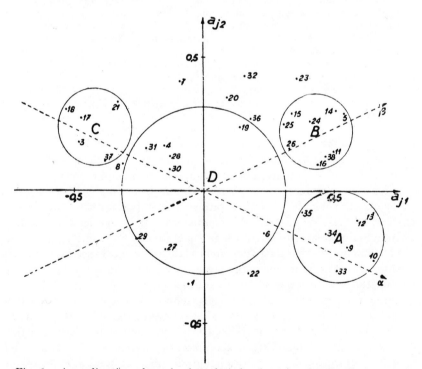

Fig. 6. An ordination of species (numbered points) in a loading hyperspace, from factor analysis of French beech forests (DAGNELIE 1962). Circles enclose four ecological groups of species; dashed lines are axes rotated to correspond to environmental factors affecting these groups.

GREIG-SMITH et. al. 1967; WHITTAKER 1967). Species may be located or ordinated in the hyperspace defined by the extracted factors, and samples may be located or ordinated in a closely related hyperspace. Fig. 6 arranges species from French beech forests. The species are primarily scattered, not clustered; nevertheless it is possible to recognize ecological groups of species which tend to occur together in the same communities in response to the same environmental factors. Circles A, B, and C enclose three ecological groups; circle D encloses a number of more widespread companion species. As in Fig. 5, the axes are oblique in relation to the environmental gradients to which these ecological groups seem most closely related. Fig. 7 represents an ordination of samples. The types of communities (which are defined by representation of the ecological groups in the new classification on the right) occupy different areas of the hyperspace, as the community-types do in relation to the environmental gradients in Figs. 3 and 4. The samples in Fig. 7 are, however, scattered, not clustered.

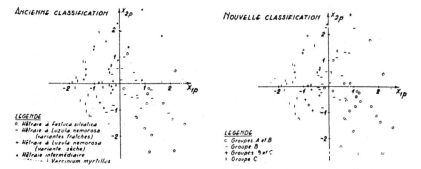

Fig. 7.   An ordination of samples in a factor-value hyperspace, from factor analysis of French beech forests (DAGNELIE 1962). Samples have been classified into community-types by undergrowth dominants, on the left, by representation of the ecological groups of Fig. 6, on the right.

It thus appears that results from direct and indirect gradient analysis converge in support of the principles of species individuality and community continuity. From both we may derive the conception of vegetation as a complex and predominantly continuous population pattern in which species have scattered positions.

DOMINANCE AND DIVERSITY

We can also „ordinate" species in relation to quite different gradients from these. It is of interest, for example, to rank the species in a community by their relative importances and ask what this ranking shows about relations between species and the nature of communities. We are in this case studying the „vertical" structure of vegetation – the manner in which species populations are organized within a given community. This organization, expressed in relative importance values or in dominance and diversity relations is related to, though different from, the more familiar vertical organization expressed in stratification and synusial relations.

There are a number of measurements expressing the relative importances of species in a community. Among these measurements coverage is easily obtained and widely useful, but productivity measurements may be most significant. Productivity values have the virtues of expressing directly the biological activity of different species, and of permitting comparison on a single scale of species widely different in size and form. For a series of plant communities in the southern Appalachian mountains I have worked out measurements or estimates of the net primary productions by their vascular plant species (WHITTAKER 1965, 1966). The species can be arranged in sequence from most productive to least productive, and plotted on a graph by this sequence and their productivities on a logarithmic scale. Seven such curves (the first of which superimposes data from three different cove forest communities) are illustrated in Fig. 8. A range of forms in these curves will

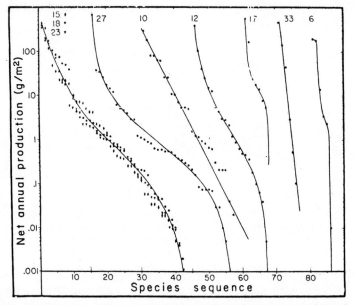

Fig. 8.   Importance-value curves for vascular plant communities in the Great Smoky Mountains (WHITTAKER 1965). Points represent species, plotted by production (logarithmic vertical scale) against the species' number in the sequence from most productive to least productive. To avoid overlap, six of the curves have been displaced to the right; zero positions for the species sequences of these are indicated by the vertical lines on the top border. Communities represented are: first three curves (15, 18, 23), superimposed, cove (valley) forests; 27, *Quercus borealis* forest; 10, *Pinus virginiana-Pinus strobus* forest; 12, *Pinus pungens* heath; 17, *Picea rubens-Rhododendron catawbiense* forest; 33, *Abies fraseri* forest; 6, mixed heath bald.

be observed – from community 33, with strong dominance by its most important species, low species diversity, and the steep, straight slope of a geometric curve, to communities 15, 18, and 23, with more mixed canopy dominance, higher species diversity, and less steep curves of sigmoid form. When such sigmoid curves are differently plotted they form lognormal distributions (PRESTON 1948; BESCHEL and WEBBER

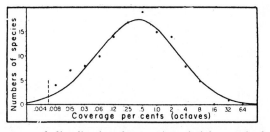

Fig. 9. A lognormal distribution for species of rich, north-slope, Sonoran desert communities of the Santa Catalina Mountains, Arizona (WHITTAKER 1965). Points are numbers of plant species, plotted for octaves or doubling-units of plant coverage, on a logarithmic scale.

1963; WHITTAKER 1965; WHITTAKER and WOODWELL 1969). Fig. 9 illustrates a lognormal distribution – a bell-shaped frequency distribution of numbers of species in octaves, or doubling units, of coverage. Such a distribution shows that a community, if it is fairly rich in species, consists of a large number of species of intermediate importances in the community, and smaller numbers of very important or dominant species and of rare species.

One may approach the interpretation of such curves through the concept of the species niche. By niche we refer to the position of the species within the community – its position in vertical (above-ground and below-ground) space, horizontal space (internal mosaic or patterning within the community), seasonal and diurnal time, community functional relations, and interactions with other species. Some of these niche characteristics form gradients by which we can arrange species. We may, for example, arrange species of a Sonoran desert community in southern Arizona by two niche axes – above-ground vertical position, and character of seasonal relations from evergreen through deciduous leaves to transitory leaves and succulence and leaflessness. The species occupy scattered positions in relation to these axes (Fig. 10).

We may conceive that the many characteristics of niches form, as ·axes, an abstract niche hyperspace (HUTCHINSON 1957). In this hyperspace the species of a given community occupy different positions; hence no two use the same resources, in the same vertical and horizontal position, at the same time, in full and direct competition with one another. We thus state the principle of GAUSE (1934) or of competitive exclusion (HARDIN 1960), the assertion that: (1) If two species are in full and direct competition in the same habitat and niche for an extended period, one must become extinct, hence (2) No two species in a given stable natural community occupy the same niche, in full competition with one another.

Niche space for a given species is related to, though not identical with, resource use by that species, and consequently to the level of productivity the species is able to achieve in the community. For understanding of the importance value curves of Fig. 8 we may ask how niche space and resources may be divided among species, to produce the distributions of relative productivity that we observe. There are a number of hypotheses

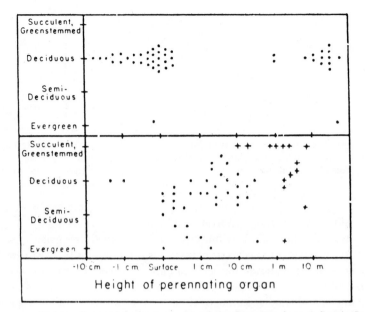

Fig. 10. Below: Perennial plant species of the Sonoran desert, Santa Catalina Mountains, Arizona, plotted against gradients of leaf persistance, vertical axis, and perennating bud height, horizontal axis (WHITTAKER and NIERING 1965). The larger woody plants of the desert are spiny, as indicated by crossed points. Above: Species of a southern Appalachian cove forest plotted in the same way. Since these species have evolved toward clustering around two life-forms (broadleaf-deciduous trees, and hemicryptophytes), niche differentiation must be assumed to affect other plant characteristics.

on division of niche space and form of importance-value curves (MOTO-MURA 1932; FISHER et. al. 1943; PRESTON 1948; MACARTHUR 1960), but two interpretations may be most appropriate for communities of vascular plants (WHITTAKER 1965; WHITTAKER and WOODWELL 1969):

1) Geometric series. When the number of species in the community is small, there may be a tendency for the first or dominant species to occupy some fraction, say k, of niche space, and the second species to occupy a similar fraction k, of niche space that has not been taken or pre-empted by the first, and the third species to occupy a similar fraction of the space not occupied by either of the two preceding species. The fraction k need not be a constant; but if the k values and the ratios of importance values for successive species $(1 - k)$ do not vary too widely, the importance values will approach a geometric series (Fig. 11A).

2) Lognormal distribution. If the number of species is larger, and the manner in which they relate to niche space is more complex, more complex curves of importance values will be formed (Fig. 11B and C). Such curves will include (a) a few dominants which occupy much of the niche space and use most of the community resources, (b) a larger number of species of intermediate importances, variously fitting themselves into the community by different patterns of niche re-

lationships and resource use, and (c) a smaller number of rare species, mostly rather narrow specialists utilizing restricted, distinctive, niche positions and resources. These curves will be of sigmoid form and, when the numbers of species and of factors affecting their relative importances are large, the importance values of these species will approach the lognormal distribution.

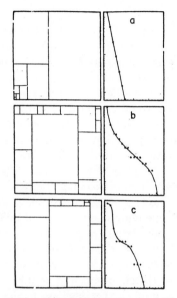

Fig. 11.   Models for species and niche relations which may underlie importance-value curves (WHITTAKER 1965). The squares in each case represent a niche space which is divided among species in the community, represented by rectangles. Sizes of the rectangles for species represent their shares of niche space and environmental resources, as expressed in their productivities. In the curve to the right of each square, the areas of the species are plotted in the manner of Fig. 8. In the top square, *a*, each species occupies 0.6 of the niche space not already occupied by more successful species. In squares *b* and *c* the species occupy, by more complex rules, niche spaces around those occupied by the most important species or dominants.

We judge that importance-value curves for vascular plants in communities (Fig. 8) are of a range of intergrading forms from geometric series to lognormal distributions. We judge further that the species in a given community evolve away from direct competition and toward difference in niche. Since all the vascular plants in a community require some of the same resources – light, water, above-ground and soil space, and nutrients – they are to some degree in competition. But they evolve toward partial competition, toward niche positions and resource uses that differ in part. It is because of this partial differentiation of niche that many vascular plant species are able to co-exist in the same community. The number of species in the community, its species diversity, is affected by relative favorableness vs. severity of environment, and by evolutionary time during which niche differentiation has evolved among the species in that environment (WHITTAKER 1965). The importance-value curves we have illustrated express the consequence of that evolution as environmental favorableness and evolutionary time affect the richness in species of the community and the manner in which niche space and resources for production are divided among the species.

We have observed species populations to be twice-scattered – in relation to environmental gradients and patterns, and in relation to niche characteristics and hyperspaces. This may be a most significant dual observation and convergence of results from gradient analysis and dominance-diversity studies. I would draw these conclusions on this observation, its meaning as I would interpret it, and its implications:

1) A common theme – evolution toward reduction of competition – underlies the scattering of adaptive centers of species in relation to both habitats and niches. Species which are in competition within a given community evolve toward niche differentiation; by this differentiation they become partial competitors and are able to survive in one another's presence. Species which are in partial competition along an environmental gradient evolve toward different locations of their population centers along the gradient. By this habitat differentiation also competition is reduced; the species utilize (in their population centers or modes, at least) the resources of different parts of the environmental gradient.

2) The latter process implies that species do not evolve toward the formation of groups of associates with closely similar distributions. A first judgment on evolution would suggest that species should evolve toward the formation of natural groups, with the species in each adapted to occurrence together by niche differentiation and other accomodations. These natural groups would appear as distinct clusters of species, separate from other clusters, in the results of gradient analysis. But the logic of evolution in response to competition supports the observations from gradient analysis: Vascular plant species evolve away from the formation of clusters of associates, toward scattering of their population centers in relation to environmental gradients, by which the intensity of competition in their distributional centers is reduced (WHITTAKER 1967).

3) We may thus state a conception of the population structure of vegetation. A plant community is a system of interacting, niche differentiated, partially competitive species. These species have evolved toward scattering of their population centers along environmental gradients. Because the species are only partial competitors, their bell-shaped distributions may overlap broadly, as observed in gradient analysis. Along environmental gradients undisturbed communities consequently, in most areas, intergrade continuously. In relation to patterns of environmental gradients, communities form complex and largely continuous population patterns. Relative discontinuities may occur in some undisturbed vegetation, par-

ticularly in some communities of more rigorous environments, steeper environmental gradients, and strong single-species dominance (WHITTAKER 1956, 1967; DAHL 1957; BEALS 1969). Discontinuities in vegetation may also result from parent material and topographic discontinuity and from disturbance. In general, however, when disturbance is not too extensive, the flora of a landscape is organized into a complex and largely continuous population pattern corresponding to the pattern of environments; such is the population structure of the vegetational mantles that we observe.

4) Associations and other community-types are man-made class concepts, abstractions from the intergrading complexities of vegetation in the field (WHITTAKER 1962). This comment on the nature of classification is not a criticism of the process of classification.

The implication of gradient analysis is not that plant communities cannot be classified, only that the classification of intergrading communities, in which species are differently distributed from one another, encounters difficulties in practice – as we all know. The relation between gradient analysis and classification is not one of antagonism, but one of complementarity and potential partnership. Many, in fact most, studies in gradient analysis use some community classification to present their results. Results of community classification, on the other hand, can often be used for gradient analysis with the community-types arranged in ecological series or community patterns. I think of classification and gradient analysis as alternative modes of abstraction, which may often be used together in a given research project in a way which enhances the effectiveness of each. Such work as that of HANSEN (1930, 1932); ELLENBERG (1950, 1952, 1963); POORE (1956, 1962); GROENEWOUD (1965); and VAN DER MAAREL (1966, VAN DER MAAREL and LEERTOUWER 1967) represents this synthesis, as does that of CURTIS (1959 and myself (WHITTAKER 1956, 1960; WHITTAKER and FAIRBANKS 1958; WHITTAKER and NIERING 1965, 1968; FRYDMAN and WHITTAKER 1968).

Gradient analysis clarifies some of the practical difficulties of classification. It may also show reason for the success of detailed floristic analysis of vegetation as applied by BRAUN-BLANQUET (1964) and TÜXEN (1955) and others of my audience in the intensively occupied and intensively studied landscape of Europe. The system of BRAUN-BLANQUET makes maximum use for purposes of classification, and environmental indication based on classification, of the information available from the distributional relations of species. It permits classifications finer in detail and expression of environmental difference, than many English-language classifications by physiognomy and dominance. Understanding from gradient analysis gives altered perspective on the meaning of such concepts as the association and other units, fidelity and character-species, differential species, ecological group, ecological series, typical and mixed or transitional community, and community complex and pattern, without reducing the value of these concepts.

I thus feel that results from gradient analysis in English-language ecology and classification in Continental phytosociology will to some extent converge in the end, and I hope for the reduction by mutual understanding of the „ecological distance" between our traditions.

SUMMARY

My purpose is to discuss certain relations between three areas of study – gradient analysis, dominance-diversity, and classification.

Species evolve toward niche differentiation, toward different positions and kinds of function, within a given plant community. The community is a system of interacting, niche-differentiated species. These species are only partial competitors, and their population distributions along gradients consequently can overlap broadly. They do not generally exclude one another at sharp boundaries.

Species evolve also toward habitat differentiation, toward having their population centers at different positions along environmental gradients. By this means also they reduce competition in their centers of distribution. The result of this is the kind of population distribution we see in gradient analysis. Species populations form bell-shaped curves, whose centers are scattered along the gradient and which overlap broadly. All the species together along the gradient form a complex, flowing continuum of populations. The communities we see mostly intergrade along such continua. In some areas the continuity is much interrupted by disturbance, but the vegetation of a landscape may be conceived as a complex and largely continuous population pattern corresponding to the pattern of environmental gradients.

Direct and indirect gradient analysis seek to deal with this population pattern by means of transects of gradients and coordinate systems. Direct gradient analysis studies population distributions along known environmental gradients. Indirect gradient analysis, such as the school of Wisconsin approach and factor analysis, seeks to arrange or ordinate samples and species in an abstract hyperspace, the axes of which represent directions of compositional variation in the vegetation pattern. Samples and species in most cases have scattered positions in the hyperspace; this scattering expresses community continuity and species individuality.

Classification, as an alternative approach, groups samples into community-types on the basis of physiognomy, dominance, species composition, or other criteria. The most widely successful system is that of BRAUN-BLANQUET, which uses various criteria but is based primarily on groups of character- and differential-species. The system represents a most effective use for classification of those complex distributional relations among species which are shown by gradient analysis.

Gradient analysis and classification are not antagonistic, but complementary approaches. Gradient analysis provides the means of understanding some of the problems of classification. Classification provides an essential means of presenting some of the results of gradient analysis. Vegetation units derived from classification can often be related to one another through gradient analysis. In many cases the two approaches can be combined in a given study, to the increased effectiveness of both. It is thus to be hoped that understanding of the population structure of vegetation may contribute to convergences of interest and method between ecologists and phytosociologists.

ZUSAMMENFASSUNG

Meine Absicht ist, gewisse Beziehungen zwischen drei Forschungsbereichen zu diskutieren: Gradient-Analyse, Dominanz-Unterschiede und Klassifikation.

Die Arten entfalten in einer gegebenen Pflanzengesellschaft Nischen-Unterschiede, verschiedene Stellungen und Wirkweisen. Die Gesellschaft ist ein System von sich gegenseitig beeinflussenden, nach Nischen unterschiedenen Arten. Diese Arten sind nur teilweise Konkurrenten und ihre Populations-Verteilung kann sich infolgedessen an Gradienten breit überlappen. Sie schließen einander im allgemeinen nicht mit scharfen Grenzen aus.

Die Arten entwickeln auch eine Standorts-(habitat)Unterscheidung, in dem ihre Populationen Zentren in verschiedenen Lagen entlang der Standortsgradienten besitzen. Durch diese Mittel verringern sie auch den Wettbewerb in ihren Verbreitungszentren. Das Ergebnis davon ist die Art der Populations-Verteilung, die wir in der Gradienten-Analyse sehen. Die Arten-Populationen bilden glockenförmige Kurven, deren Scheitelpunkte mit breiten Überlappungen entlang des Gradienten verstreut sind. Alle Arten bilden längs dem Gradienten ein komplexes, fließendes Kontinuum von Populationen. Die Gesellschaften sehen wir meist entlang solcher Kontinua angeordnet. In einigen Gebieten ist das Kontinuum durch Störungen unterbrochen, aber die Vegetation einer Landschaft kann als ein komplex und ausgedehntes ununterbrochenes Populations-Muster (pattern) aufgefaßt werden, das dem Muster des Standorts-Gradienten entspricht.

Direkte und indirekte Gradient-Analysen versuchen diese Populations-Muster mit Hilfe von Gradienten-Transekten und Koordinaten-Systemen zu behandeln. Die direkte Gradienten-Analyse studiert die Populations-Verteilung entlang' bekannter Standorts-Gradienten. Die indirekte Gradienten-Analyse sucht wie die Wisconsin-Schule mit Annäherungs- und Faktoren-Analyse, Probeflächen und Arten in einem abstrakten Überraum (hyperspace) zu gruppieren oder zu ordnen, dessen Achsen Richtungen der Anordnungs-Unterschiede im Vegetations-Muster darstellen. Probebestände und -Arten haben in den meisten Fällen zerstreute Stellungen in dem Überraum (hyperspace). Diese Streuung ist der Ausdruck für die Kontinua der Gesellschaft und die Individualität der Arten.

Die Klassifikation gruppiert als die andere Möglichkeit der Forschung Probebestände zu Gesellschaftstypen auf der Grundlage der Physiognomie, der Dominanz, der Artenkombination oder anderer Kriterien. Das bei weitem erfolgreichste System ist das von BRAUN-BLANQUET, das verschiedene Kriterien verwendet, aber ursprünglich auf Gruppen von

Charakter- und Differentialarten begründet ist. Das System ist von größtem Nutzen für die Klassifikation der komplexen Verteilungsbeziehungen unter den Arten, die von der Gradient-Analyse gezeigt werden.

Gradient-Analyse und Klassifikation sind keine gegensätzlichen sondern sich ergänzende Forschungszweige. Die Gradient-Analyse liefert Möglichkeiten einige Probleme der Klassifikation zu verstehen. Die Klassifikation gibt wesentliche Mittel um einige Ergebnisse der Gradient-Analyse darzustellen. Durch die Klassifikation gewonnene Vegetations-Einheiten können oft durch Gradient-Analyse zueinander in Beziehung gesetzt werden. In vielen Fällen können die beiden Forschungsrichtungen in einer gegebenen Untersuchung zu größerer Wirksamkeit beider vereinigt werden. So dürfen wir hoffen, daß das Verständnis der Gesellschaftstruktur der Vegetation dazu beitragen möge, Interessen und Methoden zwischen Ökologen und Pflanzensoziologen einander näher zu bringen.

LITERATURE

Austin, M. P. and Orloci, L.: Geometric models in ecology. II. An evaluation of some ordination techniques. – J. Ecol. **54**: 217–227, 1966.
Ayyad, M. A. G. and Dix, R. L.: An analysis of a vegetation-microenvironmental complex on prairie slopes in Saskatchewan. – Ecol. Monogr. **34**: 421–442. 1964.
Beals, E. W.: Vegetational change along altitudinal gradients. – Science, N.Y. **165**: 981–985. 1969.
—, and Cottam, G.: The forest vegetation of the Apostle Islands, Wisconsin. – Ecology **41**: 743–751. 1960.
Beschel, R. E. and Webber, P. J.: Bemerkungen zur log-normalen Struktur der Vegetation. – Ber. Naturwiss.-Med. Vereins Innsbruck **53** (Festschr. Gams): 9–22. 1963.
Braun-Blanquet, J.: – Pflanzensoziologie: Grundzüge der Vegetationskunde. 3rd. ed. – Wien. 865 pp. 1964.
Bray, J. R. and Curtis, J. T.: An ordination of the upland forest communities of southern Wisconsin. – Ecol. Monogr. **27**: 325–349. 1957.
Brown, R. T. and Curtis, J. T.: The upland conifer-hardwood forests of northern Wisconsin. – Ecol. Monogr. **22**: 217–234. 1952.
Clements, F. E.: Nature and structure of the climax. – J. Ecol. **24**: 252–284. 1936.
Curtis, J. T.: A prairie continuum in Wisconsin. – Ecology **36**: 558–566. 1955.
— The vegetation of Wisconsin: An ordination of plant communities. – University of Wisconsin Press, Madison. 657 pp. 1959.
— and McIntosh, R. P.: An upland forest continuum in the prairie-forest border region of Wisconsin. – Ecology **32**: 476–496. 1951.
Dagnelie, P.: Contribution a l'étude des communautés végétales par l'analyse factorielle. (Engl. summ.) – Bull. Serv. Carte Phytogéogr., Sér. B, **5**: 7–71, 93–195. 1960.
— L'étude des communautés végétales par l'analyse des liaisons entre les espèces et les variables écologiques. – Inst. Agron. de l'État, Gembloux. 135 pp. 1962.
Dahl, E.: Rondane: Mountain vegetation in South Norway and its relation to the environment. – Skr. Norske Vidensk-Akad., Mat.-Naturv. Kl., 1956, (3): 1–374. 1957.
Ellenberg, H.: Unkrautgesellschaften als Maß für den Säuregrad, die Verdichtung und andere Eigenschaften des Ackerbodens. - Ber. Landtechn. **4**: 130–146. 1948.

— Landwirtschaftliche Pflanzensoziologie. I. Unkrautgemeinschaften als Zeiger für Klima und Boden. – Stuttgart. 1950. 141 pp.
— Landwirtschaftliche Pflanzensoziologie. II. Wiesen und Weiden und ihre standörtliche Bewertung. – Stuttgart. 143 pp. 1952.
— Vegetation Mitteleuropas mit den Alpen in kausaler, dynamischer und historischer Sicht. *In*: Einführung in die Phytologie, by H. Walter, *IV*. Grundlagen der Vegetationsgliederung, Pt. 2. – Stuttgart. 943 pp. 1963.
Fisher, R. A., Corbet, A. S. and Williams, C. B.: The relation between the number of species and the number of individuals in a random sample of an animal population. – J. Anim. Ecol. **12**: 42–58. 1943.
Frydman, I. and Whittaker, R. H.: Forest associations of southeast Lublin Province, Poland. (Germ. summ.) – Ecology **49**: 896–908. 1968.
Gause, G. F.: The struggle for existence. – Williams and Wilkins, Baltimore. 163 pp. 1934.
Gleason, H. A.: The individualistic concept of the plant association. – Bull. Torrey. Bot. Club **53**: 7–26. 1926.
Goff, F. G. and Cottam, G.: Gradient analysis: the use of species and synthetic indices. – Ecology **48**: 793–806. 1967.
Goodall, D. W.: Objective methods for the classification of vegetation. III. An essay in the use of factor analysis. – Austral. J. Bot. **2**: 304–324. 1954a.
— Vegetational classification and vegetational continua. (Germ. summ.) – Angew. Pflanzensoziologie (Wien), Festschr. Aichinger **1**: 168–182. 1954b.
— The continuum and the individualistic association. (French summ.) – Vegetatio **11**: 297–316. 1963.
Greig-Smith, P., Austin, M. P. and Whitmore, T. C.: The application of quantitative methods to vegetation survey. I. Association-analysis and principal component ordination of rain forest. – J. Ecol. **55**: 483–503. 1967.
Groenewoud, H. van: Ordination and classification of Swiss and Canadian coniferous forests by various biometric and other methods. (Germ. summ.) – Ber. Geobot. Inst. ETH, Stiftg. Rübel, Zürich 1964, **36**: 28–102. 1965.
Hansen, H. Mølholm: Studies on the vegetation of Iceland. *In*: The Botany of Iceland, ed. L. K. Rosenvinge and E. Warming **III**, Pt. 1, No. 10. Copenhagen. 186 pp. 1930.
— Norholm Hede, en formationsstatistisk Vegetationsmonografi. (Engl. summ.) – K. Danske Vidensk. Selsk. Skr., Naturv. Math. Afd., Ser. 9, 3 (3): 99–196. 1932.
Hardin, G.: The competitive exclusion principle. – Science, N.Y. **131**: 1292–1297. 1960.
Hutchinson, G. E.: Concluding remarks. – Cold Spring Harbor Symp. Quant. Biol. **22**: 415–427. 1957.
Loucks, O. L.: Ordinating forest communities by means of environmental scalars and phytosociological indices. – Ecol. Monogr. **32**: 137–166. 1962.
Maarel, E. van der: Rapport inzake de vegetatie van het duingebied van de Stichting ,,Het Zuid-Hollands Landschap'' bij Oostvoorne. – Report ZHL Delft. 1960.
— Over vegetatiestructuren, -relaties en -systemen in het bijzonder in de duingraslanden van Voorne (Engl. summ.) – Thesis, Utrecht. 1966.
— and Leertouwer, J.: Variation in vegetation and species diversity along a local environmental gradient. – Acta Bot. Neerl. **16**: 211–221. 1967.
MacArthur, R. H.: On the relative abundance of species. – Am. Nat. **94**: 25–36. 1960.
Maycock, P. F. and Curtis, J. T.: The phytosociology of boreal conifer-hardwood forests of the Great Lakes region. – Ecol. Monogr. **30**: 1–35. 1960.
McIntosh, R. P.: The continuum concept of vegetation. – Bot. Rev. **33**: 130–187. 1967.
Motomura, I.: A statistical treatment of associations. (Japanese). – Japan. J. Zool. **44**: 379–383. 1932.
Orloci, L.: Geometric models in ecology. I. The theory and application of some ordination methods. – J. Ecol. **54**: 193–215. 1966.
Poore, M. E. D.: The use of phytosociological methods in ecological investi-

gations. IV. General discussion of phytosociological problems. – J. Ecol. **44**: 28–50. 1956.

— The method of successive approximation in descriptive ecology. – Adv. Ecol. Res. **1**: 35–68. 1962.

PRESTON, F. W.: The commonness, and rarity, of species. – Ecology **29**: 254–283. 1948.

RAMENSKY, L. G.: Die Grundgesetzmäßigkeiten im Aufbau der Vegetationsdecke. (Russian). – Woronesh. Wjestn. opytn. djela 1924. 37 pp. 1925. – Bot. Cbl. N.F. **7**: 453–455, 1926.

ROWE, J. S.: Uses of undergrowth plant species in forestry. – Ecology **37**: 461–473. 1956.

TÜXEN, R.: Das System der nordwestdeutschen Pflanzengesellschaften. – Mitt. Flor.-soz. Arbeitsgemeinsch. (Stolzenau/Weser), N.F. **5**: 155–176. 1955.

WARING, R. H. and MAJOR, J.: Some vegetation of the California coastal redwood region in relation to gradients of moisture, nutrients, light, and temperature. – Ecol. Monogr. **34**: 167–215. 1964.

WHITTAKER, R. H.: A criticism of the plant association and climatic climax concepts. – Northwest Sci. **25**: 17–31. 1951.

— A study of summer foliage insect communities in the Great Smoky Mountains. – Ecol. Monogr. **22**: 1–44. 1952.

— Plant populations and the basis of plant indication. (Germ. summ.) – Angew. Pflanzensoziol. (Wien), Festschr. Aichinger, **1**: 183–206. 1954.

— Vegetation of the Great Smoky Mountains. – Ecol. Monogr. **26**: 1–80. 1956.

— Vegetation of the Siskiyou Mountains, Oregon and California. – Ecol. Monogr. **30**: 279–338. 1960.

— Classification of natural communities. – Bot. Rev. **28**: 1–239. 1962.

— Dominance and diversity in land plant communities. – Science, N.Y. **147**: 250–260. 1965.

— Forest dimensions and production in the Great Smoky Mountains. – Ecology **47**: 103–121. 1966.

— Gradient analysis of vegetation. – Biol. Rev. **42**: 207–264. 1967.

— and FAIRBANKS, C. W.: A study of plankton copepod communities in the Columbia Basin, southeastern Washington. – Ecology **39**: 46–65. 1958.

— and NIERING, W. A.: Vegetation of the Santa Catalina Mountains, Arizona. (II). A gradient analysis of the south slope. – Ecology **46**: 429–452. 1965.

— — Vegetation of the Santa Catalina Mountains, Arizona. IV. Limestone and acid soils. – J. Ecol. **56**: 523–544. 1968.

— and WOODWELL, G. M.: Structure, production and diversity of the oak-pine forest at Brookhaven, New York. – J. Ecol. **57**: 155–174. 1969.

AUTHOR CITATION INDEX

SUBJECT INDEX

About the Editor

ROBERT P. McINTOSH is Professor of Biology at the University of Notre Dame where he has taught since 1958. Prior to that date he taught at Middlebury College, Vermont, and Vassar College, New York.

Professor McIntosh completed his undergraduate work at Lawrence College in Wisconsin in 1942. Following service in World War II, he received his Ph.D. in 1950 at the University of Wisconsin, where he worked with J. T. Curtis. His principal phytosociological research studies have been in forest ecology, notably of Wisconsin and the Catskill Mountains of New York.

Dr. McIntosh has served on the Editorial Boards of Ecology and Ecological Monographs and from 1969–1977 as editor of the *American Midland Naturalist*. His current interests are in the history of ecology.